Essential Topics in Oceanography

Essential Topics in Oceanography

Edited by **Suzy Bullock**

R Callisto Reference

New York

Published by Callisto Reference,
106 Park Avenue, Suite 200,
New York, NY 10016, USA
www.callistoreference.com

Essential Topics in Oceanography
Edited by Suzy Bullock

International Standard Book Number: 978-1-63239-323-4 (Hardback)

Printed in the United States of America.

Contents

Preface

Oceanography, also referred to as oceanology and ocean and marine science, is a sub-topic of earth science and covers the study of oceans and seas. This field covers a large number of topics such as study of marine organisms, ecosystem dynamics-waves and currents in ocean fluids, geophysical fluid dynamics-plate tectonics and geological studies of sea bed. Oceanography also studies various chemical substances with respect to their flux, and physical as well as chemical properties of material in the ocean and nearby oceans.

If we turn the pages of history all the world over, the earliest records of oceanography can be found in ancient times, when Aristotle and Strabo recorded observations on tides. However, early exploration of the oceans was limited to its surfaces and primarily done for cartography. It was only with the 1872-76 Challenger expedition that the first true oceanographic cruise happened and was a seminal event in establishing oceanography as a discipline.

The wide range of topics encompassed in this subject, reflects the multi-disciplinary scope of this subject. It includes astronomy, biology, chemistry, climatology, geography, geology, hydrology, meteorology and physics to explain the mysterious knowledge of world oceans. Data that oceanographers acquire is further consumed in marine engineering, and also in designing and construction of oil platforms, ships, harbours, and various structures that are used in oceans by humans.

I especially wish to acknowledge the contributing authors, without whom a work of this magnitude would clearly not have been realizable. Not only do I appreciate their participation, but also their adherence as a group to the time parameters set for this publication. I hope that this book proves to be a resourceful guide for both basic and advanced concepts in Oceanography.

Editor

Is Occurrence of Harmful Algal Blooms in the Exclusive Economic Zone of India on the Rise?

K. B. Padmakumar,[1] N. R. Menon,[2] and V. N. Sanjeevan[1]

[1] Centre for Marine Living Resources and Ecology, Ministry of Earth Sciences, Kochi 37, Kerala, India
[2] School of Marine Sciences, Cochin University of Science and Technology and Nansen Environmental Research Center, Kochi 16, Kerala, India

Correspondence should be addressed to K. B. Padmakumar, kbpadmakumar@gmail.com

Academic Editor: Robert Frouin

Occurrence, increase in frequency, intensity and spatial coverage of harmful algal blooms during the past decade in the EEZ of India are documented here. Eighty algal blooms were recorded during the period 1998–2010. Of the eighty algal blooms, 31 blooms were formed by dinoflagellates, 27 by cyanobacteria, and 18 by diatoms. Three raphidophyte and one haptophyte blooms were also observed. Potentially toxic microalgae recorded from the Indian waters were *Alexandrium* spp., *Gymnodinium* spp. *Dinophysis* spp., *Coolia monotis*, *Prorocentrum lima*, and *Pseudo-nitzschia* spp. Examination of available data from the literature during the last hundred years and *in situ* observations during 1998–2010 indicates clear-cut increase in the occurrence of HABs in the Indian EEZ.

1. Introduction

The International Council for the Exploration of the Seas [1] has defined phytoplankton blooms as, those, which are noticeable, particularly to general public, directly or indirectly through their *effects* such as visible discolouration of the waters, foam production, and fish or invertebrate mortality or toxicity to humans. Among around 5000 species of marine phytoplankton, around 300 species including diatoms, dinoflagellate, raphidophytes, prymnesiophytes, cyanophytes, and silicoflagellates can at times cause algal blooms. Only a few dozen of these species have the ability to produce potent toxins. Harmful Algal Blooms (HABs) are global phenomenon, reported from over 30 countries including India. Number of HABs, economic losses from them, the types of resources affected, and number of toxins and toxic species have also increased dramatically in recent years [2–4].

The first recorded observation on algal blooms in Indian waters is by Hornell in 1908 [5]. He witnessed massive fish mortality, largely of sardines floating on the dark yellow coloured water that contained plankton. In 1916, he found *Euglena* and *Noctiluca* species to be responsible for such episodes. Subrahmanyan [6] identified this organism as *Hornellia marina*, and later in 1982 Hara and Chihara [7] reclas-sified this as the raphidophyte *Chattonella marina*. Since then, there have been several reports on various algal blooms dominated by blooms of *Noctiluca scintillans* (= *N. miliaris*) and *Trichodesmium* spp. from the EEZ of India. Episodic observations on algal blooms were reported from 1917 onwards.

First record of Paralytic Shellfish Poisoning (PSP) was during 1981 from coastal Tamilnadu, Karnataka, and Maharashtra [8]. In 1981, PSP resulted in death of 3 persons and hospitalization of 85 people due to consumption of affected mussel *Meretrix casta* in Tamilnadu. An outbreak of PSP has occurred in Kumbla near Mangalore following consumption of clams in 1983 [9]. Godhe and Karunasagar [10] reported *Gymnodinium catenatum*, a potent PSP producing alga in both plankton samples and cysts in sediments from the coastal waters of Mangalore. In September 1997, an outbreak of PSP was reported from Vizhinjam, Kerala, resulting in the death of seven persons and hospitalization of over 500 following consumption of mussel, *Perna indica* [11]. Recently, in September 2004, an unusual nauseating smell emanating from the coastal waters was recorded from Kollam to Vizhinjam in the southwest coast of India. More than 200 persons, especially children, complained of nausea and breathlessness for short duration due to the smell. This coincided with massive fish kills in this region. The causative organism was *Cochlodinium polykrikoides*. Several publications are available dealing with different types of blooms in the Indian EEZ. The frequency and extend of HABs are reported to be increasing in the Indian waters. These blooms adversely affect the living resources, destroy coastal aquaculture and can even be fatal to humans. Taking these into consideration, Centre for Marine Living Resources and Ecology (CMLRE), Ministry of Earth Sciences commenced a national programme on monitoring HABs since 1998. This programme envisages extensive monitoring of HABs in the Indian EEZ, identification of causative toxic/harmful microalgal species, and dynamics of bloom formation, spread and crash and its ecological consequences on marine ecosystems.

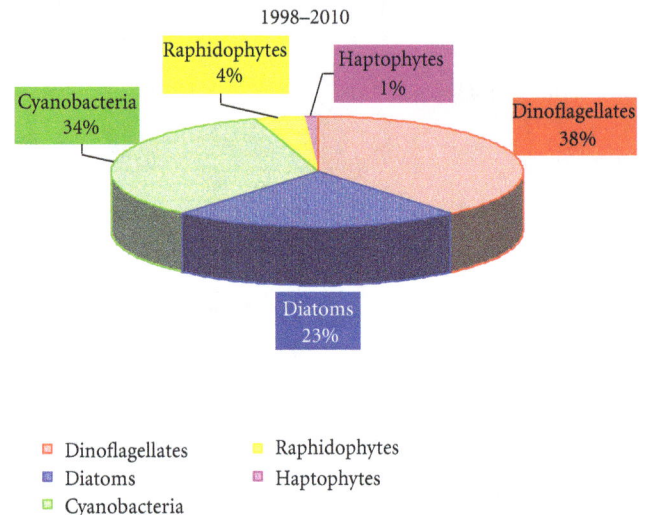

FIGURE 1: Groups of microalgae in percentage, contributing to the algal blooms during 1998 to 2010.

2. Materials and Methods

Ministry of Earth Sciences has been conducting regular oceanic cruises for the surveillance, identification, enumeration, and ecology of HABs in the territorial waters and contiguous seas in the Indian EEZ. Samplings were made from 1880 stations onboard *FORV Sagar Sampada* during the period 1998–2010. From all these stations phytoplankton samples were collected by filtering ~50 litres of surface water through 20 μm bolting silk. The filtrates were preserved in 1–3% neutralized formaldehyde solution. Microalgal samples were also collected from bloom areas and non-bloom patches. Quantitative and qualitative analysis of micro algae were carried out by using Sedgewick Rafter counting cell under a Nikon Eclipse microscope following standard identification keys [12]. Area of the bloom and mortality of fishes if any was estimated on the basis of visual observation. Vertical profiling of parameters, such as temperature, salinity, and density was done using Conductivity-Temperature-Depth profiler (CTD Seabird 911 plus). Sea Surface Temperature (SST) was measured by a bucket thermometer. Chlorophyll *a* was measured spectrophotometrically using a double beam UV-Visible spectrophotometer [13]. Major nutrients like nitrite, nitrate, phosphate, and silicate were analysed onboard using a segmented flow Auto Analyzer (SKALAR) by following standard procedures [14]. Besides *in situ* algal bloom observations since 1998, chronological HAB events in the Indian EEZ reported by earlier researchers were also incorporated for analyzing the trend of HAB events from 1917 onwards.

3. Results and Discussion

Altogether, 422 species of microalgae were recorded from EEZ of India during the investigation. Among them, there were 219 species of diatoms, 179 species of dinoflagellates, 16 species of blue green algae, and 8 other groups (silicoflagellates, chlorophytes, coccolithophorids, raphidophytes, haptophytes, and prasinophytes). Around 35 species of harmful microalgae have been recognized during the one-decade-long study period. Among the bloom forming species, during favourable conditions only a few species formed blooms. Around 80 algal blooms were recorded from the Indian EEZ during the period. Of the eighty algal blooms, 31 blooms were formed by dinoflagellates, cyanobacteria formed 27, and diatoms in 18 blooms. Three raphidophyte and one haptophyte blooms were also observed. The percentage composition of blooms by various groups of microalgae during the period 1998–2010 is presented in Figure 1. Red *Noctiluca scintillans* (without endosymbiont *Pedinomonas noctilucae*) was the dominant and frequently occurring blooming dinoflagellate in the South Eastern Arabian Sea (SEAS) during the summer monsoon and green *Noctiluca scintillans* (with endosymbiont *Pedinomonas noctilucae*) in the North Eastern Arabian Sea (NEAS) during the winter cooling. Other commonly occurring bloom forming dinoflagellates were *Cochlodinium* sp., *Gymnodinium* sp., *Gonyaulax* sp., and *Ceratium* spp. The study reveals that the *Trichodesmium* bloom occurred annually, whereas the *Noctiluca* bloom appeared at irregular intervals.

Results obtained from the qualitative analysis of phytoplanktons over the years indicate that potentially toxic dinoflagellates were always present in the region. In particular, species responsible for several of the major shellfish poisoning syndromes have also been observed, these include *Alexandrium catenella*, *Gymnodinium* sp. (all implicated in Paralytic Shellfish Poisoning-PSP), *Dinophysis acuminata*, *D. caudata*, *D. fortii*, *D. miles*, *D. tripos* (all implicated in Diarrhetic Shellfish Poisoning (DSP) episodes), *Coolia monotis*, and *Prorocentrum lima* (causing Ciguatera poisoning). In addition, other harmful algal bloom species linked to fish and animal mortalities have been observed in Indian waters. *Chattonella marina*, *Cochlodinium polykrikoides*,

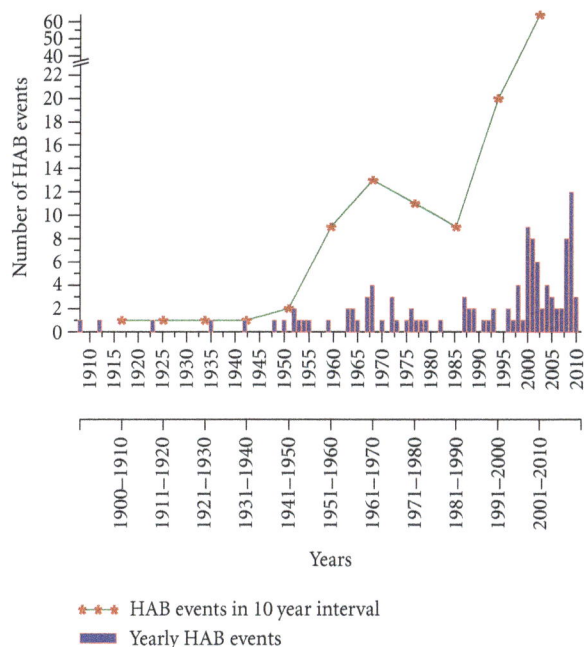

Legend:
- ★★★ HAB events in 10 year interval
- Yearly HAB events

FIGURE 2: Frequency of occurrence of harmful algal blooms during the last century.

Gonyaulax sp., *Ceratium* sp., and several *Prorocentrum* species were found abundantly in the phytoplankton population. Notwithstanding these observations, only two cases of PSP [9, 11] and one case of DSP [15] were recorded earlier from the Indian waters.

The nature of the occurrence, preponderance, and successful existences for short duration of HABs has undergone noticeable changes during the last decade in the Indian EEZ. Previous reports of algal bloom occurrence in the Indian EEZ since 1917 and *in situ* bloom observations during 1998 to 2010 were analysed to explain whether there is an increase in the HAB events, the results obtained are presented in Figures 2, 3, and 4. A major change in the HAB studies in the Indian Seas in the last two decades is the conduct of detailed ocean cruises to delineate the areas in the exclusive economic zone where the HAB events occur. This was due to the occurrence of shellfish poisoning events in the coastal communities due to the consumption of contaminated shellfish [9]. The media has become very vigilant and occurrence of toxic/harmful algal blooms is immediately reported by the local press in all coastal states especially Kerala.

Atmospheric nitrogen fixing *Trichodesmium erythraeum* is the only bloom forming cyanophycean member in the Indian waters. Presence of *Trichodesmium* blooms along the southwest coast during the premonsoon indicates the nitrogen recycling process in the ocean. Several reports [16–22] explained the regular occurrence of *Trichodesmium* blooms in the Arabian Sea during the period from January to June. From the present investigation it has come to light that the occurrence and intensity of *Trichodesmium* blooms has increased in the recent past in Indian EEZ especially in the Arabian Sea. According to Naqvi and Sen Gupta [23] Arabian Sea became nitrate deficient due to denitrifications. Nitrogen

limitation in upwelling regions is a natural consequence of nitrogen gas loss through N_2 production [24–26]. Therefore, the regions of intense upwelling generate oxygen depleted waters with N_2 production in the subsurface waters. Hence the primary production in surface waters at these areas is often nitrogen limited probably triggering blooms of cyanobacteria species capable of fixing atmospheric nitrogen. According to Prasannakumar et al. [27] there is an increase in the SST of Arabian Sea (ca 0.2°C) and is attributed to climate change and global warming. Generally the bloom of this filamentous algae occurred during hot weather with brilliant sunlight and stable high salinity [28, 29]. It is clear from these reports and our *in situ* studies that the Arabian Sea has become both nitrogen deficient and warmer, causing recurrent blooming of *Trichodesmium* in the area.

Reports on diatom blooms were meagre in the Indian EEZ during the period 1917–1997 [30–35]. In the present investigation (1998–2010) diatom blooms were observed in a number of occasions in the coastal waters and these were caused by *Coscinodiscus* spp., *Proboscia alata*, *Rhizosolenia* spp., *Fragilaria oceanica*, *Thalassiosira* spp., *Chaetoceros* spp., and *Asterionella japonica*. Upwelling, formation of mud banks, estuarine discharge, and land run off during southwest and northeast monsoons cause eutrophication in the coastal waters triggering the flowering of diatoms which responds faster to eutrophication. Toxic diatom species *Psuedo-nitzschia multiseries*, *P. australis* and *P. seriata*, collected during the investigation have the ability to produce potent neurotoxin Domoic acid. *Psuedo-nitzschia* sp. is known to produce domoic acid (DA), a potent neurotoxin that can be devastating to aquatic life via trophic transfer in the food web [36–38]. In humans, DA exposure manifests itself as amnesic shellfish poisoning (ASP) following the consumption of contaminated filter-feeding molluscs [39, 40]. Toxin production in *Pseudo-nitzschia* species has been found to show regional variations. Thus the same *Pseudo-nitzschia* species may be toxic in one part of the world but not in the other [37]. During this study *Pseudo-nitzschia* sp. was recorded from Malabar coast which is a famous shellfish harvesting ground along the west coast of India. The occurrence of toxigenic *Pseudo-nitzschia* species along our coasts is of concern, as it is possible that *Pseudo-nitzschia* spp. in the diet can result in the accumulation of domoic acid in the wild and cultured population of bivalve molluscs of the coast. Inshore waters of southwest coast of India is recently (2008 and 2009) experiencing high chlorophyll *a* concentration (10–25 mg m^{-3}), due to the localized blooms of diatoms during the southwest monsoon period from late May to October (from FORV Sagar Sampada Data centre). These are monospecific and mixed blooms by the diatom genera such as *Chaetoceros*, *Asterionella*, *Thalassiosira*, *Skeletonema*, *Thalassiothrix*, *Rhizosolenia* and *Proboscia*.

The raphidophyte flagellate, *Chattonella marina* (Subrahmanyan) Hara et Chihara is a well-known causative organism of red tides and associated mass mortality of marine fauna throughout the world oceans [41, 42]. The bloom forming harmful raphidophycean species *C. marina*, produces neurotoxin, which in nanomolecular concentrations is toxic to fishes. During the present study *C. marina* bloom

(a) 1917–1957

(b) 1958–1997

(c) 1998–2010

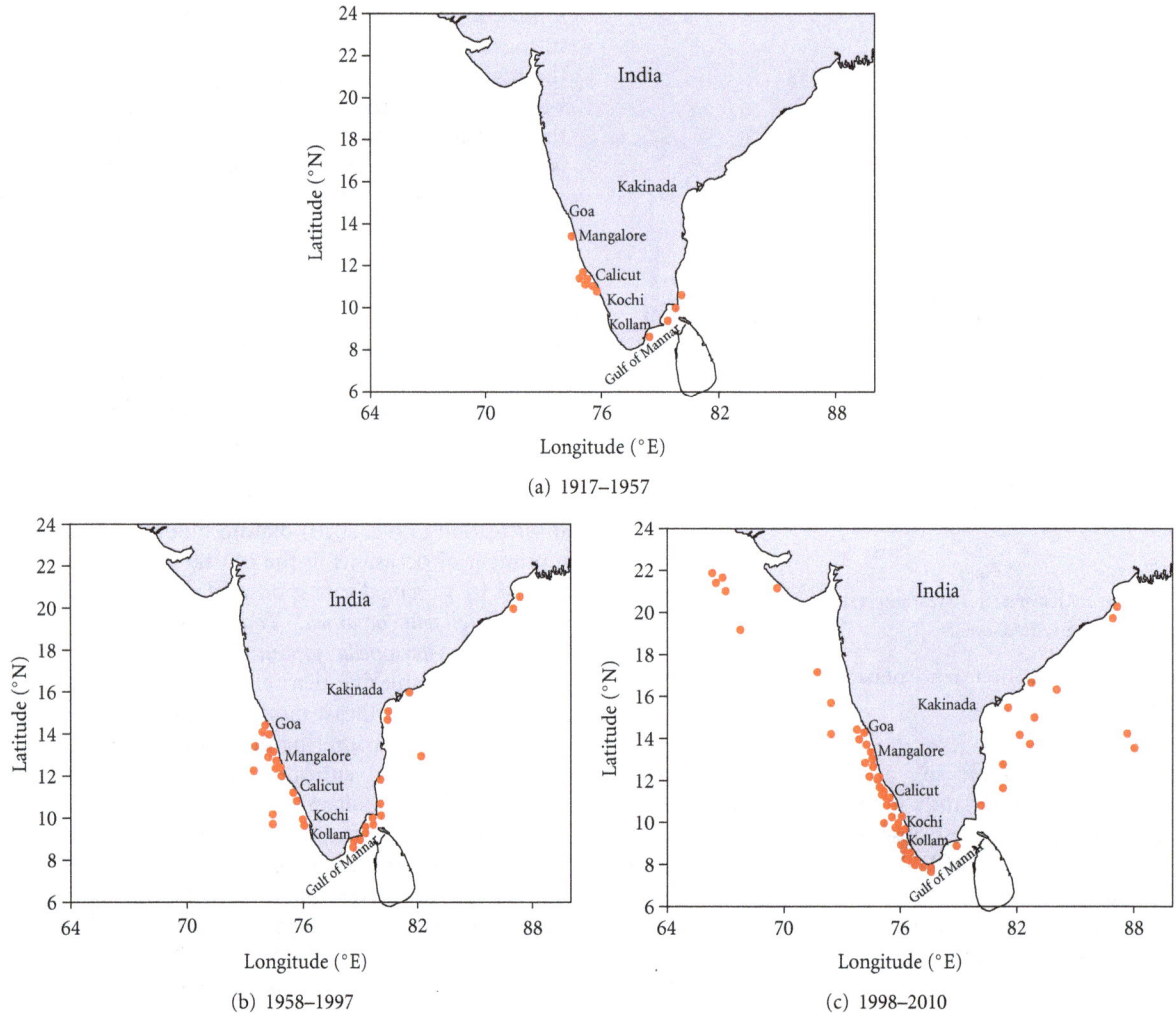

FIGURE 3: Trend figures indicating increasing incidence of harmful algal blooms in the Indian EEZ from 1917 to 2010 (red dots represent each algal blooms observed).

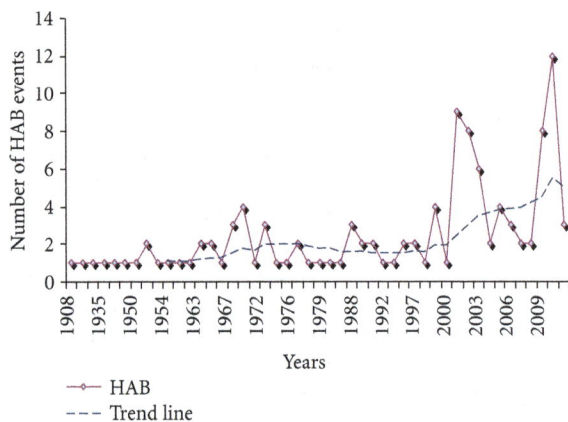

FIGURE 4: Trend line and harmful algal bloom events (actual observation) from 1908 to 2010.

and associated mortality of marine fauna was observed off Kochi during the monsoon season of 2002 and 2009. All these recorded toxic microalgae are capable of producing toxins at very low densities (100 to 1000 cells L^{-1}). The employment of scanning electron microscopy and transmission electron microscopy has been resorted to identify the species more precisely. The species capable of producing toxins are of nanoplankton size (<20 μm) and demands critical examination by electronmicroscopy. This is highly essential because the study of shellfish poisoning in the laboratory could be carried out only if the species capable of producing such toxins are identified [43]. These developments in the recognition of HABs in the Indian waters have created very good awareness about this phenomenon among the coastal community. Around 3.5 million fisherfolks depend on the fish and fish products for their livelihood in the region.

The dinoflagellate genus *Pyrophacus* has three species with comparable cellular morphology and cosmopolitan distribution. One of the species *Pyrophacus steinii* (Schiller) Wall and Dale is reported from Indian EEZ in sparse quantities. A subsurface bloom of *Pyrophacus steinii* was recorded off Mangalore during the fag end of summer monsoon (September-October, 2009). The presence of the bloom was not reflected by any conspicuous surface water discolouration. A high cell density of 7.75 × 10^5 cells L^{-1}

was recorded in the entire bloom area which was spread over a large area 10 m below surface. *P. steinii* cells are flattened, lenticular with attenuated epitheca and numerous golden brown chloroplasts. The phytoplankton population of the bloom area was composed of *Pyrophacus steinii* (94%) and other phytoplankters such as *Chaetoceros* spp., *Thalassiosira* sp., *Rhizosolenia* spp., *Pseudo-nitzschia* sp., and *Ceratium* spp. although present only in very low numbers. The chlorophyll *a* concentration of the bloom area was 26.47 μg L^{-1}. Nutrient concentration of the bloom area was 0.04 μmol L^{-1} NO$_2$-N, 0.13 μmol L^{-1} NO$_3$-N, 1.72 μmol L^{-1} PO$_4$-P, and 3.90 μmol L^{-1} SiO$_4$. There is no report about the harmful effects of this species. This observation on *Pyrophacus steinii* bloom in the Arabian Sea is unique in character and is the first record of its kind from the world oceans.

An examination of the trend line clearly depicts around 15% increase in the frequency of algal blooms in the Indian seas during the last twelve years. Further, the occurrence of algal blooms in the Arabian sea and Bay of Bengal shows that Arabian sea experiences more number of algal blooms. Bay of Bengal has recorded bloom by and large during the northeast monsoon period when cyclonic storms occur in this region. This probably is indicative of the fact that should the wind effects affect the circulation pattern of the upper ocean layers, global warming and the resultant storminess could influence the frequency of bloom formation in the Indian seas. Qualitative studies from 1998–2010 have shown that at least two species hitherto not recorded as bloom formers have staged their appearance. This could be an instance of blooming of innocuous species which could be due to eutrophication of coastal waters or introduction of bloom formers in the open sea due to ballast water discharge.

4. Conclusion

Phytoplankton blooms are known from ancient times but it requires an effort and expertise to properly investigate and accurately identify the bloom forming species. Harmful algal blooms have become a global epidemic. Increase in harmful algal bloom events in the Indian waters could be due to increase in anthropogenic enrichment of the coastal environment or other eutrophication such as increased coastal aquacultural farming, discharge of ballast water drawn from coastal ports, and so forth. The types of resources affected and the quantity of toxins and toxic species encountered have increased. The results clearly indicate that occurrence of HABs is on the rise in quality and quantity in the recent years. The time and duration that HABs occur annually may also expand as a result of climate change. Water temperature when on the rise can promote growth of toxic algae especially dinoflagellates. No serious attempts have been made so far, to find out the influence of ballast water discharge on the occurrence of HABs, but for a record by Madhu et al. [44]. The extensive occurrence of a single species both spatially and temporally necessitates investigation on the possibility of physiological races of the same species occurring in a wider area [43]. In the case of

the west coast of India, considerable increase in the number of blooms that appeared during the last decade clearly points towards the necessity for closer monitoring of the hydrology of the region which is one of the richest habitat for pelagic fishery.

Dinoflagellates are capable of movement and hence can seek deeper layers for nutrients to overcome surface nutrient impoverisation. This may lead to increased number of HABs as a result of climate change when upper stratification of surface layers occurs. Of late epidemiological studies to demonstrate the effects of algal toxins on human health have been undertaken. However, laboratory experiments to study the toxicity of bloom forming algae on shellfish from Indian waters is totally lacking and requires immediate attention as shellfish culture along the Indian coast is increasing rapidly. It is necessary to accurately identify and establish early monitoring programmes in order to decrease the negative impact of HAB. The effects of these blooms in seafood quality, economic liabilities on fisherfolks, correct assessment of factors triggering blooms, and subsequent effects in the trophodynamics are areas which need elaborate studies.

Acknowledgments

The authors are thankful to all the participants of FORV Sagar Sampada cruises from 1998 to 2010. We are also grateful for the comments and suggestions made by anonymous reviewers who helped to improve the manuscript. This observation was made under the Marine Living Resources Programme funded by the Ministry of Earth Sciences, Government of India, New Delhi.

References

[1] ICES, "Report of the ICES special meeting on the causes, dynamics and effects of exceptional marine blooms and related events," International Council Meeting Paper 1984/ E, 42, ICES, 1984.

[2] T. J. Smayda and A. W. White, "Has there been a global expansion of algal blooms? If so is there a connection with human activities?" in *Toxic Marine Phytoplankton*, E. Granelli, Ed., p. 516, Elsevier, 1990.

[3] G. M. Hallegraeff, "A review of harmful algal blooms and their apparent global increase," *Phycologia*, vol. 32, no. 2, pp. 79–99, 1993.

[4] F. M. Van Dolah, "Marine algal toxins: origins, health effects, and their increased occurrence," *Environmental Health Perspectives*, vol. 108, no. 1, pp. 133–141, 2000.

[5] J. Hornell, "A new protozoan cause of widespread mortality among marine fishes," in *Madras Fisheries Investigations Bulletin*, vol. 11, pp. 53–56, 1917.

[6] R. Subrahmanyan, "On the life history and ecology of *Hornellia marina* gen. et sp. Nov., (Chloromonadineae), causing green discoloration of the sea and mortality among marine organisms off the Malabar coast," *Indian Journal of Fisheries*, vol. 1, pp. 182–203, 1954.

[7] Y. Hara and M. Chihara, "Ultra structure and taxonomy of *Chattonella* (class Raphidophyceae) in Japan," *Japanese Journal of Phycology*, vol. 30, pp. 47–56, 1982.

[8] V. P. Devassy and S. R. Bhat, "The killer tides," *Science Reporter*, vol. 28, no. 5, pp. 16–19, 1991.

[9] I. Karunasagar, H. S. V. Gowda, M. Subhurah, M. N. Venugopal, and I. Karunasagar, "Outbreak of paralytic shellfish poisoning in Mangalore, west coast of India," *Current Science*, vol. 53, p. 247, 1984.

[10] A. Godhe and I. Karunasagar, "*Gymnodinium catenatum* on west coast of India," *Harmful Algae News*, vol. 15, p. 1, 1996.

[11] I. Karunasagar, B. Joesph, K. K. Philipose, and I. Karunasagar, "Another outbreak of PSP in India," *Harmful Algae News*, vol. 17, p. 1, 1998.

[12] C. R. Tomas, *Identifying Marine phytoplankton*, Academic press, New York, NY, USA, 1997.

[13] T. R. Parsons, Y. Maita, and C. M. Lalli, *A Manual of Chemical and Biological Methods for Seawater Analysis*, Pergamon Press, New York, NY, USA, 1984.

[14] UNESCO, "Protocols for the Joint Global Ocean Flux Study (JGOFS), core measurements," in *IOC Manual and Guides*, vol. 29, p. 170, 1994.

[15] R. Santhanam and A. Srinivasan, "Impact of dinoflagellate *Dinophysis caudata* bloom on the hydrography and fishery potentials of Tuticorin Bay, South India," in *Harmful and Toxic Algal Blooms*, T. Yasumoto, Y. Oshima, and Y. Fukuyo, Eds., pp. 41–44, IOC-UNESCO, Sendai, Japan, 1996.

[16] N. K. Panikar, "Indian fisheries," *Current Science*, vol. 28, no. 12, pp. 53–54, 1959.

[17] A. K. Nagabhushanam, "On an unusually dense phytoplankton bloom around Minicoy Island (Arabian Sea), and its effect on the local tuna fisheries," *Current Science*, vol. 36, no. 22, pp. 611–612, 1967.

[18] S. Z. Qasim, "Some characteristics of a *Trichodesmium* bloom in the Laccadives," *Deep Research-II*, vol. 17, no. 3, pp. 655–660, 1970.

[19] R. K. Sarangi, P. Chauhan, and S. R. Nayak, "Detection and monitoring of *Trichodesmium* blooms in the coastal waters off Saurashtra coast, India using IRS-P4 OCM data," *Current Science*, vol. 86, no. 12, pp. 1636–1641, 2004.

[20] R. K. Sarangi, P. Chauhan, S. R. Nayak, and U. Shreedhar, "Remote sensing of *Trichodesmium* blooms in the coastal waters off Gujarat, India using IRS-P4 OCM," *International Journal of Remote Sensing*, vol. 26, no. 9, pp. 1777–1780, 2005.

[21] A. A. Krishnan, P. K. Krishnakumar, and M. Rajagopalan, "*Trichodesmium erythraeum* (Ehrenberg) bloom along the Southwest coast of India (Arabian Sea) and its impact on trace metal concentrations in seawater," *Estuarine, Coastal and Shelf Science*, vol. 71, no. 3-4, pp. 641–646, 2007.

[22] K. B. Padmakumar, B. R. Smitha, L. C. Thomas et al., "Blooms of *Trichodesmium erythraeum* in the South Eastern Arabian Sea during the onset of 2009 summer monsoon," *Ocean Science Journal*, vol. 45, no. 3, pp. 151–157, 2010.

[23] S. W. A. Naqvi and R. Sen Gupta, "NO, a useful tool for the estimation of nitrate deficits in the Arabian Sea," *Deep Sea Research Part A*, vol. 32, no. 6, pp. 665–674, 1985.

[24] R. Dugdale, J. Goering, R. Barber, R. Smith, and T. Packard, "Denitrification and hydrogen sulfide in the Peru upwelling region during 1976," *Deep Sea Research*, vol. 24, no. 6, pp. 601–608, 1977.

[25] C. Mantoura, R. F. C. Fauzi, C. S. Law et al., "Nitrogen biogeochemical cycling in the Northwestern Indian Ocean," *Deep Sea Research Part II*, vol. 40, no. 3, pp. 651–671, 1993.

[26] A. G. Davies and C. E. Morales, "An appraisal of the stoichiometry of dissolved oxygen/nutrient inter-relationships in the upwelling system off Northern Chile," *Journal of the Marine Biological Association of the United Kingdom*, vol. 78, no. 3, pp. 697–706, 1998.

[27] S. Prasannakumar, R. P. Roshin, J. Narvekar, P. K. Dineshkumar, and E. Vivekanandan, "Response of the Arabian Sea to global warming and associated regional climate shift," *Marine Environmental Research*, vol. 68, no. 5, pp. 217–222, 2009.

[28] S. Suvapepant, "*Trichodesmium* blooms in gulf of Thailand," in *Marine Pelagic Cyanobacteria: Trichodesmium and Other Diazotrophs*, E. J. Carpenter, D. G. Capone, and J. G. Rueter, Eds., pp. 343–348, Kluwer Academic publisher, 1992.

[29] K. G. Sellner, "Physiology, ecology, and toxic properties of marine cyanobacteria blooms," *Limnology and Oceanography*, vol. 42, no. 5, pp. 1089–1104, 1997.

[30] D. V. S. Rao, "*Asterionella japonica* bloom and discolouration off Waltair, Bay of Bengal," *Limnology and Oceanography*, vol. 14, pp. 632–634, 1969.

[31] A. Subramanian and A. Purushothaman, "*Hemidiscus hardmanianus* bloom and fish mortality," *Limnology and Oceanography*, vol. 30, no. 4, pp. 910–911, 1985.

[32] K. K. Satpathy and K. V. K. Nair, "Occurrence of phytoplankton bloom and its effect on coastal water quality," *Indian Journal of Marine Sciences*, vol. 25, no. 2, pp. 145–147, 1996.

[33] S. B. Choudhury and R. C. Panigrahy, "Occurrence of bloom of diatom *Asterionella glacialis* in near shore waters of Gopalpur, Bay of Bengal," *Indian Journal of Marine Sciences*, vol. 18, pp. 204–206, 1989.

[34] R. C. Panigrahy and R. Gouda, "Occurrence of bloom of the diatom *Asterionella glacialis* (Castracane) in the Rushikulya estuary, East Coast of India," *Mahasagar*, vol. 23, no. 2, pp. 179–182, 1990.

[35] S. Mishra and R. C. Panigrahy, "Occurrence of diatom blooms in Bahuda estuary, east coast of India," *Indian Journal of Marine Sciences*, vol. 24, pp. 99–101, 1995.

[36] L. Fritz, M. A. Quilliam, J. L. C. Wright, A. M. Beale, and T. M. Work, "An outbreak of domoic acid poisoning attributed to the pennate diatom *Pseudonitzschia australis*," *Journal of Phycology*, vol. 28, no. 4, pp. 439–442, 1992.

[37] S. S. Bates, D. L. Garrison, and R. A. Horner, "Bloom dynamics and physiology of domoic acid- producing *Pseudo-nitzschia* species," in *Physiological Ecology of Harmful Algal Blooms*, D. M. Anderson, A. D. Cembella, and G. M. Hallegraeff, Eds., p. 267, Springer, Heidelberg, Germany, 1998.

[38] C. A. Scholin, F. Gulland, G. J. Doucette et al., "Mortality of sea lions along the central California coast linked to a toxic diatom bloom," *Nature*, vol. 403, no. 6765, pp. 80–83, 2000.

[39] S. S. Bates, C. J. Bird, A. S. W. de Freitas et al., "Pennate diatom *Nitzschia pungens* as the primary source of domoic acid, a toxin in shellfish from eastern Prince Edward Island, Canada," *Canadian Journal of Fisheries and Aquatic Sciences*, vol. 46, no. 7, pp. 1203–1215, 1989.

[40] V. L. Trainer, W. P. Cochlan, A. Erickson et al., "Recent domoic acid closures of shellfish harvest areas in Washington State inland waterways," *Harmful Algae*, vol. 6, no. 3, pp. 449–459, 2007.

[41] Y. Onoue and K. Nozawa, "Separation of toxin from harmful red tides occurring along the coast of Kagoshima Prefecture," in *Red Tides-Biology, Environmental Science and Toxicology*, T. Okaichi, D. M. Anderson, and T. Nemeto, Eds., pp. 371–374, Elsevier, New York, NY, USA, 1989.

[42] T. Oda, A. Ishimatsu, S. Takeshita, and T. Muramatsu, "Hydrogen peroxide production by the red tide flagellate *Chattonella marina*," *Bioscience Biotechnology and Biochemistry*, vol. 58, pp. 957–958, 1994.

[43] K. B. Padmakumar, *Algal blooms and Zooplankton standing crop along the South West coast of India*, Ph.D. thesis, Cochin University of Science and Technology, Kerala, India, 2010.

[44] N. V. Madhu, P. D. Reny, M. Paul, N. Ullas, and P. Resmi, "Occurrence of red tide caused by *Karenia mikimotoi* (toxic dinoflagellate) in the Southwest Coast of India," *Indian Journal of Geo-Marine Sciences*, vol. 40, no. 6, pp. 821–825, 2011.

Predicting Sea Surface Temperatures in the North Indian Ocean with Nonlinear Autoregressive Neural Networks

Kalpesh Patil,[1] M. C. Deo,[1] Subimal Ghosh,[1] and M. Ravichandran[2]

[1] Indian Institute of Technology, Bombay, Mumbai 400 076, India
[2] Indian National Centre for Ocean Information Services, Hyderabad 500090, India

Correspondence should be addressed to M. C. Deo; mcdeo@civil.iitb.ac.in

Academic Editor: Stefano Vignudelli

Prediction of monthly mean sea surface temperature (SST) values has many applications ranging from climate predictions to planning of coastal activities. Past studies have shown usefulness of neural networks (NNs) for this purpose and also pointed to a need to do more experimentation to improve accuracy and reliability of the results. The present work is directed along these lines. It shows usefulness of the nonlinear autoregressive type of neural network vis-à-vis the traditional feed forward back propagation type. Neural networks were developed to predict monthly SST values based on 61-year data at six different locations around India over 1 to 12 months in advance. The nonlinear autoregressive (NAR) neural network was found to yield satisfactory predictions over all time horizons and at all selected locations. The results of the present study were more attractive in terms of prediction accuracy than those of an earlier work in the same region. The annual neural networks generally performed better than the seasonal ones, probably due to their relatively high fitting flexibility.

1. Introduction

The temperature of water at around 1 m below the ocean surface, commonly referred to as sea surface temperature (SST), is an important parameter to understand the exchange of momentum, heat, gases, and moisture across air-sea interface. Its knowledge is necessary to explain and predict important climate and weather processes including the summer monsoon and El-Nino events. SST predictions are sought after by the users of coastal communities dealing with fishing and sports. Like the air above it SST changes significantly over time, although relatively less frequently due to a high specific heat. The changes in water temperature over a vertical are high at the sea surface due to large variations in the heat flux, radiation, and diurnal wind near the surface, and hence SST estimations involve considerable amount of uncertainty.

There are a variety of techniques for measuring SST. These include the thermometers and thermistors mounted on drifting or moored buoys and remote sensing by satellites. In case of satellites the ocean radiation in certain wavelengths of an electromagnetic spectrum is sensed and related to SST. Microwave radiometry based on an imaging radiometer called the moderate resolution imaging spectroradiometer is also popularly used to record SST.

In order to predict SST physicallybased as well as data driven methods are practiced. The latter type is many times preferred when site specific information is required and considering the convenience. The data driven schemes include a seasonally varying Markov model (Wu, [1]), an analogue method that searches for a similar time progression from the past (Xue and Leetmaa, [2]), empirical models designed through canonical correlation analysis (Agarwal et al., [3]), a regression model based on a lagged relationship (Laepple et al. [4]), models based on hurricane numbers (Neetu et al., [5]), and genetic algorithms and empirical orthogonal functions (Wu et al., [6]). One of the most popular methods in modern data driven approaches however is neural network (NN), also called artificial neural network. Some investigators have in recent past applied this technique to predict the SST as described below.

2. SST Predictions Using Neural Networks

Tripathi et al. [7] produced seasonal predictions of SST over a certain region in the tropical Pacific using an NN that had the input of seven modes of the initial wind stress empirical orthogonal function and also that of the SST anomalies themselves. Over a lead time of 6 months the NN models yielded predictions with a similar level of accuracy as that of the El Nino Southern Oscillation (ENSO) based models. The NN worked satisfactorily even up to a 12-month prediction horizon in ENSO predictions. A comparison of NN with canonical correlation analysis and a sophisticated version of linear statistical regression in predicting equatorial Pacific SST was made by Tanvir and Mujtaba [8], who did not find significant difference in prediction skills probably because of the linearity of the dynamics at seasonal scales. The study of Pozzi et al. [9] indicated usefulness of NN as a complementary tool to more conventional approaches in SST and paleoceanographic data analysis. Wasserman [10] predicted SST anomalies over the tropical Pacific using NN. Authors derived the SST principal components over a 3- to 15-month lead time using the input of SST anomalies and sea level pressure. Garcia-Gorriz and Garcia-Sanchez [11] developed many NN based correlations for predicting the temperature elevation of sea water during desalination.

SST data for a certain region in the Indian Ocean were analyzed by Tripathi et al. [7]. Using area-average SST twelve different NN models were developed for the twelve months in a year. Typically the model to predict SST for the month of January was based on all past observations of January. On evaluating the performance of the networks the authors found that the models were able to predict the anomalies with a good accuracy and whenever the dependence of present anomalies on past anomalies was nonlinear the NN models worked better than the linear statistical models. Collins et al. [12] used meteorological variables as input to predict targeted satellite-derived SST values in the western Mediterranean Sea. The networks trained in this way predicted the seasonal as well as the interannual variability of SST well. The impact of the heat wave that occurred during the summer of 2003 on SST was also reproduced satisfactorily. Kug et al. [13] compared prediction skills of different methods based on certain transfer functions, regressions, and NN. Authors analyzed data of radiolarian faunal abundance from surface sediments observed at the Antarctic and Pacific Oceans. The error statistics associated with the NN predictions were found to be more attractive than the other methods, although NN yielded lesser geographic trends than those of the other methods.

The prediction skills of a Bayesian NN with support vector machine and linear regression were also compared by Tang et al. [14]. Authors used SST and sea level pressure together with warm water volume across the equatorial Pacific as input. It was found that nonlinear methods were better than the linear ones, but support vector machine could not produce better overall predictions than the NN. Tanang et al. [15] used NN for identifying sources of errors in satellite-based SST estimates. The temperature of air, direction of wind, and relative humidity affected the SST derivations significantly.

A review of the readily available publications as above showed that the technique of NN is promising in predicting SST but needs more experimentation to improve on the accuracy and reliability of the results. The present work is directed along these lines. It shows usefulness of the nonlinear autoregressive type of neural network vis-à-vis the traditional feed forward back propagation type. Further, it deals with the actual numerical values of SST rather than the anomalies obtained by subtracting long-term means, since neural networks implicitly take care of such means in their training procedures and since the data used were result of reanalysis and thus bias corrected. Further, the range of data values was also not small enough (up to around 6 degrees), necessarily requiring the use of mean subtracted values. The prediction of such SST values, rather than anomalies, could have applications in planning coastal activities such as sports events and fishing expeditions and also in calibration of satellite measurements.

3. The SST Data

The data used in this study were made available by Indian National Centre for Ocean Information Services (INCOIS) and pertained to monthly mean SST at six different locations in the North Indian Ocean, code named as AS, BOB, EEIO, SOUTHIO, THERMO, and WEIO as shown in Figure 1. These data were collected from various sources such as voluntary observing ships, moorings, drifters, and Argo. The gaps in observations were filled up by suitable interpolation methods at the source level. The data involves NCEP/NCAR global reanalysis products generated through assimilation and model and uses all ship and buoy SSTs and satellite derived SSTs from the NOAA Advanced Very High Resolution Radiometer (AVHRR). The duration ranged from January 1945 to December 2005. There were thus 732 values over the period of 61 years.

Basic data statistics, namely, the minimum, maximum, and mean values, standard deviation, kurtosis, and skewness can be seen in Table 1. As can be seen from the table the temperatures varied from 23.61°C to 30.67°C at the different locations. All of the locations had comparable maximum values while the locations near the land boundary (AS, BOB, and THERMO) had lower minimum values than those in the open ocean and near the equator (EEIO and WEIO). Expectedly the sites of EEIO and WEIO that are closer to the equator had higher means. While the minimum temperatures showed relatively large changes (from 23.61°C to 27.34°C) the standard deviations at locations AS, BOB, and THERMO were higher than those at sites EEIO, SOUTHIO, and WEIO, which could be due to their coastal proximity and shallower depths. Most of the data with the exception of EEIO were associated with negative skewness as well as kurtosis values indicating concentration of data to the right of the mean in the frequency distributions and that such distributions were flatter than the normal distribution with wider peaks and spreads around the mean, respectively. These complexities

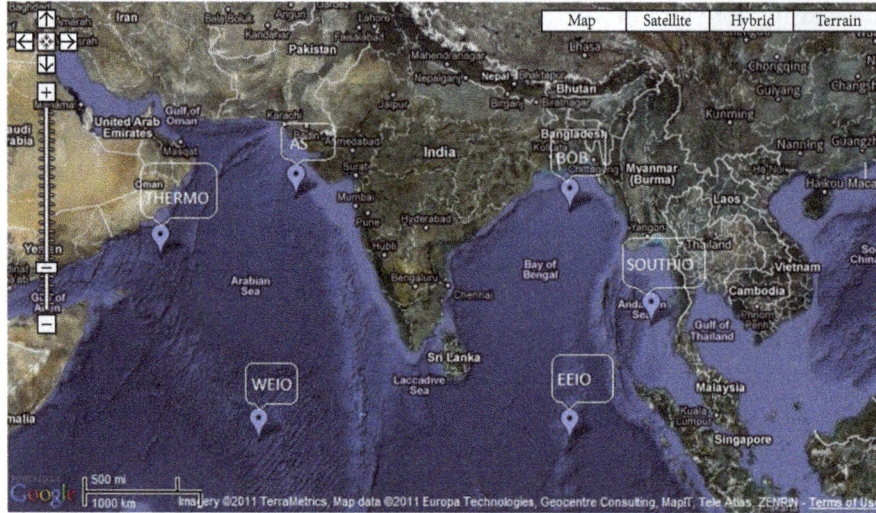

FIGURE 1: The study locations around India (http://itouchmap.com/latlong.html/) (latitudes and longitudes in degrees, respectively THERMO (16S to 14S and 56E to 58E); WEIO (1S to 1N and 65E); AS (19N to 20N and 68E); BOB (18N to 19N and 90E); EEIO (1S to 1N and 90E); SOUTHIO (9S to 11S and 95E to 98E)).

TABLE 1: Statistics of SST at various locations.

Site	Minimum (°C)	Maximum (°C)	Mean (°C)	Std. Dev. (°C)	Kurtosis	Skewness
AS	24.53	30.09	27.41	1.22	−0.83	−0.39
BOB	24.26	30.45	27.98	1.38	−0.94	−0.56
EEIO	27.60	30.56	28.85	0.55	−0.16	0.34
SOUTHIO	25.87	29.95	27.93	0.77	−0.62	−0.12
THERMO	23.61	29.75	26.47	1.53	−1.30	−0.12
WEIO	27.34	30.67	28.75	0.65	−0.64	0.27

indicate the necessity of application of a nonlinear technique like neural networks to predict future values. Neural networks do not require data preprocessing as a precondition for training and hence the same was not performed. Incidentally the presence of statistically significant trend in the observations was as well not seen.

4. Networks and Training

A neural network in general consists of interconnected neurons, each acting as an independent computational element. The most common network is in the form of a multilayered perceptron with ability to approximate any continuous function (see Figure 2). Basic information on neural networks can be seen in text books such as those of Martinez and Hseih [16] and Lee et al. [17]. The primary network operation is however mentioned below.

Each neuron of the hidden and output layer sums up the weighted input, adds a bias term to it, passes on the result through a transfer function, and produces the output. (Figure 2). Mathematically the four-step procedure followed in obtaining the network output is as given below.

(1) Sum up weighted inputs, that is,

$$\text{Nod}_j = \sum_{i=1}^{\text{NIN}} \left(W_{ij} x_i \right) + \beta_j, \tag{1}$$

where Nod_j = summation for the jth hidden node; NIN = total number of input nodes; W_{ij} = connection weight ith input and jth hidden node; x_i = normalized input at the ith input node; and β_j = bias value at the jth hidden node.

(2) Transform the weighted input:

$$\text{Out}_j = \frac{1}{\left[1 + e^{-\text{Nod}_j} \right]}, \tag{2}$$

where Out_j = output from the jth hidden node.

(3) Sum up the hidden node outputs:

$$\text{Nod}_k = \sum_{j=1}^{\text{NHN}} \left(W_{jk} \text{Out}_j \right) + \theta_k, \tag{3}$$

where Nod_k = summation for the kth output node; NHN = total number of hidden nodes; W_{jk} = connection weight between the jth hidden and kth output node; and θ_k = bias at the kth output node.

(4) Transform the weighted sum

$$\text{Out}_k = \frac{1}{\left[1 + e^{-\text{Nod}_j} \right]}, \tag{4}$$

where Out_k = output at the kth output node.

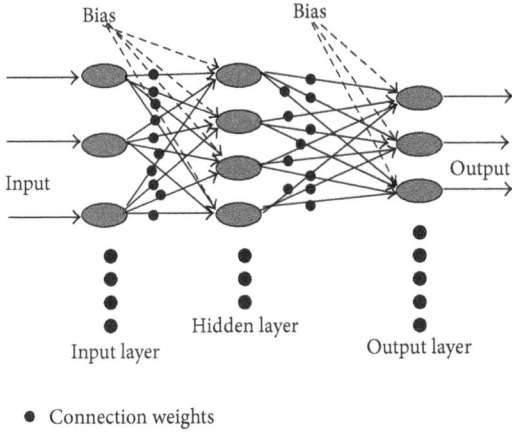

FIGURE 2: A feed forward neural network.

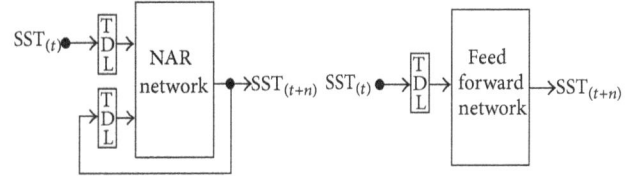

FIGURE 3: NAR architecture used for SST prediction.

Before its actual application the network has to be trained from examples or by presenting input and output pairs to the network and thereby determining the values of connection weights and bias through a training algorithm and over many epochs (presentation of complete data sets once to the network). The network is presented with the input-output pairs till the training error between target and realized outputs reaches the error goal or alternatively till no further reduction in the error is achieved despite increasing the number of epochs, as done in the present case.

In the present work SST values were predicted using a time series forecasting approach. A sequence of past values was fed as input to the network so as to enable it to recognize some hidden pattern in it and produce the forecast by moving along the time scale. The number of past values considered at a time equals the number of input neurons, while the output neuron belongs to the forecasted value over the next or any future time step. By trials aimed at getting the best outcome the past sequence length was selected as 24 months and predictions were made over a period of subsequent 12 months, but one at a time. Note that the total sample size was of 61×12 months as mentioned in the preceding section and out of it the sequence of past 24 months at every current time step (month) was used as input to the network in a sliding window manner. This is indicated mathematically as follows:

$$x(t + n)$$
$$= f[x(t), x(t-1), x(t-2), x(t-3), \ldots, x(t-23)]$$
$$\text{for } n = 1, 2, 3, \ldots, 12,$$
(5)

where $x(t + n) = $ SST at month $(t + n)$, $t = $ current month, and $n = 1, 2, 3, \ldots, 12$.

To begin with, the most common network of feed forward back propagation type (FFBP) (Figure 2) was used. However in order to see if more improved predictions were possible, another network called nonlinear autoregressive network (NAR) was also employed. The NAR is a recurrent network with feedback arrangement as shown in Figure 3 (left part), where the output is fedback to the input of the feed forward

network (right part). There is thus a feedback of the true output instead of the estimated one in the input. This makes the feed forward architecture more accurate and allows the use of only static back propagation while training. The networks were trained using the common algorithm of Levenberg-Marquardt. The feed forward network had sigmoid activation function for the hidden layer and a linear activation function for the output layer neurons. The number of input neurons was 24 while that of the output neurons was 1.

5. Testing of Networks and Results

In order to assess the performance of these networks the model predictions yielded by both FFBP and NAR were compared with actual observations for the testing period of 1997–2005, while the training was performed based on monthly SST for the period 1945–1996. As explained in the preceding section for any given time step (month) the preceding 24 months' sequence was used as network input during the training and testing exercise. The assessment of testing was done through time history comparisons and scatter diagrams, an example of which is given in Figure 4 for the location AS. The testing period ranged from 1997 to 2005. In this figure the two figures at the top show comparative predictions for the next and the next-to-next month, that is, 1 and 2 months ahead, while the bottom two figures indicate the same for the 11th and 12th months in future. These plots pertain to the application of the FFBP network. Similar comparisons based on the NAR network predictions are given in Figure 5 for the same location. Here also the testing period ranged from 1997 to 2005.

All the scatter diagrams and time history based plots as above for all the locations indicated that both FFBP and NAR were able to predict the SST values over the future time horizon varying from 1 to 12 months satisfactorily as reflected in the closeness between the observed and the predicted SST values. In order to further confirm this quantitatively and also to study the relative performance of both the networks error statistics of correlation coefficient (CC), mean absolute error (MAE), mean square error (MSE), and Nash-Sutcliffe efficiency (NSE) coefficient were worked out. The underlying expressions as well as the strengths and weaknesses of these parameters are given in the appendix.

As an example of magnitudes of the error measures of CC, MAE, MSE, and NSE and also their distribution over the lead time ranging from 1 to 12 months Table 2 can be seen. This information pertains to location AS and to the underlying network of NAR. It indicates that for all the lead times the CC values were high, varying from 0.95 to 0.99 and so also NSE

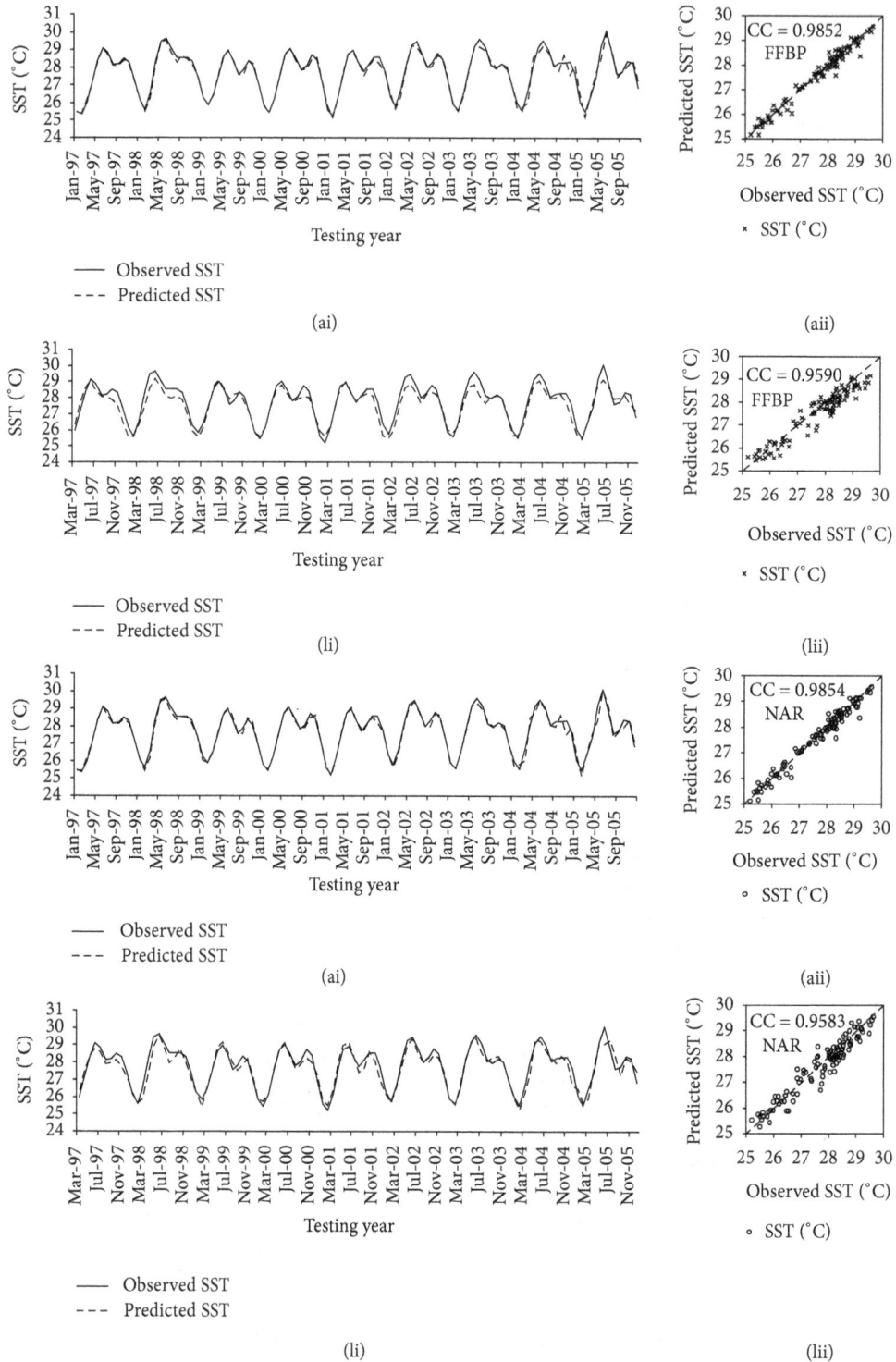

FIGURE 4: Time series of observed and predicted SST using FFBP network during testing (moving from top to bottom the prediction horizon changes as 1, 2, 11, and 12 months).

that changed from 89.20% to 96.90%. Further, the MAE and MSE were low and ranged from 0.15°C to 0.30°C and 0.04°C² to 0.15°C², respectively.

On the same lines it was noticed that for all the sites and over all prediction horizons the correlation coefficients and the NSE coefficients were high and close to 1.0, while the MAE and MSE were low. But this was true for the NAR network rather than the FFBP one. This can be seen in the example, Figure 6(a), showing the changes in MAE with increasing prediction horizons at locations AS and BOB

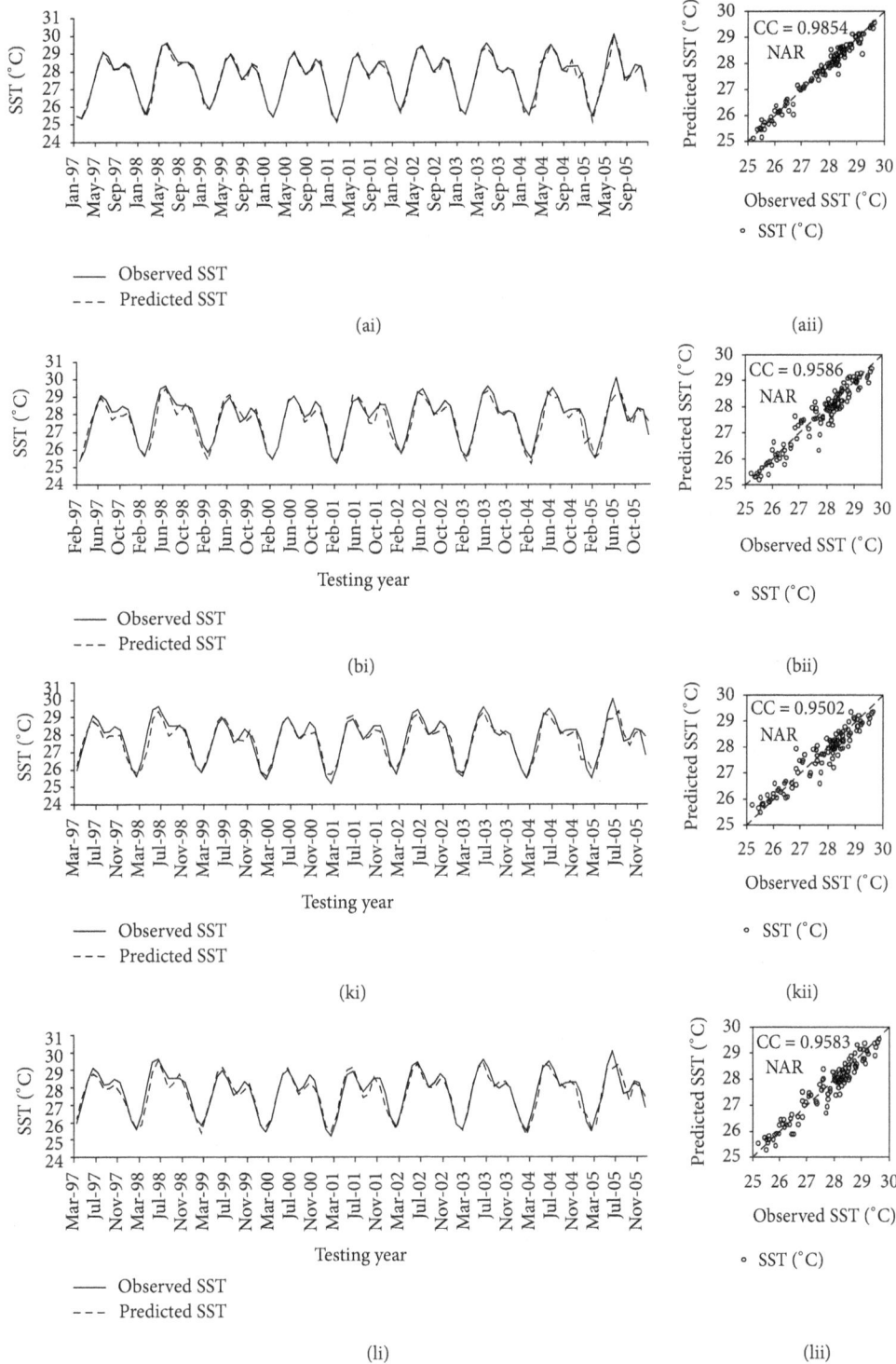

FIGURE 5: Time series of observed and predicted SST using NRA network during testing (moving from top to bottom the prediction horizon changes as 1, 2, 11, and 12 months).

and the same in NSE coefficient in Figure 6(b). The MAE gives an idea of the average error distributions and does not get influenced by their higher magnitudes. Figure 6(a) shows that it is very low in the first month of predictions but becomes high subsequently but gets confined to low values

of 0.30°C and 0.38°C at the two locations shown when the NAR network was used. The values over the next month are highly dependent on the same of the current month, and hence the MAE is high for the first month. Similar discussion is valid for the Nash-Sucliffie efficiency coefficient

TABLE 2: NAR network testing at site AS.

Prediction horizon in months	Testing phase CC	MAE in °C	MSE in °C²	NSE in %
$SST_{(t+1)}$	0.99	0.15	0.04	96.90
$SST_{(t+2)}$	0.96	0.28	0.13	90.70
$SST_{(t+3)}$	0.96	0.27	0.12	91.20
$SST_{(t+4)}$	0.95	0.29	0.15	89.50
$SST_{(t+5)}$	0.95	0.28	0.14	90.10
$SST_{(t+6)}$	0.96	0.28	0.14	90.20
$SST_{(t+7)}$	0.96	0.28	0.14	90.30
$SST_{(t+8)}$	0.96	0.30	0.14	89.90
$SST_{(t+9)}$	0.96	0.28	0.13	90.10
$SST_{(t+10)}$	0.95	0.30	0.15	89.20
$SST_{(t+11)}$	0.95	0.30	0.15	89.30
$SST_{(t+12)}$	0.96	0.28	0.12	91.00

Note: $SST_{(t+i)}$: SST at ith time (month) ahead than the present time (month) "t".

shown in Figure 6(b). The changes in the values of CC and MSE with increasing prediction horizons at sites EEIO, SOUTHIO, THERMO, and WEIO are shown in Figures 6(c) and 6(d), respectively. As regards the coefficient of correlation is concerned (Figure 6(c)) both FFBP and NAR networks showed good performance, but the NAR network were clearly advantageous than the FFBP, indicating their better capabilities to handle the data nonlinearities. The locations of EEIO and WEIO that were near the equator showed lesser prediction performance than the sites of SOUTHO and THERMO indicating high data nonlinearities that were somewhat difficult to model. Figure 6(d) further confirms the edge of NAR over FFBP through lower MSE values at all the locations. The NAR network predicted the SST values almost with equal efficiency at most of the locations.

Tripathi et al [7] had predicted monthly SST values for 12 months in advance over the Indian Ocean region using FFBP models. For this purpose the sample size was 52 years from which 12 different time series were formed, trained, and tested, unlike a sequential modeling done in this study based on 61 years' data. The testing in the present work has been done with the help of the data segment of last 10 years as against the last 5 data points of each month in the work of Tripathi et al. [7]. Table 3 shows a comparison of the present NAR based model (while testing) with this past study in terms of CC between the target and the predicted values. Although the locations were not the same generally better performance (higher CC values) of the present modeling can be noticed. The past study under reference did not indicate systematic lowering of CC with an increasing lead time as probably can be expected and so also the present one; however it had outlying CC values of 0.59 and 0.76 which were not seen in this study indicating more consistent predictions. Table 3 also shows better performances at sites AS, BOB, and THERMO where means were lower and standard deviations were higher than the other sites, that might have made the model fitting more adaptable.

The above networks pertained to annual models where a past sequence of 24 months was used to predict SST values for 12 months in advance. An alternative SST predictions were attempted based on only past season's data at the six locations. Four seasons, namely, winter (December, January, and February), summer (March, April, and May), monsoon (June, July, August, and September), and after monsoon (October and November) were considered. The monthly SST values for the next season were predicted based on the same past season and also the past month. It was however found that such seasonal predictions were not as good as those of the annual predictions described in all preceding sections. This can be seen from the example of Table 4 that shows the correlation coefficients between the predicted and target values typically at the SOUTHIO location for every calendar month based on predictions that were both season based and biannual data based. This could be due to a higher level of fitting flexibility involved when past 24 values were considered in training as against the 3 or so of the seasonal models. A larger training sequence can thus be recommended.

It is known that the climate systems of El Nino Southern Oscillation (ENSO) and Indian Ocean Dipole (IOD) are governed by higher or lower than normal SST in the Pacific and Indian Ocean, respectively. These phenomena should be discernable from the SST records. This was not attempted in the present work, since a correct way to do so would be to analyze the SST anomalies (with amplified differences) rather than the real SST values as considered here. Nonetheless a good match between the predicted and target SST values at the crests and troughs seen in the time history comparisons exemplified in Figure 5 is indicative of capturing the extremes associated with the large scale climate systems mentioned above.

It may be noted that neural networks are basically site-specific models, since they are built on data collected at a given location. Their spatial applicability would depend on the area around the particular location from where the SST values were binned.

The success of SST predictions using neural networks described in this paper may inspire application of neural networks to SST predictions over smaller time intervals such as weeks. It would also be useful to explore if other network types such as recursive networks and other training schemes like conjugate gradient decent and quasi-Newton result in better learning.

6. Conclusions

The nonlinear autoregressive (NAR) neural network was found to yield satisfactory predictions over all time horizons and at all selected locations and such predictions were superior to those based on the common FFBP type network architecture.

For most of the locations the error statistics indicated highly satisfactory performance of the NAR network with correlation coefficients between predicted and measured values lying above 0.90 and MSE, MAE less than 0.23°C and 0.38°C, respectively, and NSE coefficient of the order of 80%.

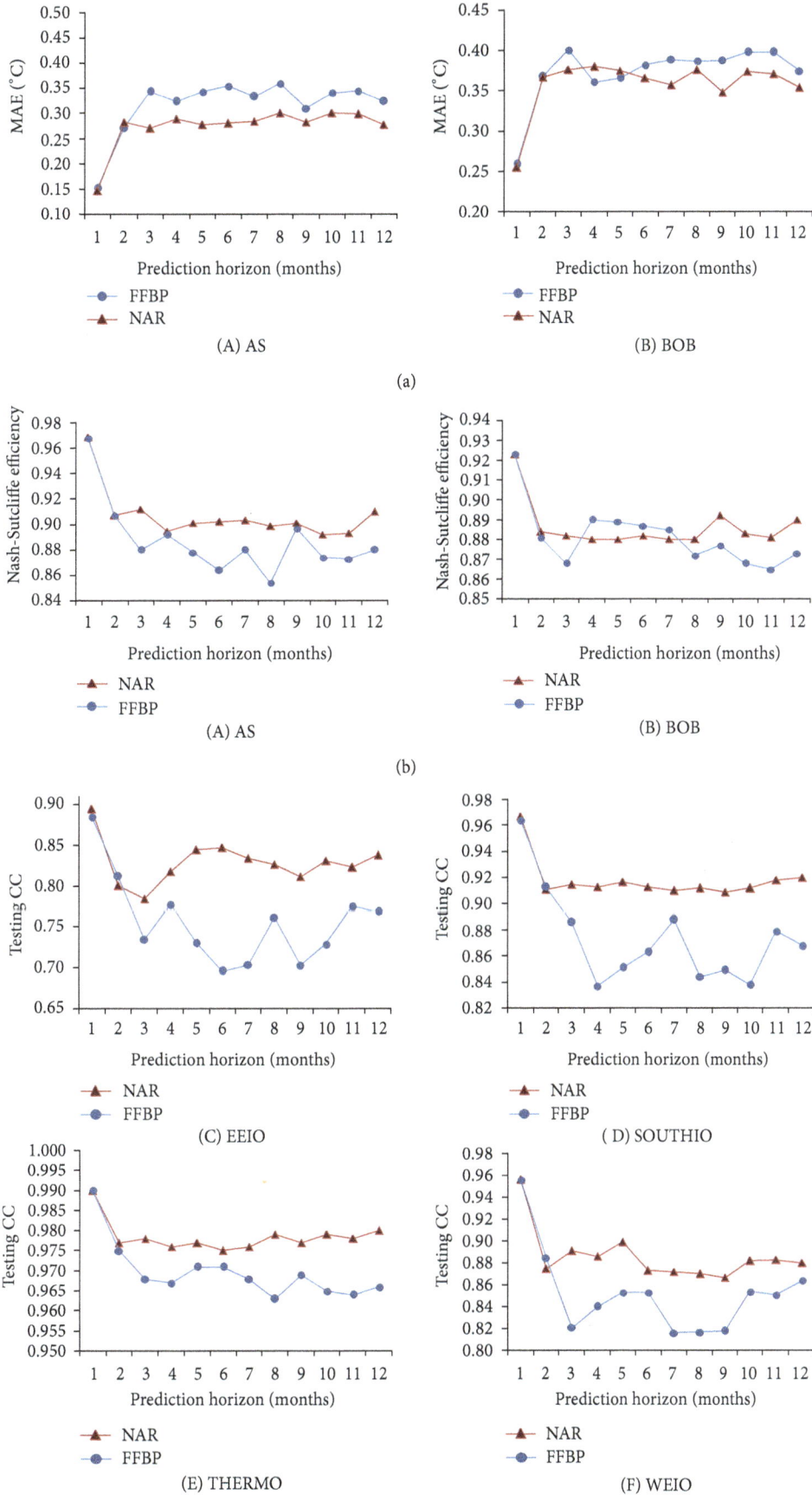

(A) AS

(B) BOB

(a)

(A) AS

(B) BOB

(b)

(C) EEIO

(D) SOUTHIO

(E) THERMO

(F) WEIO

(c)

FIGURE 6: Continued.

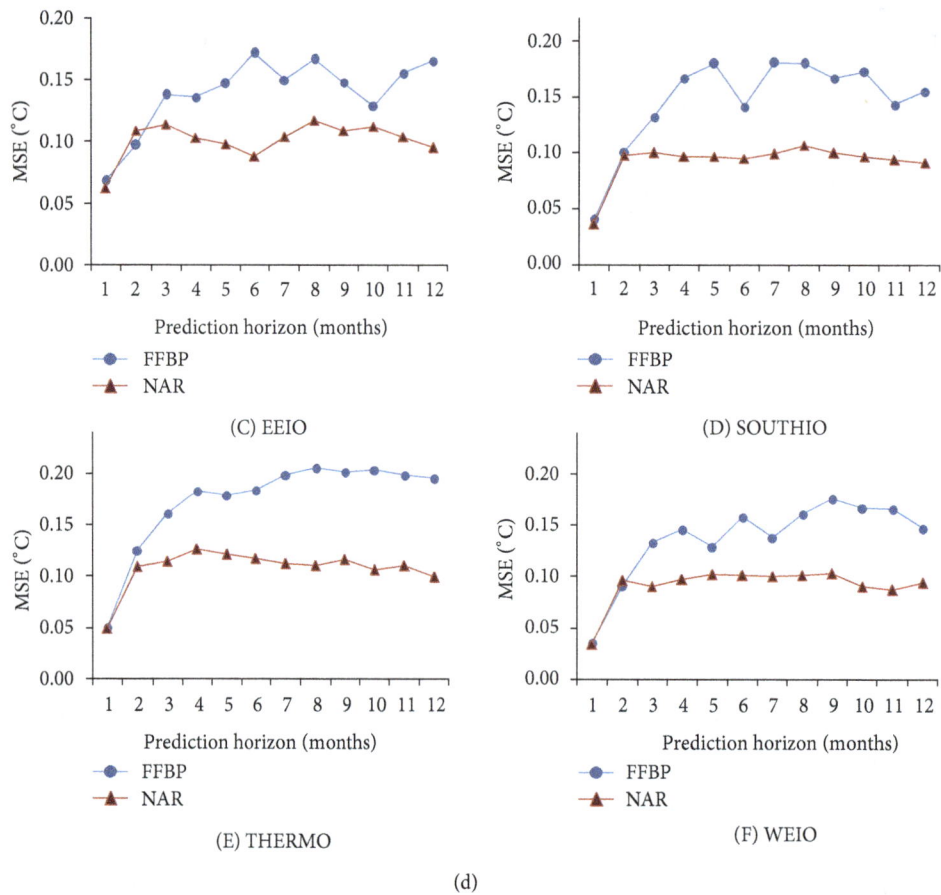

FIGURE 6: (a) MAE variations during testing at sites AS (A) and BOB (B). (b) NSE coefficient variations during testing at sites AS (A) and BOB (B). (c) CC variations during testing at sites EEIO (C), SOUTHIO (D), THERMO (E), and WEIO (F). (d) MSE variations during testing at sites EEIO (C), SOUTHIO (D), THERMO (E), and WEIO (F).

TABLE 3: Comparison with a past study.

Prediction horizon in months	CC	CC during testing for the NAR network at different locations in the present study					
	Tripathi et al. [7]	AS	BOB	EEIO	SOUTHIO	THERMO	WEIO
$SST_{(t+1)}$	0.97	0.99	0.96	0.90	0.97	0.99	0.96
$SST_{(t+2)}$	0.95	0.96	0.94	0.80	0.91	0.98	0.88
$SST_{(t+3)}$	0.59	0.96	0.94	0.79	0.92	0.98	0.89
$SST_{(t+4)}$	0.76	0.95	0.94	0.82	0.91	0.98	0.89
$SST_{(t+5)}$	0.90	0.95	0.94	0.85	0.92	0.98	0.90
$SST_{(t+6)}$	0.92	0.96	0.94	0.85	0.91	0.98	0.87
$SST_{(t+7)}$	0.77	0.96	0.94	0.83	0.91	0.98	0.87
$SST_{(t+8)}$	0.89	0.96	0.94	0.83	0.91	0.98	0.87
$SST_{(t+9)}$	0.87	0.96	0.95	0.81	0.91	0.98	0.87
$SST_{(t+10)}$	0.69	0.95	0.94	0.83	0.91	0.98	0.88
$SST_{(t+11)}$	0.76	0.95	0.94	0.82	0.92	0.98	0.88
$SST_{(t+12)}$	0.99	0.96	0.94	0.84	0.92	0.98	0.88

TABLE 4: Comparison between seasonal and annual NN models at site SOUTHIO.

Location	Month	Seasonal modeling Testing CC		Annual modeling Testing CC	
		FFBP	NAR	FFBP	NAR
Winter	Dec	0.75	0.80	0.87	0.92
	Jan	0.71	0.82	0.96	0.97
	Feb	0.79	0.84	0.91	0.91
Summer	Mar	0.40	0.42	0.89	0.92
	Apr	0.38	0.45	0.84	0.91
	May	0.36	0.41	0.85	0.92
Monsoon	Jun	0.95	0.97	0.86	0.91
	Jul	0.97	0.95	0.89	0.91
	Aug	0.92	0.96	0.84	0.91
	Sep	0.94	0.98	0.85	0.91
After monsoon	Oct	0.67	0.72	0.84	0.91
	Nov	0.66	0.73	0.88	0.92

A sequence of past 24 months' SST was found necessary to provide adequate training.

The results of the present study were more attractive in terms of prediction accuracy than an earlier work in the same region.

The network trained using the input of past 24 months' SST was found to be more beneficial than the one trained with the help of a smaller data segment.

Appendix

The Expressions of the Error Measures

Correlation coefficient (CC)

$$CC = \frac{\sum_{i=1}^{n} \left(X - \overline{X} \right)\left(Y - \overline{Y} \right)}{\sqrt{\sum_{i=1}^{n} \left(X - \overline{X} \right)^2 \left(Y - \overline{Y} \right)^2}}, \quad (A.1)$$

where X = observed SST, $\overline{X}$ = mean of X, Y = predicted SST, $\overline{Y}$ = mean of Y, and n = number of observations.

The correlation coefficient, R, shows the extent of the linear association and similarity of trends between the target and the realized outcome. It is a number between 0 and 1 such that the higher the correlation coefficients the better the model fit is. It however gets heavily affected by the extreme values.

Mean absolute error (MAE)

$$MAE = \frac{\sum_{i=1}^{n} |X - Y|}{n}. \quad (A.2)$$

The mean absolute error has the advantage that it does not distinguish between the over- and underestimation and does not get too much influenced by higher values. The lower the value of MAE is the better the forecasting performance is.

Mean square error (MSE)

$$MSE = \frac{\sum_{i=1}^{n} (X - Y)^2}{n}. \quad (A.3)$$

The mean square error is suited to iterative algorithms and is a better measure for high values. It offers a general picture of the errors involved in the prediction but is also sensitive to high values.

Nash-Sutcliffe efficiency (NSE) coefficient

$$NSE = 1 - \frac{\sum_{i=1}^{n} (X - Y)^2}{\sum_{i=1}^{n} \left(X - \overline{X} \right)^2}. \quad (A.4)$$

It is the ratio of sum square error to the variance of the observed values over its mean, subtracted from 1.0. It ranges from "0" ("no knowledge" model—every forecast is same as the observed mean) to "1" (perfect model). The negative values are also likely, indicating the worst model performance than the "no knowledge" model.

References

[1] K. K. Wu, *Neural Networks and Simulation Methods*, Marcel Decker, New York, NY, USA, 1994.

[2] Y. Xue and A. Leetmaa, "Forecasts of tropical Pacific SST and sea level using a Markov model," *Geophysical Research Letters*, vol. 27, no. 17, pp. 2701–2704, 2000.

[3] N. Agarwal, C. M. Kishtawal, and P. K. Pal, "An analogue prediction method for global sea surface temperature," *Current Science*, vol. 80, no. 1, pp. 49–55, 2001.

[4] T. Laepple, S. Jewson, J. Meagher, A. O'Shay, and J. Penzer, "Five-year ahead prediction of sea surface temperature in the Tropical Atlantic: a comparison of simple statistical methods," http://arxiv.org/abs/physics/0701162.

[5] Neetu, R. Sharma, S. Basu, A. Sarkar, and P. K. Pal, "Data-adaptive prediction of sea-surface temperature in the Arabian Sea," *IEEE Geoscience and Remote Sensing Letters*, vol. 8, no. 1, pp. 9–13, 2011.

[6] A. Wu, W. W. Hsieh, and B. Tang, "Neural network forecasts of the tropical Pacific sea surface temperatures," *Neural Networks*, vol. 19, no. 2, pp. 145–154, 2006.

[7] K. C. Tripathi, M. L. Das, and A. K. Sahai, "Predictability of sea surface temperature anomalies in the Indian Ocean using artificial neural networks," *Indian Journal of Marine Sciences*, vol. 35, no. 3, pp. 210–220, 2006.

[8] M. S. Tanvir and I. M. Mujtaba, "Neural network based correlations for estimating temperature elevation for seawater in MSF desalination process," *Desalination*, vol. 195, no. 1–3, pp. 251–272, 2006.

[9] M. Pozzi, B. A. Malmgren, and S. Monechi, "Sea surface-water temperature and isotopic reconstructions from nannoplankton data using artificial neural networks," *Palaeontologia Electronica*, vol. 3, no. 2, pp. 1–14, 2000.

[10] P. D. Wasserman, *Advanced Methods in Neural Computing*, Van Nostrand Reinhold, New York, NY, USA, 1993.

[11] E. Garcia-Gorriz and J. Garcia-Sanchez, "Prediction of sea surface temperatures in the western Mediterranean Sea by neural networks using satellite observations," *Geophysical Research Letters*, vol. 34, no. 11, Article ID L11603, 6 pages, 2007.

[12] D. C. Collins, C. J. C. Reason, and F. Tangang, "Predictability of Indian Ocean sea surface temperature using canonical correlation analysis," *Climate Dynamics*, vol. 22, no. 5, pp. 481–497, 2004.

[13] J. S. Kuge, I. S. Kang, J. Y. Lee, and J. G. Jhun, "A statistical approach to Indian Ocean sea surface temperature prediction using a dynamical ENSO prediction," *Geophysical Research Letters*, vol. 31, no. 9, Article ID L09212, pp. 1–5, 2004.

[14] B. Tang, W. W. Hsieh, A. H. Monahan, and F. T. Tangang, "Skill comparisons between neural networks and canonical correlation analysis in predicting the equatorial Pacific sea surface temperatures," *Journal of Climate*, vol. 13, no. 1, pp. 287–293, 2000.

[15] F. T. Tangang, W. W. Hsieh, and B. Tang, "Forecasting the equatorial Pacific sea surface temperatures by neural network models," *Climate Dynamics*, vol. 13, no. 2, pp. 135–147, 1997.

[16] S. A. Martinez and W. W. Hsieh, "Forecasts of tropical Pacific sea surface temperatures by neural networks and support vector regression," *International Journal of Oceanography*, vol. 2009, Article ID 167239, 13 pages, 2009.

[17] Y. H. Lee, C. R. Ho, F. C. Su, N. J. Kuo, and Y. H. Cheng, "The use of neural networks in identifying error sources in satellite-derived tropical SST estimates," *Sensors*, vol. 11, no. 8, pp. 7530–7544, 2011.

3

Swell and Wind-Sea Distributions over the Mid-Latitude and Tropical North Atlantic for the Period 2002–2008

Eduardo G. G. de Farias,[1] João A. Lorenzzetti,[1] and Bertrand Chapron[2]

[1] *Divisão de Sensoriamento Remoto, Instituto Nacional de Pesquisas Espaciais (INPE),*
CP 515 12227-010 São José dos Campos SP, Brazil
[2] *Laboratoire d'Océanographie Spatiale, Institut Français de Recherche pour l'Exploitation de la Mer (IFREMER),*
70 29280 Plouzané, France

Correspondence should be addressed to Eduardo G. G. de Farias, gentil@dsr.inpe.br

Academic Editor: Swadhin Behera

We present an analysis of wind-sea and swell fields for mid-latitude and tropical Atlantic for the period 2002–2008 using a combination of satellite data (altimeter significant wave height and scatterometer surface winds) and model results (spectrum peak wave period and propagation direction). Results show a dominance of swell over wind-sea regimes throughout the year. A small but clear decrease in swell energy and an associated increase in wind-sea potential growth were observed in the NE trade winds zone. A seasonal summertime increase in wind-sea energy in the Amazon River mouth and adjacent shelf region and in African coast was apparent in the results, probably associated to a strengthening of the alongshore trade winds in these regions. Albeit with a significantly smaller energy contribution of wind-seas as compared to swell energy, we could say that a kind of mixed seas is more evident in the trade winds region, with the remaining area being highly dominated by swell energy. An analysis of wave-age shows the absence of young-seas. Only ~2% of all data points was classified as wind-sea, a classification confirmed by a fit to a theoretical relation between wind speed, peak period, and significant wave height for fully developed wind-seas.

1. Introduction

Information on wave conditions is critical for human activities at sea. Among other activities, shipping, fishing, offshore industry, naval operations, coastal management and protection, can be adversely affected by wave conditions [1]. Government officials need timely wave information and forecasting results for decision-making. These information are necessary in the preparedness for and mitigation of ocean disasters, as well as sea-going rescue activities. Wave information may also be important in the calculations of ocean-atmosphere exchange of heat and momentum. It is now well established that the drag coefficient C_D, needed for calculating the wind stress and momentum fluxes, is dependent on wave height or wave age [2]. At very high winds and waves, C_D may be smaller than conventionally calculated due to formation of foam layers caused by steep wave breaking water that is sheared by the strong winds [3–5]. These results indicate that wave information should be taken into account in atmospheric modeling, particularly for cases of very strong weather systems.

Perhaps the most used lowest order classification of wave regimes is that one which separates the wave field into either, wind-sea or swell. The first regime is associated to a growing or equilibrium wave spectrum where peak period waves have phase velocity lower than the wind speed. As the wave energy grows and peak period increases, a point is reached when peak waves have phase speeds larger than the wind and will propagate out of the generation region as swell [1]. Wind-sea, as expected, should be highly correlated to the local winds. Swell fields, when generated by intense storms, can propagate out of their generation zones and travel very long distances across the ocean basins [6]. Therefore, very low correlation to the local winds should be expected for these longer waves. Sometimes a frequency of 10 seconds is used as a crude method to separate wind-sea from swell [7].

In a specific oceanic region, both wave regimes can and do coexist, but a predominance of one or the other regime is normally found. A mixture of swell and wind-sea states is normally present in the wave spectrum most of the time, making the quantification of the fraction of wind-seas and swell for large oceanic areas, or for the global ocean, a nontrivial task [8]. In coastal regions, gulfs, and bays, although mixed-seas fields are frequently observed, there is a predominance of wind-seas regime [9]. In open and large oceanic regions (as in the present study), usually there is a predominance of swell fields, which can have been generated by remote strong storms. Wind-seas, although a minor component in open oceans, are, however, always present [10]. A proper estimation of the wind-sea is needed for improving our theoretical understanding of wave growth and for validating wave models [11, 12]. Swell waves, due to their very low dissipation rate and very long distance propagation, in a similar fashion as tsunami waves, can be a potential risk for people's safety and near shore structures, and may cause sensible beach erosion [1, 13].

Different methodologies and data sets have been used to characterize swell/wind-sea regimes for the global ocean or for specific ocean basins. Some studies have used wave model results [14, 15]. Other investigations make use of satellite data, such as altimeter derived significant wave height (H_s) or Synthetic Aperture Radar (SAR) wavenumber spectra [6, 10, 16]. All these investigations show a high incidence of swell fields over the global ocean, but more enclosed ocean basins such as the Gulf of Mexico present a dominance of wind-seas [16].

In the present work we analyze the swell/wind-sea presence in the tropical/mid-latitude North Atlantic for the period 2002–2008. We compare the results obtained by the wave energy methodology presented by [10] which uses quasi simultaneous altimeter significant wave height (H_s) and scatterometer winds (U_{10}) to an analysis of wave fields taking into account the wave age and orientation of wind in relation to the phase propagation vector [12]. The second method was applied using a blend of satellite winds and wave modeling results. The main goal of this paper is to verify the consistency of the results of these two different methods, as well as to contribute to a better understanding of the climatology of swell and wind-seas present in this region.

2. Data and Methods

An important component of this study is a joint use of satellite and wave model data. The use of collocated data from independent sources has been made by different authors to study swell/wind-sea climatology [10, 16]. In principle, one could use simultaneous altimeter derived significant wave height (H_s) and surface wind (U). However, there is enough evidence about the effect of swell, which is uncorrelated with local wind, over the altimeter derived winds [17]. For the present paper, the wind data comes from scatterometer sensor, whose Bragg backscattering mechanism associated to larger incidence angles than altimeter is less sensitive to long wave effect.

For the processing, we used altimeter data (significant wave height) available in the AVISO website (http://www.aviso.oceanobs.com/) and QuikSCAT data available in Remote Sensing Systems website (http://www.ssmi.com/), for the period 2002–2008. The data set used has monthly temporal resolution and the following spatial resolution: 0.5°-QuikSCAT and 2°-Altimeter blending.

The regular QuikSCAT product consists of wind vectors (wind speed and direction) at a 25 km resolution across an 1800 km swath. The scatterometer wind vectors were derived from the radar backscatter using a semiempirical models and an inversion/ambiguity removal algorithm. The statistical interpolation used to obtain the scatterometer product is a minimum variance method related to the Kriging technique widely used in geophysical studies [18].

The altimeter product used in this work was derived using data from different satellite missions to obtain the blended altimeter product. The regular grid sampling on a global scale that allows the analysis of the climatology waves was also derived by Kriging interpolation. The following radar altimeter data were used for the generation of blended product: Topex/Poseidon, ERS-2, GFO, Jason-1, Envisat, and Jason-2 (details are available in the AVISO website http://www.aviso.oceanobs.com/).

We also used data from Wave Watch-III numerical modeling [19]. The wave model was forced with surface winds provided by ECMWF [20]. The wave fields used are provided every 3 hours at a spatial resolution of 0.5° latitude × 0.5° longitude. The products were generated for the period from 01/01/2002 to 31/12/2008. From a global grid of 1° × 1° resolution, a regional nested grid with a resolution of 0.5° × 0.5° was used for our study region. From the regional wave model output, we used peak period (T_p) and wave propagation direction data.

All databases (satellite and model) were averaged to monthly means and were interpolated to a common 2° × 2° lat/long grid. The final grid comprises the longitudes 10° E to 60° W and latitudes 10° S to 50° N, our study region.

2.1. The Wave Energy Statistical Method (WES). One of the first studies presenting global statistical indices of wind-sea and swell using only satellite data is attributed to [10]. According to that study, the degree of swell and wind-sea energy can be estimated using only satellite data: altimeter H_s and scatterometer wind (U_{10}). Total wave energy E_0 can be estimated from altimeter-derived H_s using the well known relation $H_s = 4E_0^{1/2}$. Wind-wave energy partition is estimated from a product of a probability of wind-sea (P_w) and the predicted wind-sea energy E_p, given by $E_p = H_p^2/16$. H_p is the predicted significant wave height (H_p) derived from the fully developed spectrum [10] using a given wind forcing. The fraction of the swell energy (E_s), to total energy, or the swell index (S) is estimated from

$$S = \frac{E_s}{E_0} = 1 - \left[\frac{H_p(U_{10})}{H_s}\right]^2 P_w. \qquad (1)$$

Assuming the drag coefficient is a function of wind speed as given by [21], the predicted Hp for fully developed sea can be calculated as

$$H_p = 1.614 \times 10^{-2} U_{10}^2 \quad (0 \le U_{10} \le 7.5 \text{ ms}^{-1}),$$

$$H_p = 10^{-2} U_{10}^2 + 8.134 \times 10^{-4} U_{10}^3 \quad (7.5 \le U_{10} \le 50 \text{ ms}^{-1}), \tag{2}$$

where U_{10} (ms^{-1}) is the 10 m height, neutral stratification scatterometer wind speed. If N_w and N_s are, respectively, the number of wind-sea and swell events, the probability of wind-seas, P_w, and swell events, P_s, can be estimated by

$$P_w = \frac{N_w}{N},$$

$$P_s = \frac{N_s}{N}, \tag{3}$$

where $N = N_w + N_s$. Since $N = N_s + N_w$, $P_s + P_w = 1$. The N_s and N_w are calculated by comparing H_p against H_s. Points where $H_s > H_p$ are considered swell and those for $H_s < H_p$ are taken as wind-sea. Similarly, it is possible to define a Wind Sea Index (W), which is interpreted as a growing potential for wind waves, given by

$$W = \frac{E_p - E_w}{E_p} = 1.0 - \left[\frac{H_s}{H_p(U_{10})}\right]^2 P_w. \tag{4}$$

E_w, the energy of the wind-sea component, can be estimated as $E_o \times P_w$ for the wind-sea cases. For a purely swell regime, $U_{10} = 0$ and $H_s > 0$, and S = 1. For a purely and beginning case of wind sea, $U_{10} > 0$ and $H_s = 0$ and W = 1. For a fully developed wind-sea and no swell, for a given U_{10}, H_s follows (2), and S = W = 0.

2.2. The Wave Age and Wind/Wave Direction Correlation Method (WAWD). Normally it is assumed that the mean wave propagation direction does not differ significantly from the wind direction [12]. This is a reasonable statement when there is a dominance of wind-sea. However, most of the time, waves are not strictly locally generated [1, 8]. For a wind-sea condition, the direction of wave propagation is expected not to exceed a 90° cone centered in the wind direction, that is, 45° at each side to the prevailing wind direction [12]. If wave propagation direction is out of this "cone of influence," we can assume the wave field has not been generated by local winds, a situation indicative of swell.

The separation of the wind-sea and swell fields, however, does not depend solely on the wind direction propagation of waves in relation to the wind. Even if waves are propagating inside the "cone of influence," it is necessary to consider the wave age, that is, the ratio of speed of waves at the peak of spectrum (C_p) to surface wind speed. The wave field should be considered wind-sea only if the wave propagation direction is within the "cone of influence" and the following relationship is satisfied [12]:

$$U_{10} \cos \theta_d > 0.83 C_p, \tag{5}$$

where

$$C_p = \frac{\omega}{k}. \tag{6}$$

ω is the peak angular frequency ($\omega = 2\pi/T_p$), k is the peak wave number ($k = \omega^2/g$), θ_d is the angle between wind and wave direction, and U_{10} is the local surface wind speed, in our case derived from scatterometer. Equation (5) states that the wave age (WA) for wind-seas, calculated using the wind projection in the direction of wave propagation, should be less than 1.2 [8]. C_p was estimated using peak period T_p generated from Wave Watch 3 model run forced with ECMWF reanalysis winds which also provided wave propagation direction [19].

3. Results and Discussion

Figure 1 shows the spatial distribution of the seasonal average values of swell and wind-sea indices (S and W) in color and mean surface wind field (QuikSCAT) for the period 2002–2008. The magnitudes of S and W are indicated by the color bars to the right of the figures.

As indicated by other studies, which used different methodologies or distinct periods [10, 14], our results also support the view that for winter and summer periods swell energy highly dominates the whole region (S > 0.9). Our results show two swell pools (S = 1) in the North Atlantic: one smaller and closer to the equator near the African coast, and a larger one encompassing all midlatitudes. Results obtained by [10] show a monotonic decrease of swell index, starting from the African coast to the interior of the region (see their Figures 4(b), 4(d)). Our results show the African swell pool aligned to the surface wind convergence zone associated with the ITCZ followed by a decreasing of S index (and an associated increase of W index) in the NE trade winds region, and a northward increase again of S towards the North Atlantic high pressure center region. The influence of the ITCZ over the wave climate in the western tropical Atlantic is also reported by [22].

The seasonal mean wind-sea index plots show low wind-sea potential growth (small values of W) throughout the region, indicating the lack of wind-sea recently generated. An exception to this pattern is seen, however, as a longitudinal and northward inclined band of higher W values centered between 10–15°N in the western tropical Atlantic and 15–32°N near Africa and during the summer season (JJA) over the NE Brazilian/Amazon river shelf region (around 40°W and 0°). This band of higher W values is stronger in the summer period, showing also a maximum near the African coast. The increase of W near Brazilian coast (near the African coast) in the summer period is likely related to the strengthening of the SE (NE) alongshore trade winds in these areas which is visible in QuikSCAT winds. This seasonal strengthening of regional winds during the JJA period is documented, among others, by [14, 15] based in a climatology of model results (ERA-40) for the period 1958–2001.

Our spatial pattern of two swell pools separated by a region of higher (lower) wind-sea (swell) index is also similar

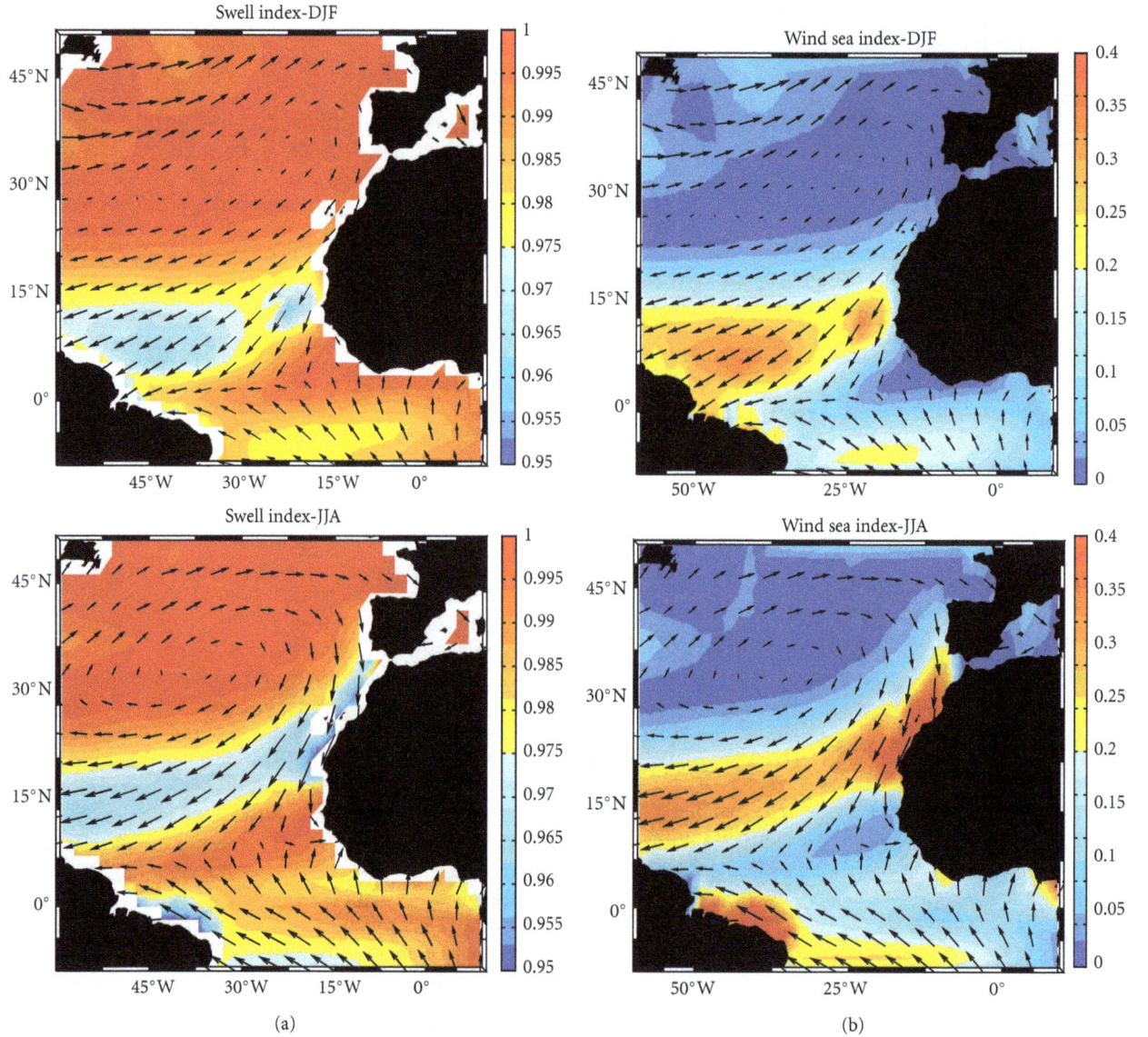

FIGURE 1: Mean seasonal swell (S) and wind-sea (W) indices (left and right, resp.) for NH winter (DJF) and summer (JJA) seasons for the period from 2002 to 2008. The black arrows: mean QuikSCAT winds for the same period.

to the inverse wave age plots given by [15], which were derived from wind and wave data from ERA-40 simulations (see their Figure 7). Also using ERA-40 data, [14] shows for our study region a double swell pool separated by a tongue of lower swell regime calculating a ratio of swell energy to total wave energy and by calculating a swell probability using a wave age (c_p/U_{10}) threshold of 1.2 (their Figures 4 and 6). In the global distribution figures of mean monthly values of peak wave period and mean wave period of [23], it is possible to see for our study region a similar two-center cell of longer period waves separated by a longitudinal band of lower period waves (see their Figures 1(c) and 1(d) for January and July).

A characterization of wave regimes by means of wave age (WA), calculated using friction velocity (u_{star}) instead of

U_{10}, is also frequently adopted [8]. The WA is in this case calculated by

$$\text{WA} = \frac{C_P}{u_{star}}, \qquad (7)$$

where the friction velocity is given by $u_{star}^2 = C_D U_{10}^2$. C_D is the drag coefficient, which can be calculated by $C_D = (0.8 + 0.065 \times U_{10})10^{-3}$ [8]. Using WA, the wind-wave regime is normally separated in two categories: (a) young wind-sea, which corresponds to a high frequency peak spectrum and waves that have just been generated by the wind, and (b) old wind-seas which are associated with a saturated spectrum and are normally encountered where weak winds blow over fast propagating swell or if a local sea is slowly decaying as a storm moves out of the region [24]. If WA, as given by (7),

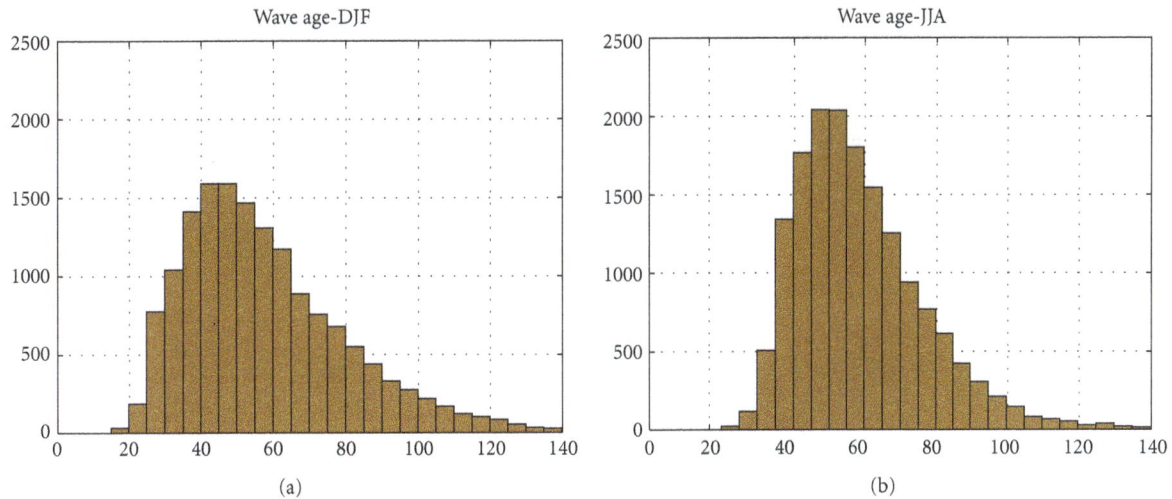

FIGURE 2: Histograms of wave age (7) for the study region. DJF: December, January, and February. JJA: June, July, and August.

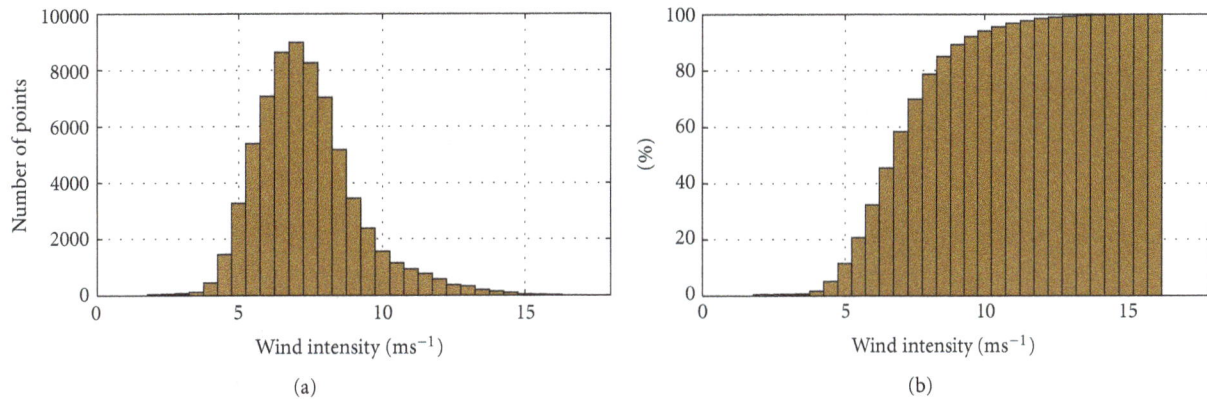

FIGURE 3: (a) Histogram of QuikSCAT wind intensity for the study region. (b) Cumulative histogram derived from QuikSCAT for our study region.

is in the range 5–10, we have young wind-seas. Old wind-seas are associated to WA values on the order of 25–30. For WA > 30 the wave field should be dominated by swell energy [8]. Histograms of WA, calculated using our data base for the whole region, for winter (DJF) and summer (JJA) are presented in Figure 2.

For both seasons, WA values in the range 40–60 dominate the WA distribution, above the old sea limit, indicative of swell dominance. A slight shift to lower values of WA peak distribution is seen for DJF. Winter-time distribution of WA has a smaller kurtosis as compared to the winter distribution. Lower values of WA in the range of 20–30, corresponding to old wind-sea, although a small fraction of total distribution, are present in both seasons. Therefore, albeit a minor contributor to the total wave energy, wind-sea near full development stage is present together with swell. Young wind-seas (5 < WA < 10) have not been observed in the distribution, which is consistent to the absence of a wind-sea index W near 1 in Figure 1.

As shown in Figure 3, the mean wind speed for the study region is about $7\,\mathrm{ms}^{-1}$, and approximately 90% of

wind intensities are below about $8\,\mathrm{ms}^{-1}$. For low wind speed oceanic areas, results of [10] show that presence of pure wind-sea fields is almost negligible (see Figure 1 of [10]). According to [10], pure wind-seas should be present in coastal regions, enclosed seas or during extreme wind events; in open ocean swell is usually present. To evaluate the proportion of swell energy over total wave energy, [14] calculated the ratio of these two quantities by integration of swell partition versus total spectral wave energy using model results. Their global results show that swell energy is everywhere above 65%, even in the extra tropics during the winter, and above 95% in the equatorial region; swell regimes were prevalent in all seasons. A mixed-seas category for our region, that still would consider a higher dominance of swell, could be associated with lower S and higher W indices, which are observed (Figure 1) in the ITCZ region and near the Brazilian north coast and Africa specially during the summer.

In order to compare the results obtained using the WES and WAWD methods, all our data bases of 91 140 set of points (H_s, U_{10}, C_p and wave direction quadruplets) were tested for direction and wave age as indicated in Section 2.2. From

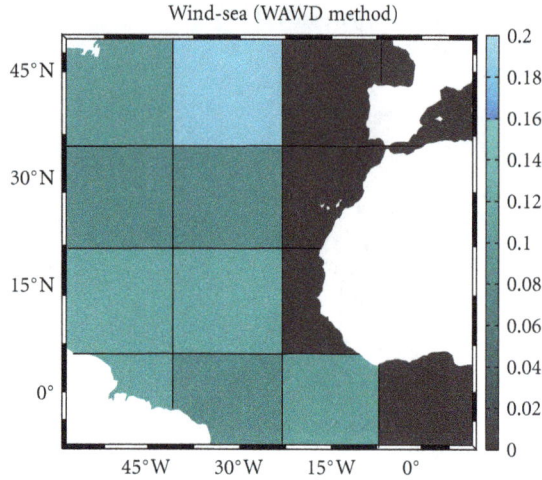

FIGURE 4: 2002–2008 average spatial distribution of the wind-sea points given by WAWD method (yellow points depicted in Figure 5).

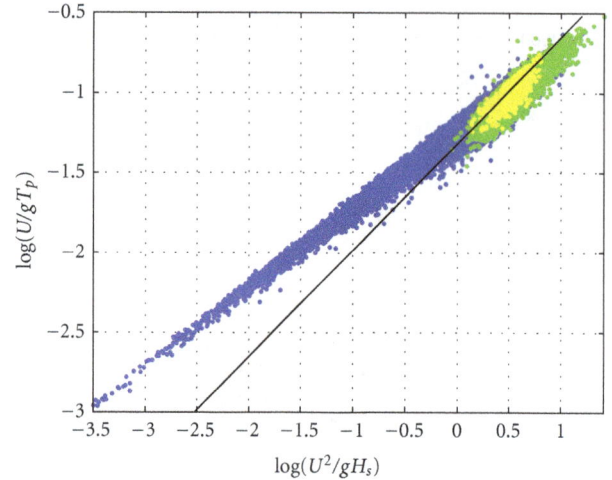

FIGURE 5: $\log(U/gT_p)$ versus $\log(U^2/gH)$. Solid black line represents the theoretical wind-sea relation given by (8) [16, 25].

all data points, 47947 satisfied the criteria of being inside the 90° cone of influence, that is, the waves were more or less propagating in the wind direction. However, from this subset, only 1686 data points passed the wave age criteria (5) corresponding to about 2% of total dataset. This extremely low fraction of purely wind-sea regime in the region given by the WAWD method agrees very well with results presented in Figure 1.

Considering the reduced number of data points, it was not possible to make a plot of the spatial mean distribution of the wind-sea points given by the WAWD method with the same resolution as the wind-sea index of Figure 1. A tentative plot is, however, presented in Figure 4 in a coarse $15° \times 15°$ resolution and an average plot for the whole period. In each resolution cell is indicated the percentage of wind-sea points in the cell. The color bar to the right shows the distribution of the wind-sea in percentage. The eastern portion of the domain is practically void of wind-sea points. The increase in wind-sea density seen centered in 15°N on the ITCZ region and near the Amazon River mouth at the equator matches a similar pattern (in higher resolution) in Figure 1.

Another way of assessing the fraction of our data points that corresponds to wind-sea condition is the following: for a wind forced ocean wave systems [25, 26] show the existence of close correlations between the dimensionless peak frequency ($f_p^* = f_p U/g$) and wave energy ($E^* = g^2 E/U^4$) to the dimensionless fetch ($x^* = xg/U^2$). As shown by [16], it is possible to eliminate x^* from these relations and derive the following equation between the wind speed (U), peak period (T_P), and significant wave height (H):

$$\frac{U}{gT_p} = 4.8 \times 10^{-2} \left(\frac{U^2}{gH}\right)^{0.66}. \tag{8}$$

Figure 5 shows the scatter plot between U/gT_p and U^2/gH calculated from our data set (U from scatterometer, H from altimeter, and T_p from WW-III model). The solid black line represents the theoretical curve for full-developed wind sea

given by (8). In Figure 5, the yellow region corresponds to points which passed the full WAWD test (~2%). The green points only passed the cone of influence part of WAWD method (~52%). The blue points did not pass the WAWD method. It is possible to note that the blue points are far off the theoretical fully developed wind-sea curve. These points correspond to wide range of swell fields found in the study region.

A fit between U/gT_p and U^2/gH just for the yellow points of Figure 5 is presented in Figure 6 (dashed line). The best fit equation to these data points is given by

$$\frac{U}{gT_p} = 5.2 \times 10^{-2} \left(\frac{U^2}{gH}\right)^{0.62}. \tag{9}$$

The close similarity between the theoretical fully developed wind-sea relation given by (8) and fitted relation using only points which passed the WAWD method also confirms first that these are in fact wind-sea cases, and secondly that wind-sea condition represents only a very small fraction of the total wave regime in our study region.

4. Conclusion

In this paper we have used satellite data and wave model results to analyze the space and seasonal variability of wind-sea and swell wave regimes in the mid-latitude and tropical North Atlantic for the period 2002–2008. The swell/wind-sea separation was first done using an energy approach suggested by [10] that uses exclusively satellite data; significant wave height was provided by radar altimetry and surface winds derived from scatterometer sensor. The separation between swell and wind-sea is indicated by a swell index and a wind-sea growth index. The results of this methodology show very clearly the high dominance of swell over wind-sea throughout the study region and for both winter and summer periods. A small but clear decrease in the swell energy and an associated increase in the potential for wind-sea growth were observed in the tropical North Atlantic

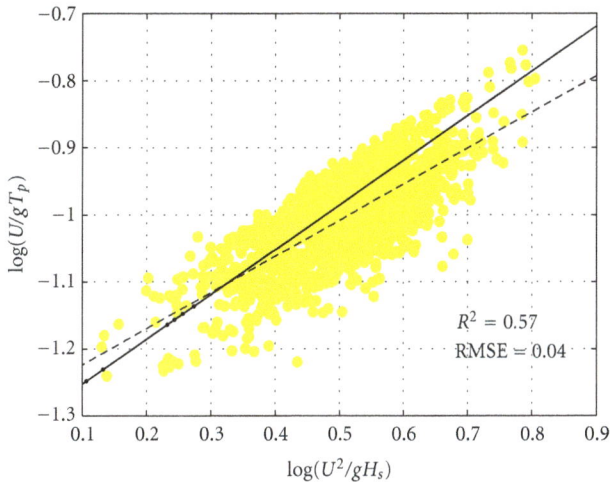

FIGURE 6: $\log(U/gT_p)$ versus $\log(U^2/gH)$ for data points that passed by the WAWD method (yellow points in Figure 5). Solid black line is the theoretical fully developed wind-sea line given by [16], and dashed line is the best fit line to the data (9).

NE trade winds zone. This spatial pattern of swell/wind-sea distribution is similar to those presented by different authors using distinct data sets, methods, or periods [14, 15, 23]. Also, of notice is a seasonal summertime increase in wind-sea index in the Amazon River mouth and adjacent shelf region and in Africa. This seasonal increase in wind-sea energy is probably associated with a strengthening of the alongshore SE and NE trade winds in these regions in the summertime. Albeit with a significantly smaller energy contribution of wind-seas as compared to swell energy, we could say that a kind of mixed seas is more evident in the region in the trade winds, with the remaining area being highly dominated by swell energy. The distribution of WA index indicates very clearly the absence of young-seas in the region.

The second method for assessing the swell/wind-sea contributions used scatterometer winds and wave spectrum peak period and wave propagation direction from wave model results. Wind-sea cases were determined from a maximum ± 45° deviation of wind propagation direction from local winds and a maximum wave age of 1.2 (5). From a set of about 91 K data points, only about 2% were considered wind-sea by this method; a result in agreement to the first method. An indication of wind-sea increase in the NE trade winds region and near the Amazon shelf is present in the spatial distribution of wind-sea data points (Figure 3).

A further analysis of the swell/wind-sea regimes was done by observing the correlation between U/gT_p and U^2/gH. As expected from the analysis using the WES and WAWD methods, the majority of data points strongly deviate from the theoretical wind-sea fully developed spectrum as given by (8). A scatter plot of data points that had passed the WAWD method (Figure 5) shows that a good fit can be adjusted with a very similar correlation law (9).

A particular difficulty in using the presented methods is associated with the very different spatial and temporal resolutions of model results and satellite data. In our study,

model results were available in a grid of 1° × 1° every 3 hours, while altimeter data is given every 7 km along the satellite track [16] and scatterometer winds in a spatial resolution of 25 km. This study shows that the use of interpolated and blended satellite data and model results provided reliable estimates of theoretical and statistical indices of wave field present in the region. We argue here that the consistency observed between different approaches, using the same data set, is indicative of the reliability of results obtained.

References

[1] L. H. Houlthuijsen, *Waves in Oceanic and Coastal Waters*, Cambridge University Press, Cambridge, UK, 2007.

[2] C. W. Fairall, E. F. Bradley, J. E. Hare, A. A. Grachev, and J. B. Edson, "Bulk parameterization of air-sea fluxes: updates and verification for the COARE algorithm," *Journal of Climate*, vol. 16, no. 4, pp. 571–591, 2003.

[3] M. D. Powell, P. J. Vickery, and T. A. Reinhold, "Reduced drag coefficient for high wind speeds in tropical cyclones," *Nature*, vol. 422, no. 6929, pp. 279–283, 2003.

[4] I. J. Moon, I. Ginis, and T. Hara, "Impact of the reduced drag coefficient on ocean wave modeling under hurricane conditions," *Monthly Weather Review*, vol. 136, no. 3, pp. 1217–1223, 2008.

[5] O Saetra, J. Albretsen, and P. A. E. M. Janssen, "Sea-State-dependent momentum fluxes for ocean modeling," *Journal of Physical Oceanography*, vol. 37, no. 11, pp. 2714–2725, 2007.

[6] F. Collard, F. Ardhuin, and B. Chapron, "Monitoring and analysis of ocean swell fields from space: new methods for routine observations," *Journal of Geophysical Research C*, vol. 114, no. 7, Article ID C07023, 2009.

[7] W. Pierson and L. Moskowitz, "A proposed spectral form for fully developed wind seas based on the similarity theory of s.a. Kitaigordskii," *Journal of Geophysical Research*, vol. 69, pp. 5181–5190, 1964.

[8] G. J. Komen, L. Cavaleri, M. Donelan, K. Hasselmann, S. Hasselmann, and P. Janssen, *Dynamics and Modelling of Ocean Waves*, Cambridge University Press, Cambridge, UK, 1994.

[9] P. A. Hwang, H. García-Nava, and F. J. Ocampo-Torres, "Dimensionally consistent similarity relation of ocean surface friction coefficient in mixed seas," *Journal of Physical Oceanography*, vol. 41, pp. 1227–1238, 2011.

[10] G. Chen, B. Chapron, R. Ezaraty, and D. Vandemark, "A global view of swell and wind sea climate in the ocean by satellite altimeter and scatterometer," *Journal of Atmospheric and Oceanic Technology*, vol. 19, pp. 1849–1859, 2002.

[11] F. Ardhuin, B. Chapron, and F. Collard, "Observation of swell dissipation across oceans," *Geophysical Research Letters*, vol. 36, no. 6, Article ID L06607, 2009.

[12] M. A. Donelan, W. H. Hui, and J. Hamilton, "Directional spectra of wind-generated waves," *Philosophical Transactions of the Royal Society A*, vol. 315, no. 1534, 1985.

[13] G. Dodet, X. Bertin, and R. Taborda, "Wave climate variability in the North-East Atlantic Ocean over the last six decades," *Ocean Modelling*, vol. 31, no. 3-4, pp. 120–131, 2010.

[14] A. Semedo, K. Suselj, A. Rutgersson, and A. Sterl, "A global view on the wind sea and swell climate and variability from ERA-40," *Journal of Climate*, vol. 24, pp. 1461–1479, 2011.

[15] K. E. Hanley, S. E. Belcher, and P. P. Sullivan, "A Global climatology of wind-wave interaction," *Journal of Physical Oceanography*, vol. 40, no. 6, pp. 1263–1282, 2010.

[16] P. A. Hwang, W. J. Teague, G. A. Jacobs, and D. W. Wang, "A statistical comparison of wind speed, wave height, and wave period from satellite altimeters and ocean buoys in the Gulf of Mexico region," *Journal of Geophysical Research*, vol. 103, pp. 451–468, 1998.

[17] R. E. Glazman and S. H. Pilorz, "Effects of sea maturity on satellite altimeter measurements," *Journal of Geophysical Research*, vol. 95, no. 3, pp. 2857–2870, 1990.

[18] NASA Quick Scatterometer, "QuikSCAT science data product, user's manual, overview & geophysical data products," 2000, Version 2.0-Draft, Jet Propulsion Laboratory, California Institute of Technology, Doc. D-18053.

[19] H. Tolman, "User manual and system documentation of WaveWatch 3 version 3.14," Tech. Rep. NOAA/NWS/NCEP/ MMAB, 2009.

[20] D. P. Dee, "collaborators The ERA-Interim reanalysis: configuration and performance of the data assimilation system," *Quarterly Journal of the Royal Meteorological Society*, vol. 137, pp. 553–597, 2011.

[21] S. Hasselmann, "The WAM model: a third generation ocean wave prediction model," *Journal of Physical Oceanography*, vol. 18, pp. 1775–1810, 1988.

[22] C. Pianca, P. L. F. Mazzini, and E. Siegle, "Brazilian offshore wave climate based on NWW3 reanalysis," *Brazilian Journal of Oceanography*, vol. 58, no. 1, pp. 53–70, 2010.

[23] I. R. Young, "Seasonal variability of the global ocean wind and wave climate," *International Journal of Climatology*, vol. 19, no. 9, pp. 931–950, 1999.

[24] Edson J. et al., "The coupled boundary layers and air-sea transfer experiment in low winds," *Bulletin of the American Meteorological Society*, vol. 88, pp. 341–356, 2007.

[25] Hasselmann K. and al et, "Measurements of wind-wave growth and swell decay during the joint north sea wave project(JONSWAP)," Tech. Rep. Ergänzungsheft zur Deutschen Hydrographischen Zeitschrift, 1973.

[26] Hasselmann K., B. Ross D., Miller P., and Sell W., "A parametric wave prediction model," *Journal of Physical Oceanography*, vol. 6, pp. 200–228, 1976.

4

Analysis of Interfering Fully Developed, Colinear Deepwater Waves

J. P. Le Roux

Geology Department, Faculty of Physical and Mathematical Sciences, University of Chile/Andean Geothermal Center of Excellence, Post-office Box 13158, Santiago, Chile

Correspondence should be addressed to J. P. Le Roux, jroux@ing.uchile.cl

Academic Editor: Lakshmi Kantha

The sea surface is normally irregular as a result of dissimilar waves generated in different areas. To describe such a sea state, various methods have been proposed, but there is no general consensus as to the best characterizing parameters of the interwaves. Three simple methods are proposed here to calculate a characteristic interwave period, length, and height for fully developed, colinear deepwater waves. The results of this study indicate that the interwave period and length are equal or very close to the period and length of the dominant component wave, irrespective of the periods of the subordinate waves. In cases where the dominant wave period is double or more than double the periods of the subordinate waves, the wave period, length and height are within 4% of the dominant wave parameters, so that such interfering, irregular waves have virtually the same characteristics as monochromatic waves. Secondary, individual interwaves propagate at the velocity of the component wave with the shortest period, that is, slower than the primary interwaves which have the same celerity as the dominant component wave.

1. Introduction

Waves are generated by wind in different areas and may be dissimilar in period, length, height, and celerity. Where such waves converge, interference takes place so that their crests are reinforced or diminished depending on whether crest-crest or crest-trough interference occurs. For ocean and coastal engineers, it is important to understand such interactions, because the design of structures naturally has to consider the peak conditions arising from them. Wave interference also influences sediment transport as well as coastal and coral reef erosion.

Complex wave fields arise because of different stages of wave development and water depths (both of which result in significant changes in the individual wave profiles), as well as the fact that waves may come from different directions. Modelling all the possible combinations resulting from these differences would be a very demanding task, so that this paper focuses instead on the relatively simple situation of interfering colinear waves with Airy characteristics. Such sinusoidal waves are typical of deepwater conditions where the wind has blown for a sufficient period of time over a long enough fetch for them to become fully developed. This would be the case for synoptic-scale (100–1000 km) pressure systems developed over the deep ocean, where there is no interference from bathymetric and land effects and where waves generated upwind have longer wavelengths than downwind waves, thus overtaking the latter. For example, a 13 s wave would require a wind speed and duration of $20\,\mathrm{m\,s^{-1}}$ and 32 hours, respectively, over a minimum fetch of about 600 km, whereas an 8 s wave would be generated by a $12.5\,\mathrm{m\,s^{-1}}$ wind blowing for at least 24 hours over a fetch of about 300 km [1]. Because the wavelength and height of fully developed waves are fixed by their respective periods (see (2) and (3)), changing the latter (T) automatically adapts the corresponding lengths and amplitudes.

Currently two approaches are used to characterize the undulating surface formed by interfering waves, hereafter referred to as the interwave surface or simply interwaves. Spectral methods (e.g., [2, 3]) are based on the Fourier transform of the sea surface [4], whereas wave train analysis uses direct measurements of the sea surface elevation at set intervals (typically every second) at a specific locality

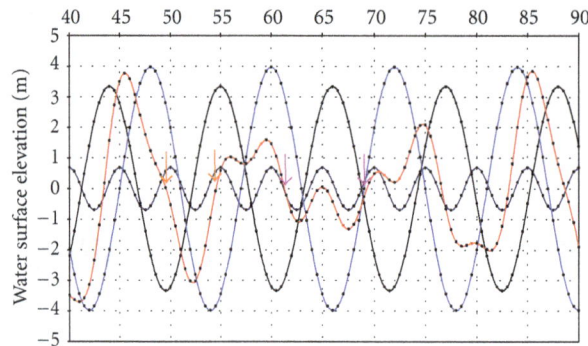

FIGURE 1: Definition of prominent troughs on t/η diagram. The width-depth value A for the trough between the orange arrows is $3.0218(54.5 - 49.5) = 15.109$, which is more than 50% of $\overline{A} = 0.5(29.6735 - 10.3164) = 9.6786$. The two troughs between the pink arrows have A-values less than 9.6786, but in combination their A-value of $1.3138(69.5 - 61.5) = 10.5104$ exceeds this cutoff, so that they are counted as one trough. Blue line: dominant component wave; black lines: subordinary component waves; red line: resultant interwaves.

or measuring station [5]. The main drawback of this method is that it cannot distinguish the wave direction [4]. Nevertheless, if criteria can be developed to identify individual waves in a record of colinear waves, the way they interact can be used for prediction and design purposes.

The approach followed here was to generate Airy waves of different periods on the accompanying Excel spreadsheet (INTERWAVES) and to examine the ways in which they interact. This reveals useful relationships between the characterizing interwave parameters.

2. Simulation of Interfering Colinear Fully Developed Deepwater Waves

In the discussion below, the subscript 1 refers to the dominant wave (i.e., the component wave with the longest period), and the subscripts 2 and 3 to the subordinate waves, 2 indicating a component wave with a longer period than 3. The subscript i refers to interwaves.

In deep water, interfering colinear Airy waves can be modeled by the standard equation [4]

$$\eta_i = \eta_1 + \eta_2 + \eta_3 = \frac{H_1}{2}\cos\left(\frac{2\pi X}{L_1} - \frac{2\pi t}{T_1}\right)$$
$$+ \frac{H_2}{2}\cos\left(\frac{2\pi X}{L_2} - \frac{2\pi t}{T_2}\right) + \frac{H_3}{2}\cos\left(\frac{2\pi X}{L_3} - \frac{2\pi t}{T_3}\right),$$
$$(1)$$

where η_i is the surface elevation of the interference wave, $\eta_1 + \eta_2 + \eta_3$ are the elevations of the respective wave components, X is the distance from the measuring station in the direction of wave propagation, L_1, L_2, L_3, and T_1, T_2, T_3 are the lengths and periods of the components waves, respectively, and t is the time in seconds. In Figures 1 and 2, the blue and black lines represent the component wave shapes and the red lines the resultant interwaves.

In the spreadsheet setup, two-component and three-component interwaves were simulated using different combinations of T_1, T_2, and T_3. For each period, the wavelength L was obtained from [6]

$$L = \frac{gT^2}{2\pi},$$
$$(2)$$

and the deepwater fully developed wave height H from [7, 8]:

$$H = \frac{gT^2}{18\pi^2}.$$
$$(3)$$

Equation (3) is based on data collected since 1967 during the Joint North Sea Wave Project (JONSWAP) as graphically summarized in [9].

The spreadsheet was designed to calculate the water surface elevation at intervals each representing 1/240th of the time required for 10 dominant wavelengths to pass the observation point, and also at 240 points over a distance of 10 dominant wavelengths. This was found to be sufficient in establishing interference patterns and relationships among the various parameters. Two graphs were plotted, one showing the water surface elevation η over time t at any specific point, and the other showing η against distance X in the direction of wave propagation at any specific moment in time. The different wave periods and heights were determined using the t/η graph, whereas the X/η graph was employed for the wavelengths. At the graph origin ($x = 0; t = 0$), full crest interference among the different wave components was modeled, but the spreadsheet was designed so that different wave phases could also be simulated at this point if required. The different combinations of wave periods used are shown in the headings of Table 1. For every combination of waves, the different interwave parameters described below were calculated using the water surface elevation as plotted on the graphs, with the results summarized in Table 1.

3. Expression of Interwave Parameters

The parameters used to describe interwaves differ from those of sinusoidal, monochromatic waves. For example, where

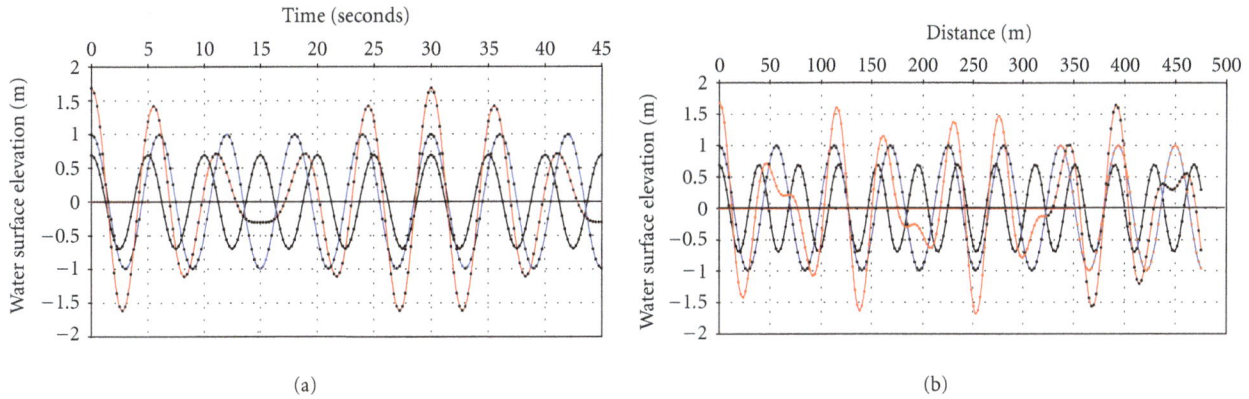

FIGURE 2: (a) Wave "groups" on a t/η diagram (a) for 6, 5 s interwave. This is the sea surface elevation recorded at any specific station over a time period of 45 seconds. (b) On an X/η diagram, the "groups" are much less prominent, that is, an observer looking at the sea surface over a distance of 500 m at any specific moment in time would hardly be able to discern groups. Blue line: dominant component wave; black lines: subordinate component waves; red line: resultant interwave.

the wave period (T) for the latter is defined as the time interval (in seconds) between the arrival of two successive crests or troughs (which gives the same result), in the case of interwaves the mean time interval between neighboring crests will not necessarily be the same as that between neighboring troughs.

One way to determine the interwave period is to calculate the average time interval between the arrival of all successive crests (T_{ic}) or troughs (T_{it}) on the t/η record. The most commonly used method, however, describes the interwave period as the time interval between successive crossings of the mean or still water level (SWL) by the water surface in a downward or upward direction, respectively, which are known as the zero downcrossing (T_{izd}) and zero upcrossing (T_{izu}) periods [5].

Here, a rapid method is proposed to obtain the interwave period T_{ix}, which consists of dividing the time lapse t_t between the first and last prominent troughs on the t/η record by the number of prominent troughs (N_t) between them (excluding the first but including the last trough). Troughs were used here because by design only half of the first and last crests are displayed on the t/η record, which might lead to ambiguity in the case of double-crested waves.

A potential problem in using this method concerns the definition of "prominent" troughs. Under certain conditions of wave interference, a whole series of crests and troughs of different sizes are generated, and while in the case of two-component interwaves there should be no problem in identifying them, it may not be so easy where three or more waves with at least one period close to that of the dominant wave interfere. In this case, the following method was found to give consistent results. For each clearly prominent trough on the record, the distance between its zero downcrossing and upcrossing point is measured and multiplied with its maximum depth below the SWL. This value is here referred to as the trough width-depth value (A). The mean $(\overline{A})$ and standard deviation (As) of all the prominent trough width-depth values are then calculated, where the standard deviation provided in Excel is given by $s = \sqrt{(n \sum x^2 - (\sum x)^2)/n(n-1)}$.

For any disputed trough to be counted as a prominent trough, its A-value must be at least 50% of $(\overline{A} - sA)$. Figure 1 shows the case of a 12, 11, 5 s wave, where a series of undulations around the SWL is produced between 40 and 90 seconds on a t/η diagram. The trough between the orange arrows has an A-value of 15.109, which is more than $0.5(\overline{A} - As) = 9.6786$. None of the two troughs between the pink arrows reaches the 50% cutoff, but together they have a value of 10.5104. They are thus counted as one trough.

The peak interwave period T_{ip} is defined as the time interval between maximum crest interference at the same locality on the t/η record.

Interwave length is commonly described as the average distance between successive zero down- or upcrossing points (L_{izd}, L_{izu}). Here, a characteristic interwave length L_{ix} was calculated by dividing the distance (X_t) between the first and last prominent wave trough on the X/η record by the number of prominent troughs (N_t) between them, including one of the end troughs. "Prominent" is defined as for the wave period. The peak interwave length (L_{ip}) is the distance between crests of maximum interference at any specific moment on the X/η record.

Interwave heights are also characterized by different methods, for example, the mean height (H_{im}) that averages every successive crest-trough elevation difference in the record and the root-mean-square height (H_{irms}), in which the square root is obtained after squaring, summing, and averaging these individual differences. The commonly used significant wave height (H_{is}) is the mean height of the highest one-third of all waves on the record, whereas the maximum wave height H_{imax} refers to the largest difference between any recorded adjacent trough and crest.

In modeling monochromatic waves, it is found that the standard deviation of the sea surface ηs can be related to the wave height as obtained from (3), by multiplying ηs by 2.8168. For example, a value of 1.3805 m is obtained for a 5 s wave by calculating the standard deviation of the water surface elevation for 240 s, compared to 1.3805 m as determined from (3). This relationship was therefore used

TABLE 1: Wave parameters as determined from t/η and X/η diagrams for different combinations of component wave periods.

(a) Two-component interwaves

Wave parameters	6, 5 s	6, 4 s	6, 3 s	6, 2 s	6, 1 s	12, 11 s
T_{ic}	6.00	6.00	6.00	6.00	6.00	12.00
T_{it}	6.00	6.11	6.00	6.00	6.00	12.22
T_{izd}	6.03	6.08	6.00	6.00	6.00	12.11
T_{izu}	6.03	6.08	6.00	6.00	6.00	12.11
T_{ix}	6.06	6.11	6.00	6.00	6.00	12.11
T_{ip}	30.00	12.00	6.00	6.00	6.00	132.00
L_{izd}	56.2124	57.2489	56.2080	56.2080	56.2080	230.6821
L_{izu}	55.5054	55.9478	56.2080	56.2080	56.2080	230.6821
L_{ix}	56.2080	55.6876	56.2080	56.2080	56.2080	230.6821
L_{ip}	1405.18	224.832	56.2080	56.2080	56.2080	27204.3
H_{im}	2.2394	2.1672	1.9880	2.2088	1.9880	9.2315
H_{irms}	2.4736	2.2117	1.9880	2.2088	1.9880	10.0715
H_{is}	3.2124	2.6077	1.9880	2.2088	1.9880	14.0973
H_{imax}	3.3007	2.6077	1.9880	2.2088	1.9880	14.4979
H_{ix}	2.4296	2.1821	2.0531	2.0020	1.9892	9.8970
C_{ip}	9.3680	9.3680	9.3680	9.3679	9.3679	17.9550
C_{isc}	7.8067	6.2452	4.6839	3.1226	1.5613	17.9550

(b) Three-component interwaves

Wave parameters	6, 5, 4 s	6, 5, 3 s	6, 5, 2 s	6, 4, 3 s	6, 4, 2 s	6, 3, 2 s	12, 11, 10 s
T_{ic}	5.00	5.00	4.29	6.00	6.00	6.00	10.91
T_{it}	5.00	5.00	4.17	6.17	6.08	6.00	10.95
T_{izd}	5.00	6.06	6.08	6.06	6.08	6.00	10.95
T_{izu}	4.98	6.16	6.03	6.06	6.19	6.00	10.90
T_{ix}	6.11	6.08	6.06	6.17	6.08	6.00	12.17
T_{ip}	60.00	30.00	30.00	12.00	12.00	6.00	660.00
L_{izd}	39.4794	45.9884	56.2080	56.9886	57.2478	56.2080	185.4824
L_{izu}	38.7331	45.9884	56.4682	55.9478	56.2080	56.2080	184.5457
L_{ix}	57.2489	56.2080	56.4682	55.9478	55.4273	56.2080	231.8531
L_{ip}	5620.72	1405.18	1405.18	224.8320	224.832	56.208	680106.6
H_{im}	2.2017	2.1213	1.7129	2.3535	2.3117	2.2088	9.7358
H_{irms}	2.4517	2.3611	2.1374	2.3946	2.3535	2.2088	11.4729
H_{is}	3.3410	3.2423	3.1588	2.9752	2.7534	2.2088	17.8234
H_{imax}	3.9894	3.3595	3.4893	2.9752	2.7534	2.2088	19.8847
H_{ix}	2.5947	2.4854	2.4421	2.2432	2.1956	2.0671	11.6291
C_{ip}	9.3679	9.3679	9.3679	9.3679	9.3679	9.3679	17.0628
C_{isc}	6.2452	4.6839	3.1226	4.6839	3.1226	3.1226	17.0628

to calculate H_{ix}, here termed as the characteristic interwave height, in Table 1.

As used here, the primary interwave celerity C_{ip} refers to the propagation velocity of prominent or primary wave crests (C_{ipc}) or troughs (C_{ipt}) within the wave train. Some of these larger wave forms may be similar to the concept of wave groups as currently used. Individual secondary crests and troughs superimposed on the primary interwave forms propagate at speeds different from C_{ipc} or C_{ipt}, which is referred to as the secondary interwave crest or trough celerity (C_{isc} or C_{ist}), respectively. All these celerities can be measured on the X/η diagram by monitoring, for example, the propagation of any specific secondary crest over the distance of 10 primary wavelengths while changing t in (1). On the t/η diagram this cannot be done because the graph only depicts the water surface elevation at a specific station over time.

4. Two-Component Interwaves

Experimenting with different two-wave combinations, the following general observations were made.

4.1. Interwave Period. If the dominant wave period T_1 is double or more than double the period of the subordinate wave T_2, all interwave periods including the characteristic interwave period (T_{ix}) and peak interwave period (T_{ip}) are equal to T_1.

If the dominant wave period is less than double that of the subordinate wave, all interwave periods except the peak interwave period are very close to that of the dominant wave, differing by less than 2%.

The peak interwave period for all two-component interwaves is given by

$$T_{ip} = \frac{T_1 T_2}{\text{LCF}}, \qquad (4)$$

where LCF is the largest common factor by which T_1 and T_2 can be divided. For example, for a 6, 2 s wave $T_{ip} = (6)(2)/2 = 6$ s, for a 6, 3 s wave $T_{ip} = (6)(3)/3 = 6$ s, and for a 6, 5 s wave $T_{ip} = (6)(5)/1 = 30$ s.

4.2. Interwave Length. If the dominant wave period is double or more than double the period of the subordinate wave, all interwave lengths are exactly equal to the wavelength L_1 of the dominant wave. When the dominant wave period is less than double that of the subordinate wave, all interwave periods except the peak interwave period are very close to that of the dominant wave, differing by less than 3%. This means that L_{ix} can be obtained directly from the t/η record together with (2), substituting T_{ix} for T.

The peak interwave length for all two-component waves is also given by (2), in this case using T_{ip} instead of T_{ix}. This is the case even though T_{ip} is determined from the t/η diagram and L_{ip} from the X/η diagram. For example, for a 6, 4 s interwave, $T_{ip} = 12$ s and $L_{ip} = 224.832$ m as read from the X/η diagram, the small discrepancy (0.0034 m) being due to slight inaccuracies caused by line smoothing. This relationship is useful because waves are usually monitored at a single station and the results displayed on a t/η diagram where the peak interwave length will not be visible but can still be determined from (4) and (2).

4.3. Interwave Height. If the dominant wave period is double or more than double the period of the subordinate wave, the wave height as determined by H_{im}, H_{irms}, H_{is}, and H_{imax} has the same value in all four cases because the same wave form is repeated. In some combinations, this height coincides with the height of the dominant wave H_1, but in other cases it can be more than 10% larger than the latter. For example, the four different methods calculate a height of 1.988 m for both 6, 3 and 6, 1 s interwaves, agreeing closely with that of the dominant 6 s wave (1.9879 m), whereas a 6, 2 s interwave is calculated to be 2.2088 m high.

In these cases, the characteristic interwave height H_{ix} is 2.0531, 2.002, and 1.9892 m for 6; 3, 6; 2, and 6; 1 s interwaves, respectively. This is considered to better represent the characteristic wave height than the other measures, for the following reasons: a dominant wave with superimposed smaller waves must represent higher energy conditions than the dominant wave alone, and given the fact that the interwave length is the same as the dominant wavelength,

this energy should be manifested in a higher characteristic interwave height. Furthermore, the interwave height should increase with an increase in the wave period (and height) of the subordinate wave. H_{ix} is the only one of the 5 different wave height definitions that makes this distinction. In addition, it better represents the wave height as obtained from a combination of the t/η and X/η graphs. In the case of a 6, 3 s interwave, for example, the t/η graph depicts a peaked crest and flat trough, with the former 1.2425 m above and the trough 0.7455 m below the SWL, respectively. However, on the X/η graph the crest is 1.2424 m above the SWL, and there is a double trough at a depth of 0.9851 m below the SWL, with a secondary crest 0.7455 m below the SWL in the middle thereof. Taking the average trough depth as (2 × 0.9851 + 0.7455)/3 = 0.9052 m, the wave height would be 2.1476 m. Averaging this with the 1.988 m shown on the t/η graph gives 2.0678 m for the two graphs, which agrees better with the value of 2.0531 m obtained by H_{ix} than the value of 2.2088 given by the other 4 height definitions. Lastly, in cases where the dominant wave period is less than double that of the subordinate wave, the value of H_{ix} falls between those given by H_{im} and H_{irms}. In these cases, H_{is} and H_{imax} give larger heights than H_{im}, H_{irms} and H_{ix}, which is logical as they represent the highest third and maximum wave heights, respectively.

4.4. Interwave Celerity. Interwave celerity is a somewhat controversial subject because of the concept of wave groups or "beats" as currently used. Two interfering wave trains will produce zones of reinforced crests and troughs interspersed with zones of reduced crests and troughs. The envelope curves enclosing these zones describe the wave groups, which propagate at a speed considered to be one half that of the phase velocity in deep water (e.g., [4]). However, some of these "groups" may be an artifact of the t/η diagrams commonly used to analyze wave trains. Consider the interference pattern of a 6, 5 s interwave on such a diagram (Figure 2(a)). In this case there are two clear "zones" with reinforced crests and troughs separated by a "zone" of reduced crests and troughs. However, on the X/η diagram (Figure 2(b)), which represents the real surface as it would be observed at any moment in time over a specific distance, the "groups" are far less obvious. As T_2 becomes smaller relative to T_1, the secondary interwave crests and troughs diminish in size until they can hardly be distinguished, so that the primary interwave form is almost exactly that of the dominant wave and does not form groups of any description.

It was shown above that the length and period of the primary interwaves are virtually the same as the period and length of the dominant wave for all combinations of T_1 and T_2. There seems to be no logical reason why such primary forms should be considered as "groups" when T_1 and T_2 have similar values (e.g., a 6, 5 s interwave), but not in the case of 6; 1 s interwaves, when important attributes such as period and length are shared in both cases. In this paper, therefore, all prominent wave forms produced by different combinations of T_1 and T_2 are simply considered to be primary interwaves, whereas individual smaller crests and troughs superimposed on them are described as secondary interwaves.

Analysis of the celerity of primary and secondary interwaves results in the following conclusions.

For all combinations of T_1 and T_2, the primary interwave crest (or trough) celerity C_{ipc} (or C_{ipt}) is equal to the celerity of the dominant wave C_1, except where T_2 comes to within about 15% of T_1, when

$$C_{ipc} = \frac{g(T_1 + T_2)}{4\pi}. \qquad (5)$$

For example, for a 12; 11 s interwave the observed C_{ipc} is 17.995 m s^{-1}, which agrees with $9.81(12 + 11)/4\pi$. This coincides with the wave celerity of an 11.5 s wave, that is, $(T_1 + T_2)/2$.

The celerity of any secondary interwave crest is given by

$$C_{isc} = C_1 \frac{T_2}{T_1} = C_{w2}, \qquad (6)$$

except where T_2 comes to within about 15% of T_1, when it is also given by (5). In these cases, no secondary interwave crests can be distinguished anymore because the two component waves merge completely into a single primary interwave form.

Where the dominant wave period is less than double the secondary wave period, C_{isc} is actually not constant, however, but varies along the crest trajectory. For example, the crest of a 6, 4 s wave originating at $T = 0$ and $X = 0$ (where there is maximum interference) propagates at an average celerity of 7.026 m s^{-1} during the first 6 seconds, 4.684 m s^{-1} during the next 6 seconds, and 7.026 m s^{-1} from 12 to 18 s, which is the time required to reach its next peak elevation of 1.4357 m above the SWL at a distance of 112.416 m from the origin. This time interval differs from the peak interwave period T_{ip} because the latter is reached every 12 seconds at any specific station, but represents the interference of different crests, not the cyclic decay and growth of the same crest. The mean celerity of the secondary interwave crest is 6.2453 m s^{-1} in this case, which is the same as that of the subordinate wave celerity C_{w2} given by $gT_2^3/2\pi = 9.81 \times 4/2\pi$.

Some of the interference patterns may be confused with wave groups, although the present analysis shows that they do not travel at half the phase velocity as true groups purportedly do, but at the same velocity as the dominant wave. Furthermore, the secondary crests actually propagate *slower* than the primary forms. This can be demonstrated by changing the time T incrementally on the X/η diagram, which shows that these secondary crests in fact cycle backward, thus "originating" at the front of the primary forms and "disappearing" at the back (although they simply grow and then diminish in height without actually disappearing completely).

5. Multicomponent Interwaves

In this series of virtual experiments, three Airy waves of different periods were superimposed to examine their interference characteristics.

5.1. Interwave Period. When the periods of the subordinate component waves are two-thirds or less than two-thirds of the dominant wave, the interwave period given by all the methods considered here, including T_{ix}, are the same as or very close to that of the dominant wave. However, when the subordinate wave period T_2 is more than two-thirds of the dominant wave period, the periods defined as T_{ic}, T_{it}, and in some cases also T_{izu} and T_{izd}, may be less than the dominant wave period. This results from the fact that interference in these cases produces low-amplitude undulations varying about the SWL in some parts of the spectrum. These are counted as crests and troughs by definition of the parameters above, whereas the definition of T_{ix} in this case does not consider these undulations to be prominent troughs. For example, in the case of a 12, 11, 5 s wave, there are two troughs between 61.5 and 69 s from the origin, as defined by the zero down- and upcrossing points (Figure 1). The interwave periods T_{izu} and T_{izd} would therefore be less than T_{ix}, which in all cases falls within 2% of the period of the dominant wave, T_1.

The peak interwave period T_{ip} for multicomponent waves is given by

$$T_{ip} = \frac{T_1 T_2 T_3}{\text{LCF}_1 \text{LCF}_{21}}, \qquad (7)$$

where LCF$_1$ and LCF$_2$ are the two largest common factors by which any pair between T_1, T_2, and T_3 can be divided. For example, for a 6, 4, 3 s interwave $T_{ix} = 6 \times 4 \times 3/3 \times 2 = 12$, and for a 6, 5, 4 s interwave, $T_{ix} = 6 \times 5 \times 4/2 \times 1 = 60$ s.

5.2. Interwave Length. The wavelengths determined by the down- or upcrossing method (T_{izd}, T_{izu}) are generally the same as the wavelength of the dominant wave, but are lower in some cases where the interference surface crosses the SWL in a series of small waves, for example, in the case of 6, 5, 4 s waves. L_{ix} is in all cases close to the dominant wavelength.

The peak crest interference wavelength L_{ip} is given by (5). For example, for a 6, 5, 4 s interference wave, $L_{ip} = 5620.716$ m, which is exactly divisible by 56.2072 m, 39.0327 m, and 24.981 m, the wavelengths of 6, 5, and 4 s waves, respectively.

5.3. Interwave Height. The wave heights as defined by H_{im} and H_{irms} appear to give somewhat haphazard results for three-component as compared to two-component waves. For example, a 6, 5 s wave has H_{im} and H_{irms} values of 2.2394 and 2.4736 m, respectively, but for a 6, 5, 4 s wave these values *reduce* to 2.2017 and 2.4517 m. H_{ix} in these cases yields characteristic heights of 2.4296 and 2.5947 m, respectively, which better represents the higher energy of 6; 5; 4 s waves as compared to 6, 5 s waves. For a 6, 4, 3 s interwave H_{im} and H_{irms} are also *higher* (2.3535 m and 2.3946 m) than for a 6, 5, 4 s wave, whereas H_{ix} correctly reflects a decrease in the characteristic wave height to 2.2432 m.

5.4. Interwave Celerity. The primary interwave celerity C_{ip} is in all cases, with the exception of the conditions described below, equal to the dominant wave celerity C_1. The celerity of any secondary interwave crest is given by

$$C_1 \frac{T_3}{T_1} = C_{w3}. \tag{8}$$

All secondary interwave crests therefore propagate at the celerity of the subordinate wave with the lowest wave period.

Where both subordinate waves have periods that fall within about 15% of the dominant wave period, no secondary interwaves are observed, and the interwave crest propagates at a velocity given by

$$C_{ipc} = \frac{g(T_1 + T_2 + T_3)}{6\pi}. \tag{9}$$

6. Conclusions

The following general observations can be made on the basis of the analysis presented here.

Both the interwave period T_{ix} and length L_{ix} are equal to or very close to the period T_1 and length L_1 of the dominant component wave, irrespective of the periods of the subordinate waves. This constant relationship thus allows the dominant wave to be recognized easily in any record of fully developed, interfering colinear waves. Although T_{ix} and L_{ix} generally concur with T_{izd}, T_{izu}, L_{izd}, and L_{izu} of the zero up- or downcrossing methods, the latter tend to underestimate the real interwave period and length in cases where all the component waves have similar periods because of small fluctuations of the water surface around the SWL.

The interwave height H_{ix} proposed here better characterizes the "average" wave height than H_{im} or H_{irms}, because it steadily increases with increasing energy conditions, whereas the latter two heights do not show any consistent pattern. It is also easily calculated on a spreadsheet if the water surface elevation is recorded at 1 second intervals, making use of the built-in standard deviation function (s) of Excel.

In all cases where the dominant wave period is double or more than double the periods of the subordinate waves, T_{ix}, L_{ix}, and H_{ix} are within 4% of the dominant wave parameters. In such cases, the significant and maximum wave heights (H_{is} and H_{imax}) are the same as H_{im} and H_{irms}, so that such interfering, irregular waves have virtually the same characteristics as monochromatic waves. This might explain why the nomograms in [9], which plot energy-based significant wave heights and periods against wind speed, duration, and fetch, show fairly constant relationships corresponding to Airy wave characteristics [10]. However, even if the wave parameters under these circumstances are very close to those of the dominant wave, small differences can produce substantial changes in wave-induced loadings and responses of deepwater offshore platforms and can also significantly affect the safety of ships encountering such waves in the deepwater environment.

Where the subordinate wave periods T_2 and T_3 differ by less than 15% from that of the dominant wave period, respectively, the interwave completely integrates the component waves so that no secondary interwave crests are observed. Although the interwave period and length in this case are

very close to the dominant wave period and length, the interwave height (H_{ix}) increases significantly, whereas the interwave celerity C_{ip} ($= C_{isc}$) assumes an intermediate value.

The average interwave steepness given by H_{ix}/L_{ix} varies from 0.0354 to 0.0502 for the range of conditions tested, whereas H_{is}/L_{ix} varies from 0.0354 to 0.0769 and H_{imax}/L_{ix} from 0.0354 to 0.0858. However, individual interwaves may be significantly steeper, because in the majority of cases, the highest crests occur adjacent to the deepest troughs.

Secondary interwave crests propagate at the celerity of the subordinate wave with the shortest period, whereas the primary interwaves have the same celerity as the dominant wave.

In terms of shipping and offshore design, the interrelationships outlined above can be used to forecast the highest and steepest colinear interwaves that may be expected under any particular combination of conditions (H_{imax} and H_{imax}/L_{ix}), whereas "average" conditions are better represented by H_{ix} and H_{ix}/L_{ix}. Of particular importance here is the recurrence frequency of peak conditions at any one location, as can be determined from the t/η diagram (using T_{ip}). In the case of colinear waves, such peak conditions would recur frequently at any specific location, but much less so if the waves are coming from different directions. The accompanying Excel spreadsheet (INTERWAVES) can be used to calculate the different parameters of interwaves, as well as to visualize their profiles at specific stations over time and over stretches of ocean at any moment in time.

For practical applications, it would be prudent and necessary to compare the estimates of any engineering quantity (e.g., wave forces and responses on structures or ships) derived from the methods proposed here to those based on statistical analysis of wave parameters (e.g., H_{im}, H_{rms}, and H_{max}) derived from internationally accepted standards, such as the IAHR and CEM methods. Risk and design studies should use the largest of these estimates.

Acknowledgment

The author is indebted to an anonymous reviewer for very constructive criticism, which helped to improve this paper.

References

[1] J. P. Le Roux, "Characteristics of developing waves as a function of atmospheric conditions, water properties, fetch and duration," *Coastal Engineering*, vol. 56, no. 4, pp. 479–483, 2009.

[2] B. Kinsman, *Wind Waves*, Prentice-Hall, Englewood Cliffs, NJ, USA, 1965.

[3] O. M. Phillips, *The Dynamics of the Upper Ocean*, Cambridge University Press, Cambridge, UK, 2nd edition, 1977.

[4] Z. Demirbilek and C. L. Vincent, *Water Wave Mechanics*, Coastal Engineering Manual (EM 1110-2-1100), chapter II-1, U.S. Army Corps of Engineers, Washington DC, USA, 2002.

[5] IAHR, *List of Sea State Parameters*, Supplement to Bulletin no. 52, International Association of Hydraulic Research, Brussels, Belgium, 1986.

[6] G. B. Airy, "Tides and waves," *Encyclopedia Metropolitana*, article 192, pp. 241–396, 1845.

[7] J. P. Le Roux, "A simple method to determine breaker height and depth for different deepwater wave height/length ratios and sea floor slopes," *Coastal Engineering*, vol. 54, no. 3, pp. 271–277, 2007.

[8] J. P. Le Roux, "A function to determine wavelength from deep into shallow water based on the length of the cnoidal wave at breaking," *Coastal Engineering*, vol. 54, no. 10, pp. 770–774, 2007.

[9] D. T. Resio, S. M. Bratos, and E. F. Thompson, *Meteorology and Wave Climate*, Coastal Engineering Manual, chapter II-2, U.S. Army Corps of Engineers, Washington, DC, USA, 2003.

[10] J. P. Le Roux, "An extension of the Airy theory for linear waves into shallow water," *Coastal Engineering*, vol. 55, no. 4, pp. 295–301, 2008.

5

Occurrence of Nitrogen Fixing Cyanobacterium *Trichodesmium* under Elevated $p\text{CO}_2$ Conditions in the Western Bay of Bengal

Suhas Shetye,[1] **Maruthadu Sudhakar,**[2] **Babula Jena,**[1] **and Rahul Mohan**[1]

[1] *National Centre for Antarctic & Ocean Research, Headland Sada, Goa 403 804, India*
[2] *Ministry of Earth Sciences, Prithvi Bhavan, New Delhi 110 003, India*

Correspondence should be addressed to Suhas Shetye; suhassht@gmail.com

Academic Editor: Lakshmi Kantha

Recent studies on the diazotrophic cyanobacterium *Trichodesmium* showed that increasing CO_2 partial pressure ($p\text{CO}_2$) enhances N_2 fixation and growth. We studied the *in situ* and satellite-derived environmental parameters within and outside a *Trichodesmium* bloom in the western coastal Bay of Bengal (BoB) during the spring intermonsoon 2009. Here we show that the single most important nitrogen fixer in today's ocean, *Trichodesmium erythraeum*, is strongly abundant in high ($\geq 300\,\mu$atm) $p\text{CO}_2$ concentrations. N : P ratios almost doubled (~10) at high $p\text{CO}_2$ region. This could enhance the productivity of N-limited BoB and increase the biological carbon sequestration. We also report presence of an oxygen minimum zone at Thamnapatnam. Earlier studies have been carried out using lab cultures, showing the increase in growth rate of *T. erythraeum* under elevated $p\text{CO}_2$ conditions, but to our knowledge, this study is the first to report that in natural environment also *T. erythraeum* prefers blooming in high $p\text{CO}_2$ concentrations. The observed CO_2 sensitivity of *T. erythraeum* could thereby provide a strong negative feedback to rising atmospheric CO_2 but would also drive towards phosphorus limitation in a future high CO_2 world.

1. Introduction

Climate change will significantly alter the marine environment within the next century and beyond. Future scenarios predict an increase from the current ~380 to ~750 to ~1,000 ppm CO_2 in the atmosphere towards the end of this century [1, 2]. As the ocean takes up this anthropogenic CO_2, dissolved inorganic carbon (DIC) in the surface ocean increases, while the pH decreases [3]. Rising global temperatures are likely to increase the surface ocean stratification, which may affect the light regime in the upper mixed layer as well as nutrient input from deeper waters [4]. Uncertainties remain regarding both the magnitude of the physicochemical changes and the biological responses of organisms, including species and populations of the oceanic primary producers at the base of the food web. In view of potential ecological implications and feedbacks on climate, several studies have examined the sensitivity of key phytoplankton species to $p\text{CO}_2$ [5–8]. Significant response to elevated $p\text{CO}_2$ was observed in N_2-fixing cyanobacteria [9–13], which play a

vital role in marine ecosystems by providing a new source of biologically available nitrogen to the otherwise nitrogen-limited regions [14].

Trichodesmium, a colony-forming cyanobacterium, fixes nitrogen in an area corresponding to almost half of Earth's surface [15] and is estimated to account for more than half of the new production in parts of the oligotrophic tropical and subtropical oceans [16, 17]. Future expansion of the oligotrophic subtropical provinces to higher latitudes due to surface ocean warming and increased stratification is expected to change the spatial extent of *Trichodesmium* and hence the magnitude of global N_2 fixation by this organism [18]. Recent studies focused on the impact of different environmental factors on the *Trichodesmium* species which has high abundance and forms massive blooms in tropical and subtropical areas [16, 17]. Higher $p\text{CO}_2$ levels stimulated growth, biomass production, and N_2 fixation [10, 11, 13] and affected the inorganic carbon acquisition of the cells [13]. While elevated sea surface temperatures are predicted to shift the spatial distribution of *Trichodesmium* species toward higher latitudes [18], the combined effects of pCO_2 and temperature may favour this species and increase its spatial extent even farther [10, 19]. An increase in the average light intensity, caused by the predicted shoaling of the upper mixed layer, may further stimulate photosynthesis and thus growth and N_2 fixation of *Trichodesmium* [20].

To test the hypothesis that *Trichodesmium* grows better in elevated pCO_2 conditions, we have carried out field studies within and around the *Trichodesmium* bloom in the western Bay of Bengal (BoB). In the northern Indian Ocean, studies on *Trichodesmium* bloom have been mostly in the Arabian Sea [21, 22], but N_2 fixation is a much more important source of nitrogen in BoB than in the Arabian Sea. The Bay is devoid of nutrient supply due to strong stratification, despite the nutrient supply through the rivers [23]. Satellite datasets on chlorophyll a (chl *a*), sea surface temperature (SST), wind stress (τ), and sea surface height anomaly (SSHA) were used to observe the spatial extent of bloom and to study the physical processes associated with it.

2. Methodology

2.1. In Situ Observations. Vertical profiles of CTD were collected from six different coastal regions along the east coast of India (Figure 1). At each region, 5 stations were collected based upon the water depth (50, 100, 200, 500, and 1000 m). Seawater samples were collected using a Rosette sampler with 5 litre Niskin bottles mounted on the CTD assembly. Salinity was measured with the help of an Autosal. Nutrients (silicate, phosphate, nitrate, and nitrite) were measured with a Skalar Autoanalyzer by standard colorimetric methods. Standards were used to calibrate the auto analyzer, and frequent baseline checks were made. The standard deviation for duplicates was $0.07\,\mu M$ for silicate, $0.06\,\mu M$ for nitrate, and $0.01\,\mu M$ for nitrite and phosphate. Total carbon dioxide (TCO_2) content of seawater samples was determined using a coulometer (model 5014 of U.I.C. Inc., USA). The reliability of the coulometric titration was regularly checked with certified referenced materials (CRMs, Batch number 92) provided by A. Dickson (SIO, University of California). The accuracy estimated from the CRMs values was $2\,\mu mol$ for TCO_2. The pH_T and pH_f were measured at $25°C$ by cresol red spectrophotometry [24]. The pH_T of the samples was then corrected to the *in situ* temperature following equation of Gieskes [25]. Analytical precision was ~0.002 for pH. *In situ* pCO_2 was calculated using the dissociation constants of Dickson and Millero [26].

The CO_2 exchange flux (mmol C m^{-2} day^{-1}) across the sea-air interface was calculated using equation given in Wanninkhof [27]. $F = k \times s \times \Delta pCO_2$, where k is the gas transfer velocity, s is the solubility of CO_2 gas in seawater [28], and ΔpCO_2 is the difference between surface seawater and atmospheric pCO_2. The sea-air pCO_2 difference is computed using the measured surface water pCO_2 values and the atmospheric pCO_2 from the zonal mean CO_2 concentrations reported by the GLOBALVIEW-CO_2 (2009).

Water samples were collected from patches of *Trichodesmium* blooms near the coastal Thamnapatnam (in

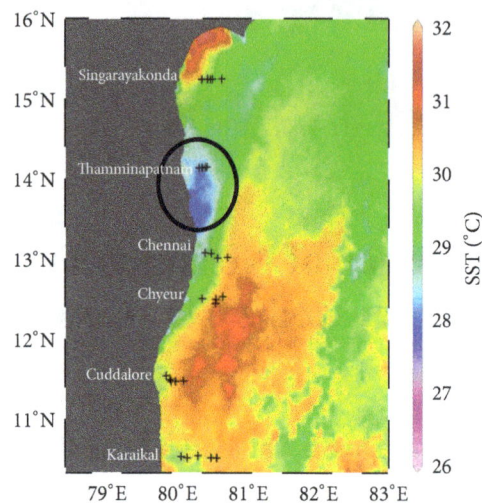

FIGURE 1: Sampling locations (+) overlaid over the NOAA-AVHRR weekly composite sea surface temperature (SST) map during 9th–16th May 2009. Circle showing the cooling feature (~27°C to ~27.5°C) associated with the *Trichodesmium* bloom near Thamnapatnam.

Tamil Nadu, India) and then fixed in a 2% formalin solution to permit the identification and counting of phytoplankton. Samples fixed in formalin were observed through a polarising microscope for species identification.

2.2. Satellite-Derived Environmental Parameters. Satellite observations were used to study the spatial extent of bloom and the physical processes associated with it. SeaWiFS (Sea-viewing Wide Field-of-view Sensor) based Level-3 global standard mapped images (SMI) of climatological chl *a* values (9 km spatial resolution) were acquired from Goddard Space Flight Centre (GSFC), for the month of May (1997–2010). Chl *a* datasets based upon SeaWiFS operational bio-optical algorithm (i.e., ocean color, OC4) developed by O'Reilly et al. [29] and later updated (OC4v6) by National Aeronautics and Space Administration (NASA) Ocean Biology Processing Group (OBPG). The previous algorithm yields a strong correlation ($r = 0.892$) with *in situ* Chl *a* on global scale that includes samples from all water types [29]. Level-3 Pathfinder SSTs dataset (4 km spatial resolution) from the Advanced Very High Resolution Radiometer (AVHRR) was obtained from NASA's Jet Propulsion Laboratory (JPL) and utilized during the bloom period. Multiple satellite altimeters (Jason-1, TOPEX/Poseidon, ERS 1/2, and GFO) merged product on sea surface height anomalies (SSHA) at a spatial resolution of $1° \times 1°$ were obtained from the NASA Physical Oceanography Distributed Active Archive Center (PODAAC). QuikScat-measured wind vector (scalar wind speed with corresponding u and v components) data files available at spatial resolution of 25 km were downloaded from http://www.ssmi.com/. The wind stress (τ) was then calculated using variable drag coefficients (C_D) given by Yelland and Taylor [30].

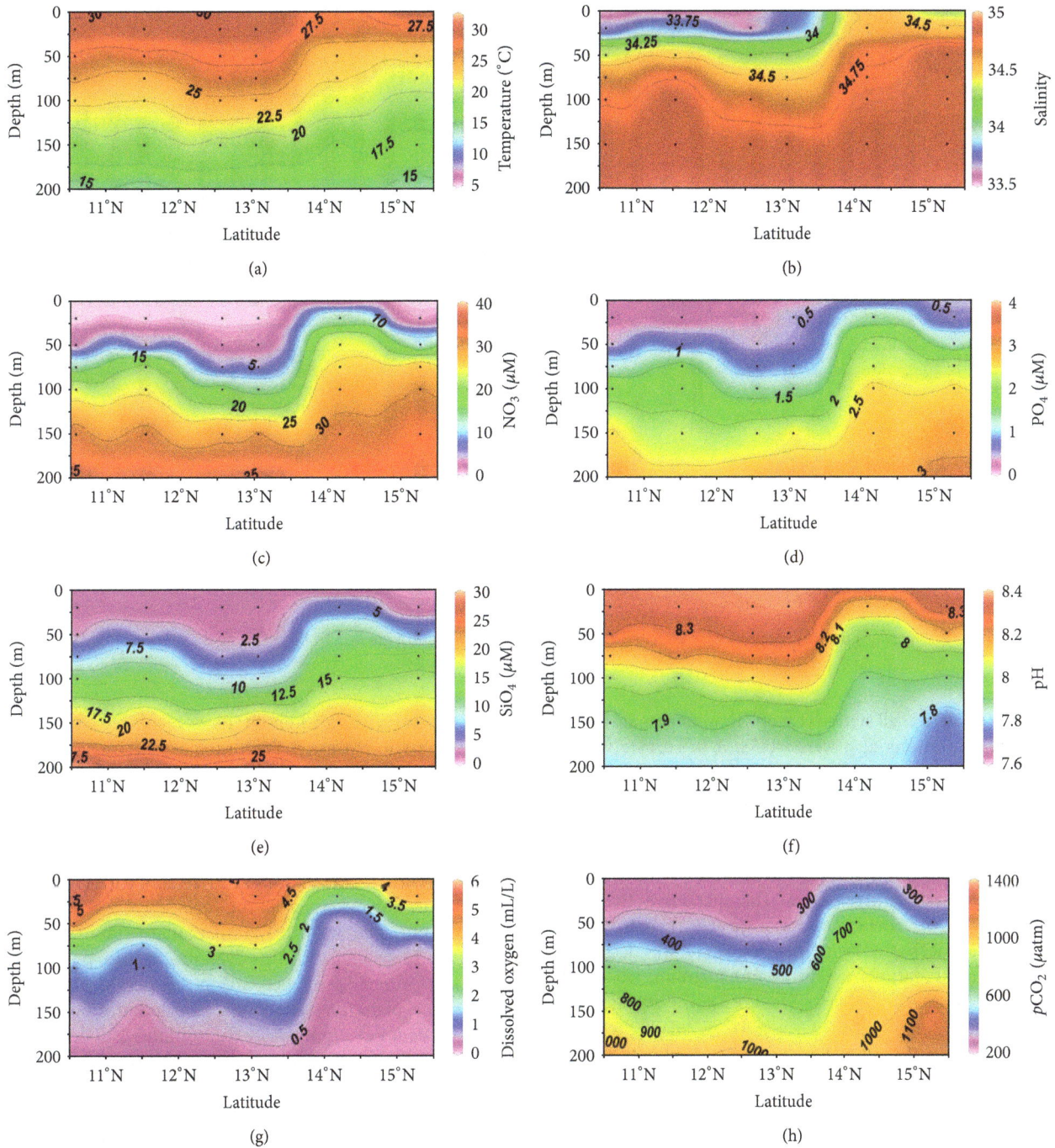

FIGURE 2: Cross-section of the zonal (80.15°E to 80.75°E) physicochemical parameters collected onboard ORV-Sagar Sampada during April-May 2009. The upliftment of thermocline, halocline, nutricline, elevated CO_2, and oxygen deficient water near 14°N (off Thamnapatnam) is due to upwelling resulted from divergence of water masses.

3. Results and Discussion

3.1. Observed Variability of Physicochemical Parameters. BoB is highly influenced by monsoons and receives large volume of freshwater from both river discharge and rainfall [31, 32] which results in low sea surface salinity (SSS). All stations had salinity < 34 and temperature > 29°C except Thamnapatnam.

At Thamnapatnam surface salinity was 34.53, and the SST was 27.2°C.

Depth profiles of SST, SSS, and nutrients (NO_3, NO_2, PO_4, and SiO_4) are shown in Figure 2. Surface NO_3 concentrations were <1 μM up to 75 m depth at all stations except Thamnapatnam, where it reached concentrations of 13 μM at 20 m and went on increasing with depth. SiO_4 concentrations were

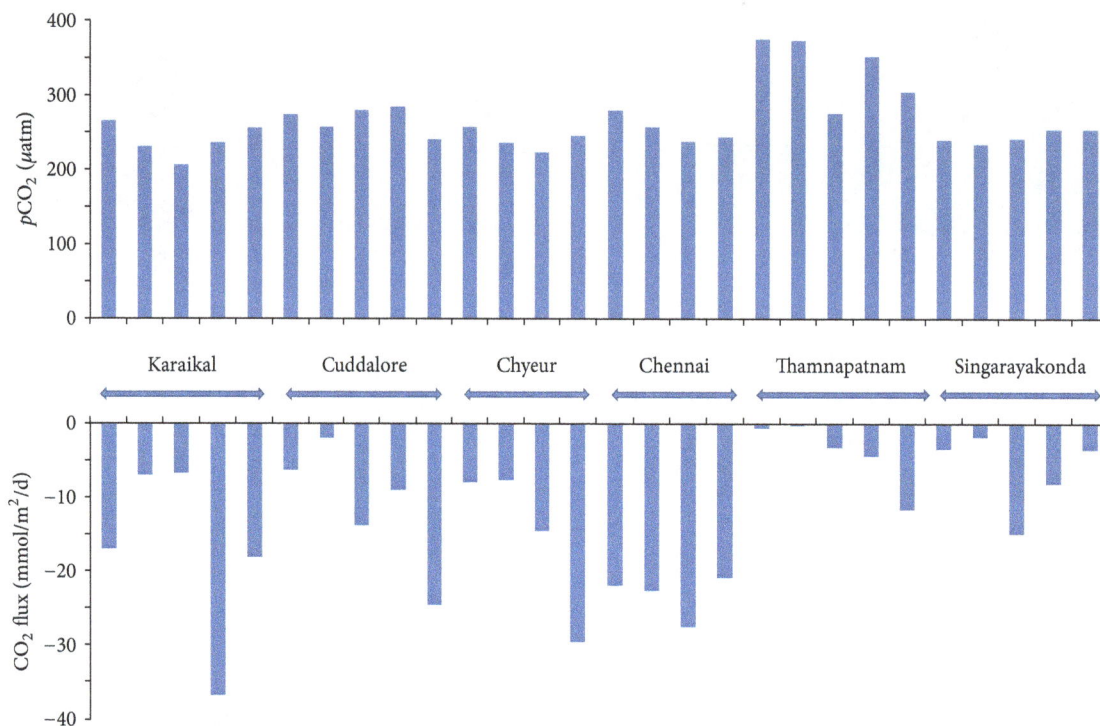

FIGURE 3: Station wise distribution of pCO_2 and carbon dioxide flux in Bay of Bengal.

<5 μM up to 20 m at all stations except Thamnapatnam. Phosphate concentrations also were <0.5 μM in surface waters, except Thamnapatnam. Dissolved oxygen concentrations were high at the surface and were above 4 mL/L at all stations. Maximum concentration was found at Chennai, reaching up to 5.89 mL/L (Figure 2). Concentration decreased rapidly with depth, and an intense oxygen minimum zone (OMZ) is seen between 150 and 500 m. At Thamnapatnam the OMZ starts from 75 m. Surface pH was greater than 8.3 at all stations and decreased with depth. pCO_2 was less than 300 uatm at all stations except Thamnapatnam, where the maximum pCO_2 was 376 uatm (Figure 3). The average CO_2 flux in Bay of Bengal was −12.8 mmol C m^{-2} day^{-1}, whereas at Thamnapatnam the CO_2 flux was −3.9 mmol C m^{-2} day^{-1} (Figure 3). The negative sign indicates that the region is a sink for atmospheric CO_2.

3.2. *Physical Forcing.* The shoaling of thermocline, halocline, nutricline, and elevated CO_2 and oxygen deficient water (Figure 2) near 14°N (off Thamnapatnam) is due to shoaling and resulted from divergence of water masses. The shoaling-induced cooling feature is well captured by NOAA AVHRR derived SST during May 12, 2009 (Figure 4(a)) and also reflecting in weekly composite product (9–16th May 2009) signifying persistency of the feature (Figure 1). Satellite observations indicated lower SST (~27°C to ~27.5°C) and associated negative SSHA ranging from −0.1 m to −0.2 m near the bloom region off Thamnapatnam (Figures 4(a) and 4(b)). These features represent an upwelling area where cold, high nutrient, and dense waters are pushed to the surface (Figure 2), leading to phytoplankton bloom. The alongshore wind stress is favourable off Thamnapatnam to induce coastal upwelling finally resulting in a phytoplankton bloom (Figure 4(c)).

3.3. *Distribution of Trichodesmium erythraeum.* Colonies of the cyanobacterium (1-2 mm size bundles of trichomes) could be seen with the naked eye in the surface water at all stations along Thamnapatnam. At Thamnapatnam stations, surface water was blooming with this cyanobacterium, possibly the dominant primary producer during this period. Analysis of the surface plankton collections revealed the presence of *Trichodesmium erythraeum* in the sample. It appeared that *T. erythraeum* was concentrated in the upper 3 meters of the water column due to its positive buoyancy [33–35] and the surface water attained a brownish color. Around 4500 filaments of *T. erythraeum* per litre seawater were observed. A colony of *T. erythraeum* collected from the Thamnapatnam consisted of ~ 3.1×10^5 cells. *T. erythraeum* comprised 100% of total cell count and showed complete dominance over other phytoplankton.

The *T. erythraeum* bloom reported here assumes considerable importance as the BoB is believed to be a region of relatively low productivity compared to the Arabian Sea. Stratification caused by freshwater influx is thought to restrict upwelling of nutrients. Stratification in the southern BoB is weaker than in the north [36, 37].

During spring intermonsoon, eddies and recirculation zones from the coastal regions of BoB due to the western BoB current (WBC) were found to enhance phytoplankton growth [38, 39]. Since the chl *a* in the upper layer of BoB is

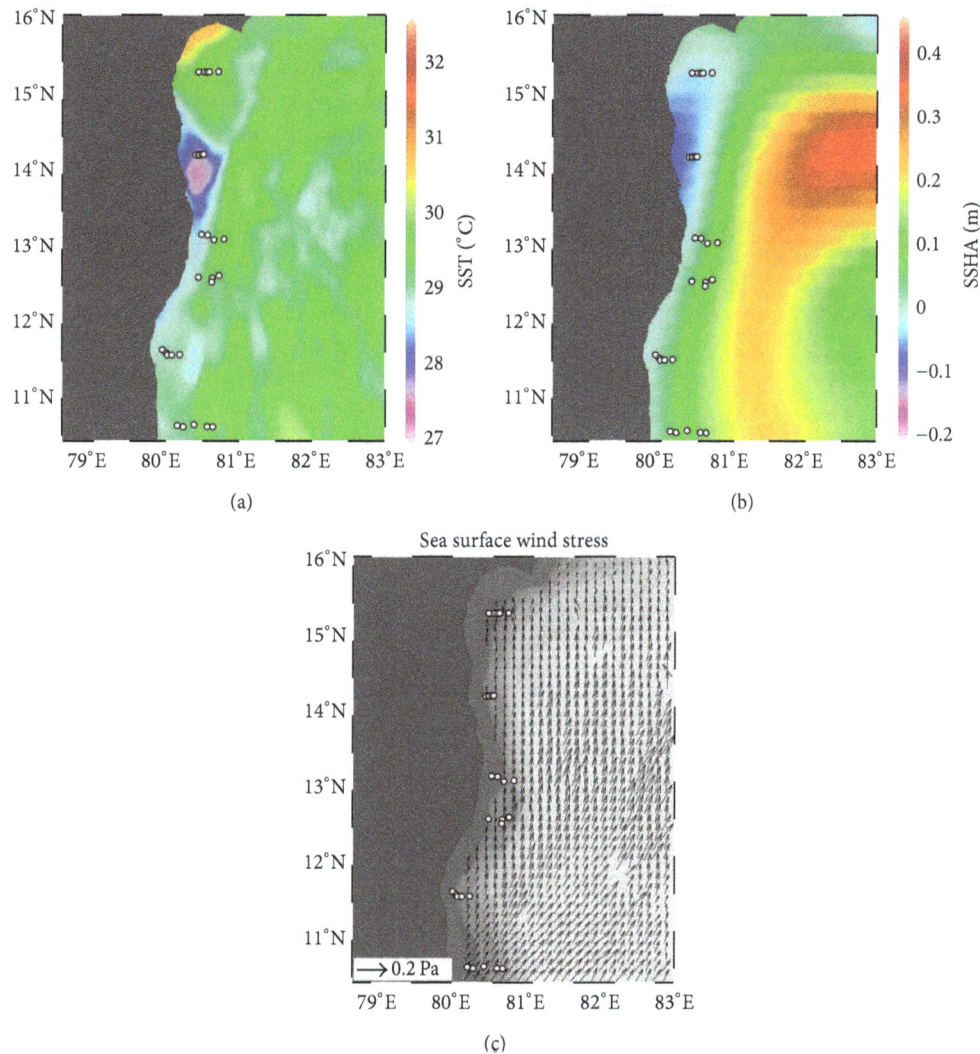

FIGURE 4: Satellite observations during May 12, 2009 illustrating (a) NOAA-AVHRR sea surface temperature in °C, (b) satellite altimetry-based sea surface height anomaly (SSHA) in meter, and (c) QuikSCAT-measured wind stress vectors (Pa).

limited by the availability of nutrients, such oceanic processes that can bring nutrients into the euphotic zone are of prime importance. Nutrients are brought to surface waters by coastal upwelling driven by alongshore winds (Figure 4(c)). In order to identify the physical process that could cause the high chl a blooms described previously, we have examined the relevant physicochemical oceanographic data, and the results corresponding to May 12, 2009 are shown in Figure 2. The winds have a large alongshore component (Figure 4(c)) and strong offshore component, thus resulting in coastal upwelling.

Since the surface water temperature was >26°C, it would have favored the *Trichodesmium* bloom as marine cyanobacteria exhibit temperature optima somewhere in the range of 25–35°C [40]. Studies by Suvapepun [41] and Sellner [42] have reported that cyanobacteria require higher temperature optima for growth than other phytoplankton and that the

temperature has been the most important factor contributing to cyanobacterial dominance. Buoyancy regulation by cyanobacteria plays a key role in this phenomenon, as soon as the wind abates; cyanobacteria float rapidly towards the surface due to their positive buoyancy [42]. The surface accumulations are quickly dispersed over the water column by wind-induced mixing at wind speeds over 6–8 m s^{-1} [43].

While effects of CO$_2$-related seawater acidification have been demonstrated for a variety of marine microalgae and cyanobacteria [44], mainly focusing on carbon acquisition and concentrating mechanisms [45], little is known about its impact on marine diazotrophs. Earlier studies by Hutchins et al. [10], Ramos et al. [9], Levitan et al. [11], and Kranz et al. [13] reported from lab experiments that *T. erythraeum* prefers high CO$_2$ conditions. This trend is predominantly attributed to changes in cell division [10, 11] but also altered elemental ratios of carbon to nitrogen [11] or nitrogen to

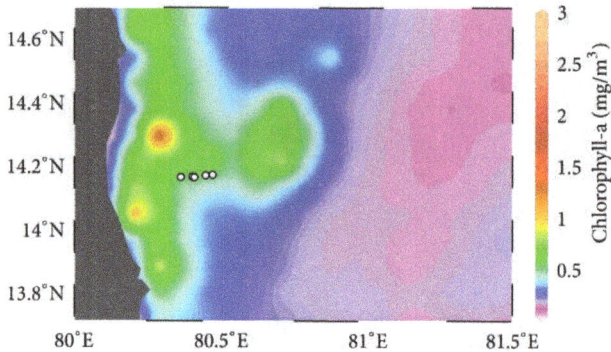

FIGURE 5: SeaWiFS observed climatological chl *a* concentration (1997–2010) for the month of May at the *Trichodesmium* bloom location in western Bay of Bengal.

phosphorus [9]. A first step toward a mechanistic understanding of responses in *T. erythraeum* has been taken by Levitan et al. [11], focusing on pCO_2 dependency of nitrogenase activity and photosynthesis. We tried to investigate the same hypothesis in natural environment, and we found *T. erythraeum* blooms at Thamnapatnam station which had the highest pCO_2 concentrations. At Thamnapatnam all the 5 stations had pCO_2 concentrations $\geq 300\,\mu$atm. In *T. erythraeum*, photosynthetically generated energy (ATP and NADPH) is primarily used for the fixation of CO_2 in the Calvin-Benson cycle. Cyanobacterial Rubisco possesses one of the lowest CO_2 affinities among phytoplankton [46], their Rubisco is confined to multiple capsid-like carboxysomes and helps to minimize CO_2 leakage [47] thus giving them advantage over other phytoplankton groups.

Nitrogen fixation by *Trichodesmium* also alters the N : P ratio and could drive towards phosphate limitation. N : P ratio was 10 at Thamnapatnam, whereas in Chennai it was 5.3, clearly indicating phosphorus limitation in *Trichodesmium* blooms. The average N/P ratio upto 200 m at all stations was less than 10, indicating large deviations from the standard Redfield's N/P ratio of 16. Analysis of climatological SeaWiFS chlorophyll a image for the month of May off Thamnapatnam (Figure 5) indicated elevated chl a concentration up to $\sim$3 mg/m^3. Jyotibabu et al. [48] have reported primary productivity as high as 2160 mg C m^{-2} d^{-1} with in the bloom in BoB.

Upwelling brought high pCO_2 water and nutrients from subsurface to surface and led to blooming of *Trichodesmium* at surface. The underlying processes responsible for the strong CO_2 sensitivity in this important diazotroph are currently unknown. Our results clearly indicate that with rise in CO_2 in future, the abundance of *Trichodesmium sp.* and N_2 fixation could have potential biogeochemical implications, as it may stimulate productivity in N-limited oligotrophic regions and thus provide a negative feedback on rising atmospheric CO_2 levels.

4. Conclusion

This study has been carried out by conjunctive analysis of *in situ* and satellite observations within and outside a

Trichodesmium bloom in the western Bay of Bengal during the spring intermonsoon 2009. Oceanic environmental parameters such as SST, chl *a*, wind, and SSHA measured from remote sensing satellites are useful to study the bloom due to its synoptic, high spatial resolution, and repetitive characteristics. In this study we show that *Trichodesmium* is strongly abundant in high pCO_2 ($>$300 μatm) concentrations. N : P ratios almost doubled ($\sim$10) at this high CO_2 region. We also report presence of an oxygen minimum zone in the Bay of Bengal. Earlier studies have been carried out using lab cultures, showing the increase in growth rate of *Trichodesmium* in high pCO_2 conditions, but our study proves for the first time that in natural environment also *Trichodesmium* bloom in high pCO_2 waters. The observed CO_2 sensitivity of *Trichodesmium* could thereby provide a strong negative feedback to rising atmospheric CO_2 but could also lead towards phosphate limitation. Our study also proves that Thamnapatnam is a site for coastal upwelling, OMZ, and *Trichodesmium* bloom formation. However, metabolic processes must be studied in detail to understand the responses of *Trichodesmium* to changes in environmental conditions.

Acknowledgments

The authors would like to thank the Ministry of Earth Sciences for providing the necessary ship time. The authors thank the captain, crew, participants, and technical staff of the FORV Sagar Sampada for their invaluable assistance. The authors also thank the chief scientist Dr. Rosamma Philip for her support. SeaWiFS dataset was provided by NASA's Goddard Space Flight Center. The AVHRR and QuikScat data were obtained from the NASA Physical Oceanography Distributed Active Archive Center (PODAAC).

References

[1] J. Raven, K. Caldeira, and H. Elderfield, *Ocean Acidification due to Increasing Atmospheric Carbon Dioxide*, The Royal Society, London, UK, 2005.

[2] M. R. Raupach, G. Marland, P. Ciais et al., "Global and regional drivers of accelerating CO_2 emissions," *Proceedings of the National Academy of Sciences of the United States of America*, vol. 104, no. 24, pp. 10288–10293, 2007.

[3] D. A. Wolf-Gladrow, U. Riebesell, S. Burkhardt, and J. Bijma, "Direct effects of CO_2 concentration on growth and isotopic composition of marine plankton," *Tellus B*, vol. 51, no. 2, pp. 461–476, 1999.

[4] S. C. Doney, "Oceanography: plankton in a warmer world," *Nature*, vol. 444, no. 7120, pp. 695–696, 2006.

[5] S. Burkhardt and U. Riebesell, "CO_2 availability affects elemental composition (C: N: P) of the marine diatom skeletonema costatum," *Marine Ecology Progress Series*, vol. 155, pp. 67–76, 1997.

[6] U. Riebesell, I. Zondervan, B. Rost, P. D. Tortell, R. E. Zeebe, and F. M. M. Morel, "Reduced calcification of marine plankton in response to increased atmospheric CO_2," *Nature*, vol. 407, no. 6802, pp. 364–366, 2000.

[7] B. Rost, U. Riebesell, S. Burkhardt, and D. Sültemeyer, "Carbon acquisition of bloom-forming marine phytoplankton," *Limnology and Oceanography*, vol. 48, no. 1, pp. 55–67, 2003.

[8] P. D. Tortell, C. D. Payne, Y. Li et al., "CO$_2$ sensitivity of Southern Ocean phytoplankton," *Geophysical Research Letters*, vol. 35, no. 4, article L04605, 2008.

[9] B. E. J. Ramos, H. Biswas, K. G. Schulz, J. Barcelose, J. LaRoche, and U. Riebesell, "Effect of rising atmospheric carbon dioxide on the marine nitrogen fixer *Trichodesmium*," *Global Biogeochemical Cycles*, vol. 21, no. 2, article GB2028, 2007.

[10] D. A. Hutchins, F. X. Fu, Y. Zhang et al., "CO$_2$ control of *Trichodesmium* N$_2$ fixation, photosynthesis, growth rates, and elemental ratios: implications for past, present, and future ocean biogeochemistry," *Limnology and Oceanography*, vol. 52, no. 4, pp. 1293–1304, 2007.

[11] O. Levitan, G. Rosenberg, I. Setlik et al., "Elevated CO$_2$ enhances nitrogen fixation and growth in the marine cyanobacterium *Trichodesmium*," *Global Change Biology*, vol. 13, no. 2, pp. 531–538, 2007.

[12] F. X. Fu, M. R. Mulholland, N. S. Garcia et al., "Interactions between changing pCO$_2$, N$_2$ fixation, and Fe limitation in the marine unicellular cyanobacterium Crosphaera," *Limnology and Oceanography*, vol. 53, no. 6, pp. 2472–2484, 2008.

[13] S. A. Kranz, D. Sultemeyer, K. U. Richter, and B. Rost, "Carbon acquisition by *Trichodesmium*: the effect of pCO$_2$ and diurnal changes," *Limnology and Oceanography*, vol. 54, no. 2, pp. 548–559, 2009.

[14] L. A. Codispoti, J. A. Brandes, J. P. Christensen et al., "The oceanic fixed nitrogen and nitrous oxide budgets: moving targets as we enter the anthropocene?" *Scientia Marina*, vol. 65, no. 2, pp. 85–105, 2001.

[15] C. S. Davis and D. J. McGillicuddy, "Transatlantic abundance of the N$_2$-fixing colonial cyanobacterium *Trichodesmium*," *Science*, vol. 312, no. 5779, pp. 1517–1520, 2006.

[16] D. G. Capone, J. A. Burns, J. P. Montoya et al., "Nitrogen fixation by *Trichodesmium* spp.: an important source of new nitrogen to the tropical and subtropical North Atlantic Ocean," *Global Biogeochemical Cycles*, vol. 19, no. 2, article GB2024, pp. 1–17, 2005.

[17] C. Mahaffey, A. F. Michaels, and D. G. Capone, "The conundrum of marine N$_2$ fixation," *American Journal of Science*, vol. 305, no. 6–8, pp. 546–595, 2005.

[18] E. Breitbarth, A. Oschlies, and J. LaRoche, "Physiological constraints on the global distribution of *Trichodesmium*—effect of temperature on diazotrophy," *Biogeosciences*, vol. 4, no. 1, pp. 53–61, 2007.

[19] O. Levitan, C. M. Brown, S. Sudhaus, D. Campbell, J. LaRoche, and I. Berman-Frank, "Regulation of nitrogen metabolism in the marine diazotroph *Trichodesmium* IMS101 under varying temperatures and atmospheric CO$_2$ concentrations," *Environmental Microbiology*, vol. 12, no. 7, pp. 1899–1912, 2010.

[20] E. Breitbarth, J. Wohlers, J. Klas, J. LaRoche, and I. Peeken, "Nitrogen fixation and growth rates of *Trichodesmium* IMS-101 as a function of light intensity," *Marine Ecology Progress Series*, vol. 359, pp. 25–36, 2008.

[21] E. Desa, T. Suresh, S. G. P. Matondkar et al., "Detection of *Trichodesmium* bloom patches along the eastern Arabian Sea by IRS-P4/OCM ocean color sensor and by in-situ measurements," *Indian Journal of Marine Sciences*, vol. 34, no. 4, pp. 374–386, 2005.

[22] N. Gandhi, A. Singh, S. Prakash et al., "First direct measurements of N$_2$ fixation during a *Trichodesmium* bloom in the eastern Arabian Sea," *Global Biogeochemical Cycles*, vol. 25, no. 4, article GB4014, 2011.

[23] A. Singh and R. Ramesh, "Contribution of riverine organic nitrogen flux to new production in the coastal northern Indian Ocean: an assessment," *International Journal of Oceanography*, vol. 11, pp. 1–7, 2011.

[24] R. H. Byrne and J. A. Breland, "High precision multiwavelength pH determinations in seawater using cresol red," *Deep Sea Research I*, vol. 36, no. 5, pp. 803–810, 1989.

[25] J. M. Gieskes, "Effect of temperature on the pH of seawater," in *Limnology Oceanography*, vol. 14, pp. 679–685, 1969.

[26] A. G. Dickson and F. J. Millero, "A comparison of the equilibrium constants for the dissociation of carbonic acid in seawater media," *Deep Sea Research I*, vol. 34, no. 10, pp. 1733–1743, 1987.

[27] R. Wanninkhof, "Relationship between wind speed and gas exchange over the ocean," *Journal of Geophysical Research*, vol. 97, no. 5, pp. 7373–7382, 1992.

[28] R. F. Weiss, "Carbon dioxide in water and seawater: the solubility of a non-ideal gas," *Marine Chemistry*, vol. 2, no. 3, pp. 203–212, 1974.

[29] J. E. O'Reilly, S. Maritorena, M. C. O'Brien et al., "Ocean color chlorophyll a algorithms for SeaWiFS, OC2, and OC4: version 4," in *SeaWiFS Postlaunch Calibration and Validation Analyses*, S. B. Hooker and E. R. Firestone, Eds., vol. 11 of *SeaWiFS Postlaunch Technical Report*, part 3, pp. 9–23, NASA, Goddard Space Flight Center, Greenbelt, Md, USA, 2000.

[30] M. Yelland and P. K. Taylor, "Wind stress measurements from the open ocean," *Journal of Physical Oceanography*, vol. 26, no. 4, pp. 541–558, 1996.

[31] P. N. Vinayachandran and T. Yamagata, "Monsoon response of the sea around Sri Lanka: generation of thermal domes and anticyclonic vortices," *Journal of Physical Oceanography*, vol. 28, no. 10, pp. 1946–1960, 1998.

[32] A. Singh, R. A. Jani, and R. Ramesh, "Spatiotemporal variations of the δ^{18}O-salinity relation in the northern Indian Ocean," *Deep-Sea Research I*, vol. 57, no. 11, pp. 1422–1431, 2010.

[33] D. G. Capone, J. P. Zehr, H. W. Paerl, B. Bergman, and E. J. Carpenter, "*Trichodesmium*, a globally significant marine cyanobacterium," *Science*, vol. 276, no. 5316, pp. 1221–1229, 1997.

[34] D. G. Capone, A. Subramaniam, J. P. Montoya et al., "An extensive bloom of the N$_2$-fixing cyanobacterium *Trichodesmium erythraeum* in the central Arabian Sea," *Marine Ecology Progress Series*, vol. 172, pp. 281–292, 1998.

[35] T. Shiozaki, K. Furuya, T. Kodama et al., "New estimation of N$_2$ fixation in the western and central Pacific Ocean and its marginal seas," *Global Biogeochemical Cycles*, vol. 24, no. 1, article GB1015, 2010.

[36] P. N. Vinayachandran, V. S. N. Murty, and V. R. Babu, "Observations of barrier layer formation in the bay of Bengal during summer monsoon," *Journal of Geophysical Research*, vol. 107, no. 12, pp. SRF 19-1–SRF 19-9, 2002.

[37] G. S. Bhat, S. Gadgil, H. P. V. Kumar et al., "BOBMEX: the bay of Bengal monsoon experiment," *Bulletin of the American Meteorological Society*, vol. 82, no. 10, pp. 2217–2243, 2001.

[38] H. R. Gomes, J. I. Goes, and T. Saino, "Influence of physical processes and freshwater discharge on the seasonality of phytoplankton regime in the bay of Bengal," *Continental Shelf Research*, vol. 20, no. 3, pp. 313–330, 2000.

[39] S. P. Kumar, M. Nuncio, J. Narvekar et al., "Are eddies nature's trigger to enhance biological productivity in the bay of Bengal?" *Geophysical Research Letters*, vol. 31, no. 7, article L07309, 2004.

[40] G. E. Fogg, W. D. P. Stewart, P. Fay, and A. E. Walsby, *The Blue-Green Algae*, Academic Press, London, UK, 1973.

[41] S. Suvapepun, "*Trichodesmium* blooms in the gulf of Thailand," *Marine Pelagic Cyanobacteria*, pp. 343–348, 1992.

[42] K. G. Sellner, "Physiology, ecology, and toxic properties of marine cyanobacteria blooms," *Limnology and Oceanography*, vol. 42, no. 5, pp. 1089–1104, 1998.

[43] M. Kahru, J. M. Leppanen, and O. Rud, "Cyanobacterial blooms cause heating of the sea surface," *Marine Ecology Progress Series*, vol. 101, no. 1-2, pp. 101–107, 1993.

[44] M. Giordano, J. Beardall, and J. A. Raven, "CO_2 concentrating mechanisms in algae: mechanisms, environmental modulation, and evolution," *Annual Review of Plant Biology*, vol. 56, pp. 99–131, 2005.

[45] S. Burkhardt, G. Amoroso, and U. Riebesell, "CO_2 and HCO_3^- uptake in marine diatoms acclimated to different CO_2 concentrations," *Limnology and Oceanography*, vol. 46, no. 6, pp. 1378–1391, 2001.

[46] M. R. Badger, T. J. Andrews, S. M. Whitney et al., "The diversity and coevolution of rubisco, plastids, pyrenoids, and chloroplast-based CO_2-concentrating mechanisms in algae," *Canadian Journal of Botany*, vol. 76, no. 6, pp. 1052–1071, 1998.

[47] M. Meyer and H. Griffiths, "Origins and diversity of eukaryotic CO_2-concentrating mechanisms: lessons for the future," *Journal of Experimental Botany*, vol. 64, no. 2, pp. 769–786, 2013.

[48] R. Jyotibabu, N. V. Madhu, N. Murukesh, P. C. Haridas, K. K. C. Nair, and P. Venugopal, "Intense blooms of *Trichodesmium erythraeum* in the open water along east coast of India," *The Indian Journal of Geo-Marine Sciences*, vol. 32, no. 2, pp. 165–167, 2003.

6

Internal Solitary Waves in the Brazilian SE Continental Shelf: Observations by Synthetic Aperture Radar

João A. Lorenzzetti and Fabian G. Dias

Divisão de Sensoriamento Remoto, Instituto Nacional de Pesquisas Espaciais (INPE), CP 515, 12227-010 São José dos Campos, SP, Brazil

Correspondence should be addressed to João A. Lorenzzetti; loren@dsr.inpe.br

Academic Editor: Robert Frouin

We present an analysis of internal solitary waves (ISWs) on the SE Brazilian continental shelf using a set of Envisat/ASAR satellite images. For the 17-month observation period, 467 ISW packets were detected. Most of observed solitons were associated to 4–6 ms^{-1} wind. The number of ISW packets shows a seasonal signal with a peak in summer, with higher concentration in the outer shelf in all seasons, followed by midshelf during the summer. Propagation direction of ISWs was predominantly onshore with packets separated by typical M_2 internal tide wavelengths (~10–40 km). The highest values of the barotropic tidal forcing F are concentrated at the shelf break between 200 and 500 m isobaths. These characteristics suggest that ISWs are formed from nonlinear disintegration of internal tides generated at the shelf break that propagate shoreward as interfacial internal waves. No significant change in the number of ISWs from spring to neap tides was observed in spite of significant tidal current variation (60%). Even not being a region of strong tides, this study shows that ISWs are a frequent and widespread feature, possibly playing a significant dynamic role, affecting biological production, sediment dispersion, and transport.

1. Introduction

Since the launch of SEASAT in 1978, numerous studies of oceanic internal solitary waves (ISWs) have been made using synthetic aperture radar (SAR) images. The availability of a great number of SAR satellite images has shown that ISWs are an ubiquitous oceanic phenomenon. They are frequently observed wherever tidal currents and stratification occur near significant seafloor topographic features, such as shelf break zones, plateaus, or sills [1]. Although a large number of remote sensing studies of ISWs have used SAR images, these features can also be observed in ocean color imagery [2] and in Sun glitter regions of visible images [3]. ISWs normally appear in SAR images as packets of 2–8 dark and bright stripes, with subsequent packets separated by distances corresponding to the wavelength of the internal tides (Figure 1) [1, 4].

At typical microwave frequencies used in SAR systems (1–10 GHz), the penetration depth of radar pulses in sea water, which is dependent on the complex permittivity, is restricted to less than about one centimeter [5]. Although practically a surface phenomenon, the backscattering of SAR electromagnetic pulses by wind-generated centimeter scale gravity-capillary surface waves (Bragg resonant waves) allows the visualization of ISWs, which are present tens of meters deep. Several physical mechanisms have been postulated as responsible for making ISWs visible in SAR images. In a pioneer work, [6] suggested as the main mechanism the hydrodynamic modulating effect of convergences/divergences of ISWs surface currents on the Bragg waves. Another mechanism was proposed by [7, 8], who showed that in low wind speed regions (less than 5 ms^{-1}), the ISW signatures in SAR images are strongly controlled by the effects of surface films or surfactants. More recently, [9] showed that wave breaking is also a major mechanism leading to the formation of surface signatures of ISWs in SAR images.

ISWs can play an important role in oceanic biological productivity. The passage of a train of ISWs can uplift the phytoplankton near the mixed-layer base, increasing light intensity and photosynthesis [10]. An enhancement in productivity can also occur by vertical mixing and upward flux of nutrients

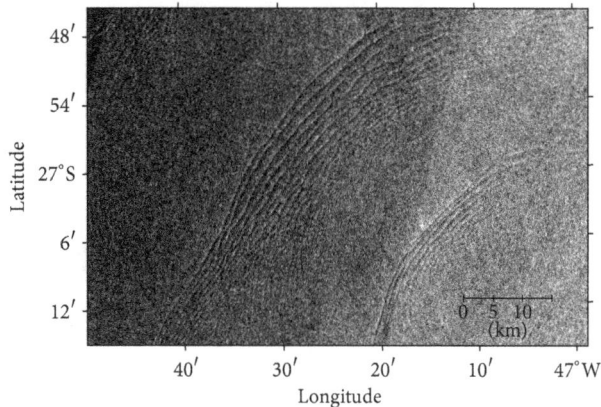

FIGURE 1: Example of two ISWs packets generated at consecutive tidal cycles observed on the Brazilian southeast coast. Image acquisition date: February 3, 2010; geographic position is indicated at insert box in Figure 2.

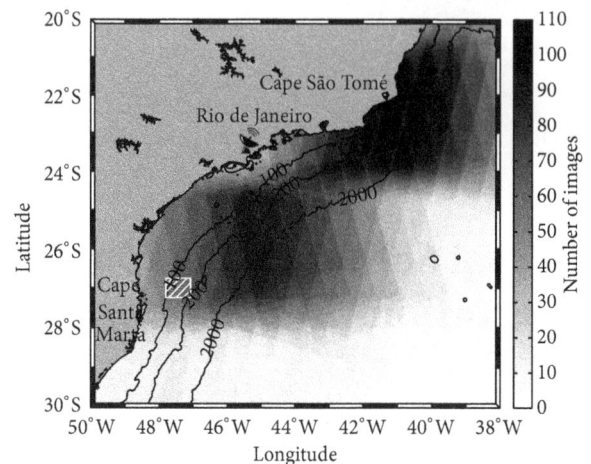

FIGURE 2: Study region with the number of ASAR scenes available in each area (see gray scale at right). Continuous black lines: isobaths in meters. White hatched rectangle represents geographic coverage of image in Figure 1. Small antenna indicates position of INPE's SAR data reception station.

during energy dissipation by internal wave breaking [11–14]. ISWs can as well affect sediment resuspension, producing turbidity signatures [15, 16] and inducing sediment transport [17]. Internal waves and solitons are reported affecting sound propagation. Anomalous sound propagation and acoustic transmission losses in the summer and in the coastal zone have been attributed to the resonance of the sound signal with ISWs [18].

ISWs have their origin in the baroclinic internal tides (ITs) which are forced by the deep ocean barotropic tide. The shelf break is the most important place for ITs generation. There, the flow of barotropic tidal currents interacts with steep bottom gradients producing large displacements of the isopycnals that start to propagate as ITs during tide reversals [19]. The most common form of ISWs generation is when IT energy, formed at the shelf break radiates horizontally as an interfacial internal wave in the thermocline/pycnocline. As the IT propagates onshore, it can disintegrate into ISWs in the form of solitons, which are able to maintain their shapes and velocities as the result of a balance between nonlinearity that tends to steepen their crests and dispersion that broaden them, much as an undular bore [20, 21]. The ISWs travel in the same direction as the generating ITs and are phase locked to their troughs [22]. In such cases, a remote generation region can normally be attributed to the observed ISWs.

Some ISW packets can appear locally, that is, well away from any specific topographic feature of origin as the shelf break. These are cases in which an IT energy ray reaches the thermocline after reflection in the bottom or at surface, producing large displacements in the thermocline. References [23, 24] reported such a case for the central Bay of Biscay, where the appearance of ISWs was associated to an IT energy beam generated at the shelf break that propagated downward, reflected at the bottom, and excited the thermocline near the location where the ISWs were found. Analyzing this local formation mechanism, [25] showed that a moderate developed thermocline is ideal for ISWs local generation.

An IT energy beam propagates almost unimpeded through a too weak thermocline, or reflects back to deeper layers on a too strong one. In both cases, no significant perturbations are produced in the thermocline and no ISWs can be locally generated.

In situ observations of internal tide activity near the shelf break over the Brazilian southeast continental shelf were reported by [26]. Using the results of a three-dimensional tidal model implemented for the region, they showed that the baroclinic M_2 energy flux is, however, weak when compared to other regions known for strong internal tides and is predominantly offshore. In a following paper, [27] modeled the influence of Brazil Current (BC)—normally near the shelf break in this region—on the propagation characteristics of ITs. Results indicated that BC acts as a barrier to onshore propagation of internal tides generated over the upper slope, tending to reflect them back toward the open ocean.

Although these first studies of ITs on the SE Brazilian shelf did not indicate that ISWs could be significant over the shelf region, using satellite data [28] showed in a global atlas of ISWs the occurrence of these features in the region. The observed ISWs had characteristics similar to ISWs of other continental shelf regions around the world. More recently, with the availability of new SAR images for the region a much more significant presence of ISWs over the shelf than previous studies indicated became clear.

The purpose of the present work is, therefore, to clarify this issue and to contribute to a better picture of ISWs occurrence in this region. This is done by an analysis of a relatively large data set of SAR images for the region. Our goal is to extract from the data set, ISWs main space and time characteristics, such as speed, orientation, and wavelength and length along the crest, and to locate the likely generation sites of internal tide and subsequent solitons in the region. To the best of our knowledge, this is the first study of its kind for the SE Brazilian shelf.

2. Study Area

The study area is located between 20–30°S and 38–50°W in the subtropical western South Atlantic (Figure 2). Despite the availability of SAR scenes for the deep oceanic basin, our focus in this study is on the continental shelf and adjacent shelf break regions, where most of the ISWs were found. The continental shelf subregion is called the Southeast Continental Shelf (SECS). It presents smooth bottom topography with the shelf break depth between 120 and 180 m [29].

The SECS can be divided into three subregions with different physical characteristics: inner, mid, and outer shelf. The inner continental shelf (ICS) is totally mixed without any vertical stratification throughout the year. It extends from the coast to about the 30 m depth during the summer and to the 60 m during the winter. The mid continental shelf (MCS) is vertically stratified during the summer, with a strong seasonal thermocline, and extends from 30 m to 80 m depth in this season. During the winter, the MCS is narrow (confined between 60 and 80 m depth), or almost absent. The outer continental shelf (OCS) extends from the 80 m depth to the shelf break (~180 m); a permanent thermocline is present there all seasons. The Brazil Current (BC), a relatively weak western boundary current, is the main dynamic feature in the OCS. It flows southward along the slope from about 10°S until the subtropical convergence region (33–38°S) [30]. A strong mesoscale circulation, with BC meandering and eddies, is present in the OCS and slope [31, 32].

From mid to end autumn (April-May) and throughout the winter (June–August), all southern Brazilian coast and the south part of SECS are affected by a northward intrusion of low salinity waters from the La Plata river estuary, the so-called Plata Plume Water (PPW). The PPW extends more than 1000 km from the estuary and induces vertical stratification over the shelf in this region [33].

The semidiurnal lunar tidal constituent (M_2) is the main tidal component in the SECS [34]. Tidal amplitudes decrease southward from 0.35 m near Cape São Tomé to 0.1 m in Cape Santa Marta [30].

3. Data Set and Methods

The radar data set consisted of 264 C-band VV polarization images from the Advanced Synthetic Aperture Radar (ASAR) on board ENVISAT satellite of the European Space Agency (ESA). ENVISAT, a sun-synchronous polar orbit satellite, flying at an orbit altitude of about 800 km with a repeat cycle of 35 days, was launched on March 1, 2002 and operated until April 2012. Images used in this investigation were acquired in January 2009 and from September 2009 to December 2010 by the National Institute for Space Research—INPE's receiving station in Brazil (22°40′58″S; 45°00′07″W; see position indicated by a small antenna in Figure 2). SAR raw data was processed at station site to Level 1b format using the ACS (Advanced Computer Systems) SAR processor. The data gap between January and September of 2009 was caused by initial technical difficulties found during implementation of station. Data analysis was limited to the end of 2010 by technical problems encountered subsequently which could not be resolved in time for the present investigation. The product used for analysis was of medium resolution (pixel size of 75 m) and wide swath of 400 km. Data set contained almost the same number of ascending (129) and descending (135) orbit images.

The data set covered predominantly the SE Brazilian continental shelf and adjacent offshore waters between 20° and 28°S (Figure 2). Between 28° and 30°S the number of images was considerably lower (about 7 times less) than on the remaining region; so the results for this area are statistically less robust. A seasonal analysis of the data set shows that winter time (JAS) images prevailed in the northern portion of the region, while images in other seasons prevail toward the south.

For each SAR image, detection of ISWs was done by visual analysis of subregions which had been digitally processed, registered to a latitude/longitude projection, and enhanced. Each detected ISW packet had its average center coordinates recorded and all main characteristics annotated. The wind speed over each observed internal wave packet was estimated using ASCAT scatterometer data at 12.5 and 25 km resolution cells obtained from [35]. A maximum acquisition time difference of 5 hours between scatterometer winds and ASAR images was guaranteed. The "climatologic" monthly mean wind speed was downloaded from [36], which is based on 10 years of QuickScat data. Bathymetry for the study region was the ETOPO 2, at a spatial resolution of 2′, which was obtained from [37].

The components of the maximum barotropic tidal velocity for M_2 constituent were obtained from the 1/4° resolution model OTIS—Oregon State University Tidal Inversion Software [38]. The velocity components were linearly interpolated onto the finer grid of ETOPO2 with resolution of 2′. The Brunt-Väisälä frequency (N) was calculated from objectively analyzed monthly temperature and salinity vertical profiles obtained from World Ocean Atlas 2009 (WOA09) [39], at a resolution of 1°.

Using the described data, the barotropic body force F [40] was calculated for the summer period (JFM). F is the forcing term of the baroclinic internal tide and is associated with the vertical motions produced by the horizontal barotropic tidal flow interacting with bottom topography in the presence of vertical stratification. The amplitude of depth-integrated body force is given by

$$F = \frac{1}{\omega} Q \cdot \nabla \left(\frac{1}{h} \right) \int_H^0 z N^2(z) \, dz, \qquad (1)$$

where ω is the tidal angular frequency (rad s^{-1}), z is the upward vertical coordinate ($z = 0$ at sea surface), $N(z)$ is the local Brunt-Väisälä frequency, Q is the barotropic flux vector $Q = (Q_x, Q_y) = (u_1 H, v_1 H)$, u_1 and v_1 are the zonal and meridional components of the barotropic tidal velocity, H is local depth, and $h(x, y)$ is the ocean bathymetry.

4. Results and Discussion

From the set of 264 SAR images, about 34% (89 images) contained some ISW signature. As a measure of comparison,

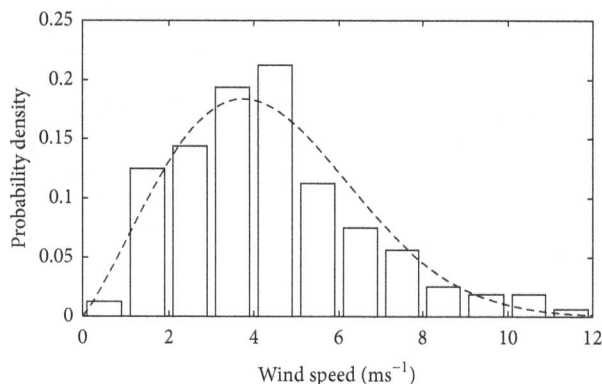

FIGURE 3: Histogram of wind speed observed over 160 ISW packets for the period 2009-2010. Dashed line: a best fit of a Weibull pdf distribution.

[41] found 5% out of 2600 images with some ISWs for the Norwegian shelf utilizing ERS1-2 and Radarsat-1 images with different swaths. In our study region, a total of 467 ISWs packets were detected for the analyzed period. The number of ISW packets was almost equally distributed among ascending (232 packets) and descending (235 packets) orbit images. Corresponding sea surface wind speed was recovered for 160 of them.

4.1. Wind Effect on ISW Detection and Seasonal Variability. Sea surface wind is an important factor for internal wave imaging according to SAR image theories. According to [42], ~3 ms^{-1} would be the weakest wind required for SAR imaging of ISWs. This low wind limit is, however, not so clearly defined. According to [7], for C band radar images, internal waves can still be seen as bright bands in a dark background in winds under 2 ms^{-1} due to a resonance mechanism of dm-waves with the ISW. As the wind becomes strong, the ISW modulation depth (the anomaly of backscatter from its mean value) is significantly reduced, making it harder to observe the ISWs in SAR images. Simulations done by [43, 44] indicate that for winds higher than about 7-8 ms^{-1} the ISW contrast between maximum and minimum signals relative to the normal sea clutter becomes very weak, making the detection difficult.

Figure 3 shows the histogram of wind intensity over 160 observed ISW packets for the study period and the Weibull probability density function (pdf) fit (shape factor $k = 2.19$ and scale factor $c = 4.96$), leading to mean and standard deviation wind values of 4.4 ms^{-1} and 2.1 ms^{-1}, respectively [45].

Approximately 97% of the observed ISW packets were associated with winds below 10 ms^{-1} although a few were detected at wind speeds of about 11-12 ms^{-1}. Similar result is reported by [46] for the northern South China Sea. Most of ISWs were observed at winds between 2.3 and 6.5 ms^{-1} (mean ± 1 sd), which according to the pdf correspond to a probability of 67% of the data set. This result is similar to the theoretical limits put by [42-44].

Summer and winter spatial distributions of wind speed for the study region based on 10 years of QuickScat data are depicted in Figure 4. As expected, wind speed decreases toward the coast and is weaker during the summer. For summer and winter, winds below 7 ms^{-1} are generally observed over the continental shelf. The seasonal average minimum winds are present during the summer (January) between Rio de Janeiro (22.9°S) and São Sebastião (23.75°S) with values close to 4 ms^{-1}; during the winter this minimum regional cell is displaced to southwest with minimum winds of 4.5 ms^{-1}. It is well known that the presence of surface films can reduce short-scale roughness and radar backscatter, inducing a negative bias in the retrieved surface wind via Geophysical Model Function used. However, by our experience in analyzing SAR images for this region, we consider that the minimum wind regions referred to above are not the result of this effect. These regional minimum wind centers are more likely the combined result of the presence of a high mountain range very close to shoreline, the orientation of coastline and the NE and E wind directions prevailing in the summer and E and SW during the winter. In the wintertime, average wind speeds above 8 ms^{-1} are observed south of 28°S. These wintertime higher winds are normally associated with the passage of cold fronts which brings S/SE average winds of 8 ms^{-1} [47]. These results indicate that typical wind speeds observed over the continental shelf at SECS are adequate for SAR observations of ISWs, and except for extreme meteorological events or after the passage of strong cold fronts, wind should not be a limiting factor.

To get a sense of the seasonal variability, we present in Figure 5 the mean number of ISWs packets observed per image and for different seasons. A clear seasonal cycle of ISWs observed in the region is present, with a dominance of wave packets during the summer as compared to winter and autumn; a small but visible increase is seen for the spring season. The high standard deviations are probably associated to the variability of conditions necessary for the generation and observation by SAR of ISWs. A fair number of images showed no ISWs whilst some had 18 ISWs packets.

4.2. Tidal Variability and ISW Generation. According to results of the OTIS tidal model, barotropic tidal currents over the shelf break in the region are about 60% stronger during spring tides. Since ISWs are generated by ITs, which are formed by the barotropic tide, it should be expected a positive correlation between observed ISWs and the fortnight tidal cycle. Surprisingly, separating the ISWs according to the phase of tidal cycle at time of image acquisition, it was verified that a few more ISWs packets were observed during neap tides than during spring tides. Considering that similar number of images was available for both tidal periods, we can conclude that no correlation exists in the region between the number of ISWs packets and the strength of tidal currents. A similar behavior is reported by [48] for the New Jersey North America east coast. The suggested explanation for this behavior was the presence of remotely generated shoaling internal tides which could induce significant changes in local internal tide generation. If the phase of shoaling IT is randomly

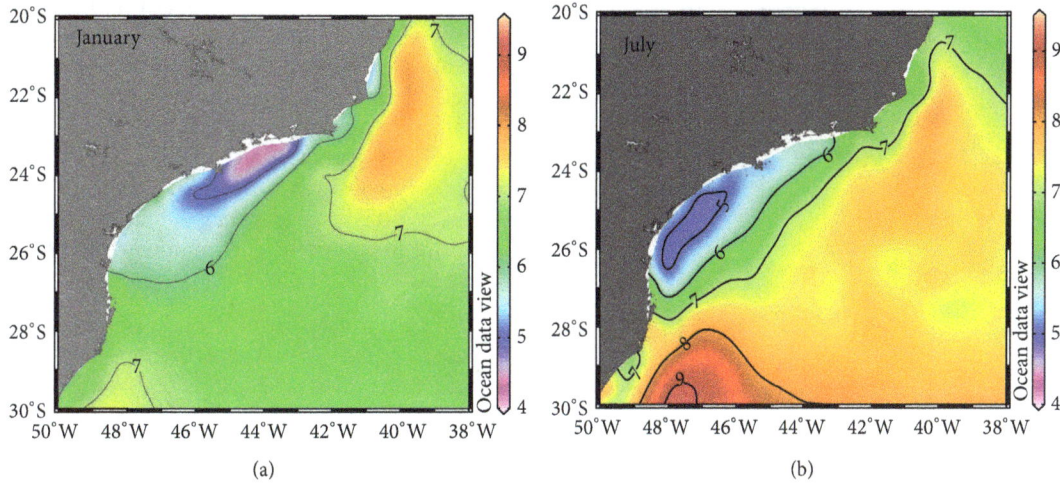

FIGURE 4: Spatial distribution of mean summer (January) and winter (July) wind intensity for the study region based on 10 years of QuickScat data.

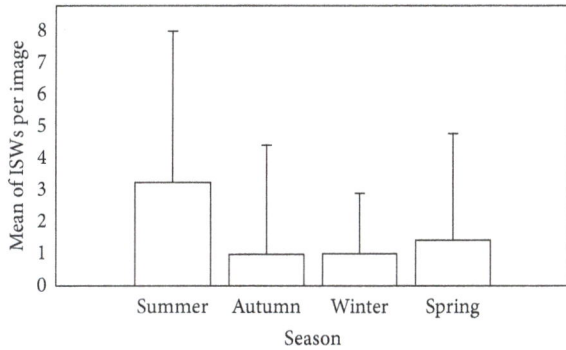

FIGURE 5: Average number of ISWs packets per scene and one standard deviation (vertical bars).

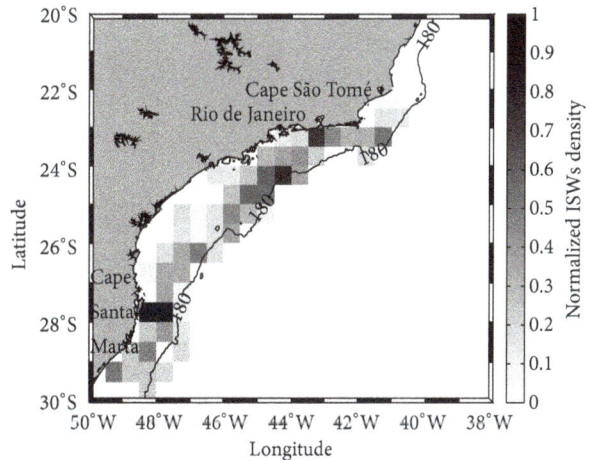

FIGURE 6: Summertime spatial distribution of ISWs. Values are normalized (see text) number of ISWs per number of scenes for the period.

changed by interaction with currents and mesoscale features, its pressure perturbations can interact constructively or destructively with barotropic tidal velocities, reinforcing or inhibiting internal tides generation. Lack of correlation of ISWs to tidal phase is also reported by [49, 50] on the Portuguese shelf. In this region, large amplitude ISWs with similar amplitudes were observed during neap and spring tides even though the barotropic tide changed by a factor of 2. Their explanation for this behavior was that the ISWs theoretical limiting amplitudes had already been achieved during the neap tides. In our case, it is difficult to say which of these theoretical explanations would be more suitable considering the limitations of the available data set.

4.3. Spatial Distribution and Direction of Propagation. The spatial distribution of observed ISWs is presented in Figures 6 and 7. The results were calculated in cells of $0.5° × 0.5°$ lat/long, corresponding to a typical length of ISWs crests. At each cell, the total number of ISWs packets was divided by the total number of satellite images covering the same area and period. The maximum number found for the region was

set to one and used for normalization so that scales vary from 0 to 1 for both figures. Although a fair number of satellite scenes covered the deep ocean (see Figure 2), almost all of the ISWs packets were observed in the continental shelf at depths lower than 200 m. For the summer period (Figure 6), except at the northernmost portion of the region, north of 22°S where no ISWs were detected, solitons were seen throughout the region. Regional maximum concentrations of ISWs were near the islands of Florianópolis (27.8°S) and São Sebastião (23.75°S) as well as near Rio de Janeiro (22.9°S). Spring season ISWs spatial distribution (not shown) has a very similar pattern.

The spatial distribution of ISWs for the grouped autumn and winter seasons is shown in Figure 7. The reason for joining these two seasons was first that they seem to have similar environmental conditions for ISWs generation and SAR observation (Figure 4) and second that the number of

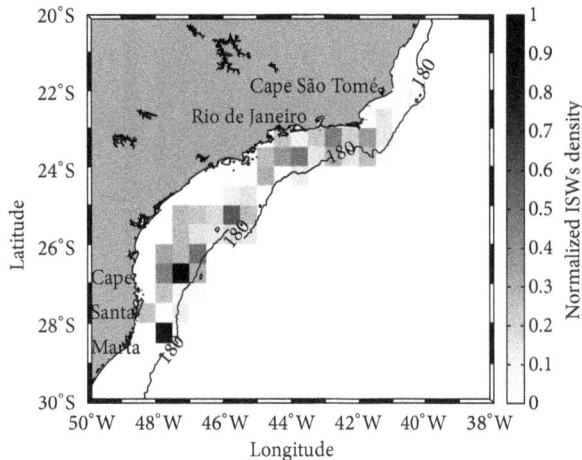

FIGURE 7: Same as Figure 6 but for grouped autumn and winter images.

ISW packets observed in winter was too low to produce a significant separate analysis.

Similar to the summer period, during autumn and winter ISW packets were observed all over the study area, but with a lower number of ISWs per scene. The seasonal decrease of observed ISW packets in the winter seems to be related to the seasonal decrease in the vertical stratification in the period (Figure 8).

The very low number of ISW packets south of 28°S can be, however, somewhat negatively biased; no winter images were available in this area and only a low coverage was obtained during autumn. Additionally, during autumn and winter the average wind speed increases, staying over $7\,ms^{-1}$ south of 28°S (see Figure 4), and as discussed before, decreasing the visibility of ISWs in SAR images. Stratification in the southern sector of this region is also singular. The PPW northward intrusion in this period causes a salinity-induced stratification and an inverted thermocline [33, 51]. Numerical experiments done by [52] indicate that vertical stability solely caused by salinity cannot produce, support, or augment the internal wave oscillations. Their results also indicate that horizontal convergence/divergence of surface currents—required for the manifestation of internal waves at the surface—are inhibited in the presence of temperature inversion. Therefore, we suppose that the PPW intrusions might also contribute to the low number of ISWs south of 28°S.

The frequency of ISW observations as a function of local depth is presented in Figure 9 for two periods: spring/summer (son/djf) and autumn/winter (mam/jja). For both periods, most of the ISW packets were observed between the 80 and 160 m isobaths at the outer shelf and shelf break regions, followed by midshelf occurrences. Inshore, the 40 m depth—at the inner shelf—the percentage of observed ISWs was very low. Toward the deep ocean, no ISWs were observed beyond the 500 m isobath.

The prevailing propagation direction of ISWs (Figure 10) was onshore at 319° (measured clockwise from north). Most

of the ISWs were observed propagating normal to the local bathymetry toward the coast as a result of wave refraction and tidal advection. Near 26°S and 23.5°S, where the 180 m isobath is oriented almost in an eastwest direction, it is possible to observe ISWs propagating northward. A very few anomalous cases of ISWs propagating offshore were observed. These cases were concentrated especially near 23.5°S and 41.5°W. A few cases of ISWs propagating southward were observed in the northern part of the domain following the topography.

4.4. Formation Region and Likely Generation Mechanism. Considering the spatial distribution and the prevailing onshore direction of propagation of ISWs, it is likely that they are being formed at the shelf break. However, the combination of local depth gradients, with dominant direction and strength of tidal currents and stratification can favor specific places as main generation sites of ITs. To estimate these probable generation sites, we calculated the depth-integrated-amplitude of the barotropic body force (F) using ETOPO2 bathymetry, the summer mean stratification, and maximum value of M_2 tidal current component (1). This variable has been extensively used as a good indicator of the source regions of internal tides [53–56]. The spatial distribution of the depth-integrated amplitude of F is shown in Figure 11.

As anticipated, the highest values of F are concentrated between the 200 and 500 m isobaths at the slope, mostly in response to the steep topography gradients of this region. This strongly indicates the shelf break as the main region for IT generation in SECS. Three areas concentrate the highest F values: one between 25 and 26°S, one between 23 and 24°S, SE of Cabo Frio, and a more extensive one between 20 and 22.5°S. Although Figure 11 suggests north of 24°S as being a favorable region for IT generation, one or more mechanisms could be acting to prevent the generation (or observation) of ISWs north of 22°S, since very few ISW packets were observed in this northern sector throughout the year (Figures 6 and 7). For the region between 20°S and 23°S, [57] show that the Brazil Current inshore thermal front is positioned in the outer shelf region, inshore the 200 m isobath. As discussed by [27], it is possible that ITs generated at the shelf break in this region are being reflected back to deep ocean by the BC before disintegrating into ISWs. Other specific characteristics of this region, such as a narrow continental shelf, a weaker stratification (Figure 8), and stronger winds (Figure 4) may be contributing to hinder ISWs generation/observation.

Considering the results presented above, it is reasonable to assume that the ITs in the SECS are being generated at the shelf break and show a clear onshore propagation. We are not excluding here offshore propagation of ITs as indicated by [27]; the point is that we only observed a very small number of solitons with offshore propagation in the region. Thus, it is very probable that the ISWs observed in the continental shelf are caused by a nonlinear disintegration of interfacial internal tides propagating shoreward.

4.5. Some ISW Characteristics. The observed solitons are phase locked to the troughs of the ITs so that sequential ISW packets must be separated by the ITs wavelength [58].

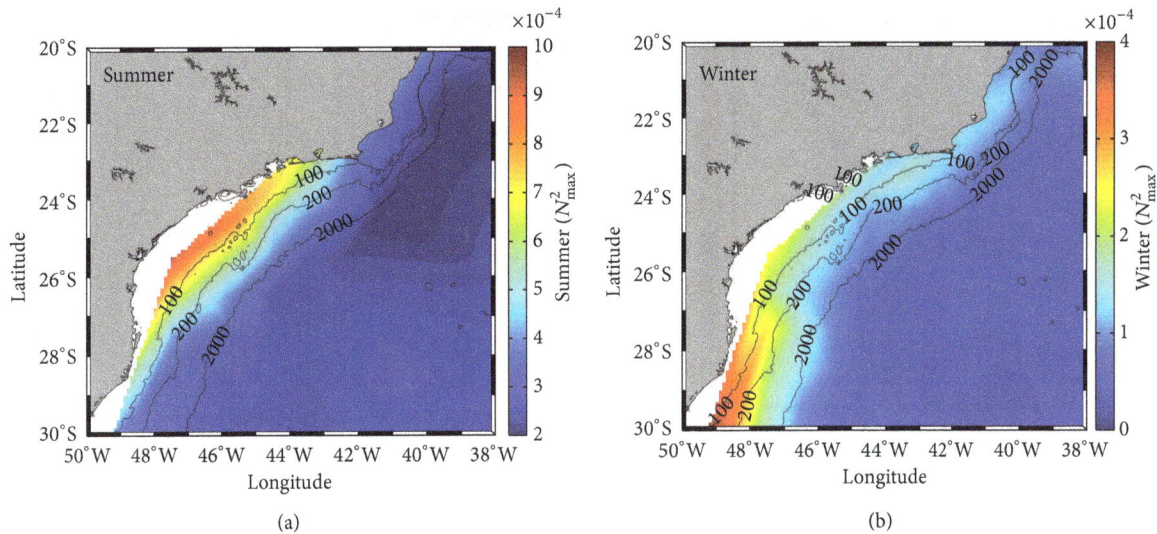

FIGURE 8: Climatological seasonal fields of maximum Brunt-Väisälä square frequency (N^2) for the study region. Data source: World Ocean Atlas 2009.

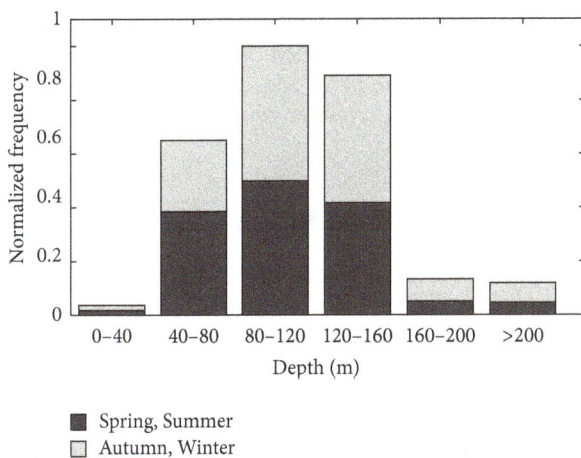

FIGURE 9: Observed seasonal occurrence of ISW packets as a function of local depth.

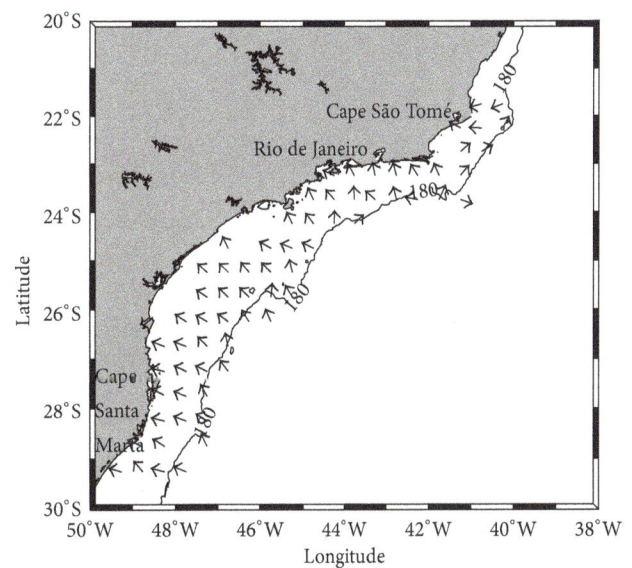

FIGURE 10: Average propagation direction (vectors) of ISWs observed in the SECS.

Sequential packets of ISWs are very similar in shape and direction of propagation and their separation can be easily measured in SAR images from the center of the first soliton of each packet. A typical IT wavelength scale can also be estimated as NHT/π, where N is the typical buoyancy frequency, H is the water depth, and T the tidal period [27, 59]. A comparison of these typical IT wavelengths, calculated using $T = 12.42$ h (M_2 tidal period), the climatological values of N, and local depth of observed solitons, against measurements of distance of sequential packets of ISWs is depicted in Figure 12. Albeit an anticipated scatter due to the simplifications assumed, a positive correlation between estimated and observed wavelengths is verified. We consider that this result goes in favor of the hypothesis that first, the ISWs observed in the region by SAR are strongly forced by the semidiurnal lunar tidal component, and second that the

N values, as given by the WOA09 data set, are a reasonable approximation of the vertical stratification for the region.

Assuming that sequential ISW packets are generated at each M_2 tidal cycle, the mean phase speed of ISWs can be estimated dividing the measured distances of sequential solitons by the M_2 tidal period [60]. The result of this simple method is presented in Figure 13. The overall mean speed was 0.64 ms^{-1} with standard deviation (std) of 0.16 ms^{-1}. The maximum phase speed did not exceed 1.04 ms^{-1}. For spring and summer, the propagation speed was about 0.66 ms^{-1} with std. of 0.16 ms^{-1}, and for autumn and winter was slightly lower, 0.57 ms^{-1} with std. of 0.08 ms^{-1}, due to the weaker stratification. Similar values of phase speed have been

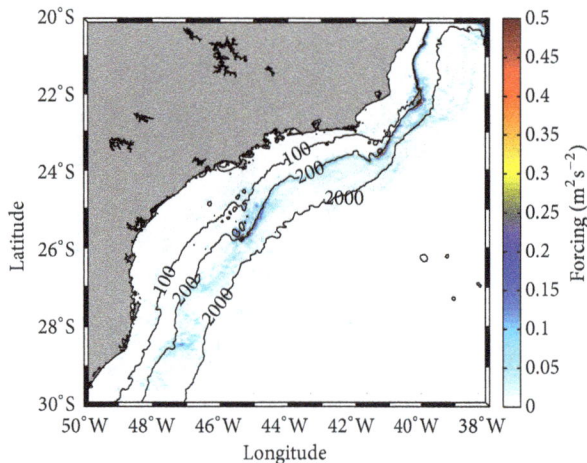

FIGURE 11: Depth integrated amplitude of M_2 barotropic tidal force field F.

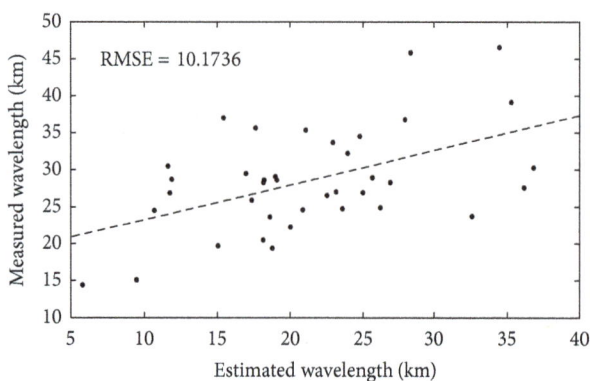

FIGURE 12: Theoretically estimated M_2 ITs wavelengths as NHT/π (horizontal axis) versus wavelength obtained from sequential ISWs packets (vertical axis).

observed in other continental shelves [1, 41, 61, 62] although much greater values, on the order of $2.5\,\mathrm{ms^{-1}}$, have been reported in deeper areas [19].

The wavelength of the ISWs (i.e., the intersoliton separation) was measured for the first three solitons of the packets. A larger number of measurements for the first soliton was done since generally for the rear of the packets the signatures were normally weaker or very noisy. The first soliton average wavelength was 1.04 km with a maximum value of 2.36 km and a std. of 0.44 km. To the rear of the packets, the wavelengths decreased. The second and third solitons showed average wavelengths of 0.9 and 0.85 km, with std. of 0.42 and 0.37 km, respectively. The amplitudes of the solitons in the ISWs packets decrease from the front to the back of the packets. Since the velocity of a soliton is directly proportional to its amplitude, distances between one wave crest and the next decreases toward the back of the packet [1]. The higher wavelength standard deviation for the first solitons is also a result of their higher degree of nonlinearity and sensibility to slight variations in the stratification [43]. The values of wavelength did not change significantly between seasons.

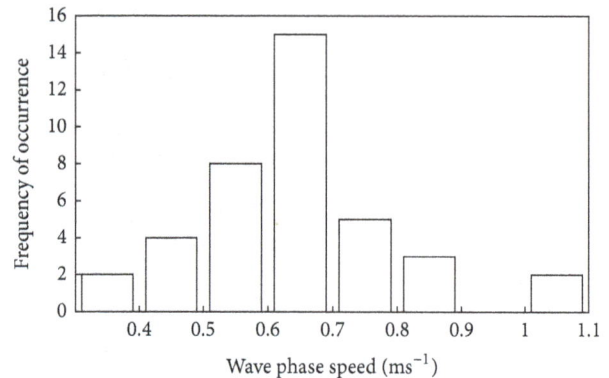

FIGURE 13: Histogram of phase speeds of ISWs estimated from the separation of sequential ISW packets.

The along-crest lengths were measured for the first soliton of the packets as this tends to be longest one. The mean observed length was 43.5 km with a standard deviation of 28 km. The maximum length was 152.5 km, but about 90% of the ISWs had along-crest lengths under 80 km. These lengths are typical of continental shelf ISWs [63]. However, much larger values have been reported in some places as in South China Sea [46, 64] and in the Mascarene ridge in the Indian Ocean [65] with along-crest lengths in excess of 200 km and 350 km, respectively.

5. Summary and Conclusions

A study of the space/time variability and main characteristics of oceanic ISWs off the Brazilian Southeast coast was carried out using a set of Envisat ASAR images. A summary of main results obtained is presented in Table 1.

Analysis of concomitant scatterometer wind data (max 5 h apart) showed that most of observed ISWs packets were associated with winds between 3 and $6\,\mathrm{ms^{-1}}$; very few solitons were detected for wind speeds above $10\,\mathrm{ms^{-1}}$ (Figure 2). Contrary to a generally assumed $3\,\mathrm{ms^{-1}}$ weakest wind threshold for SAR observations, a few solitons were still visible at winds near $1\,\mathrm{ms^{-1}}$, a fact also observed by [7]. Considering the wind speed climatology for the region (Figure 3), we can say that this variable should not be a complicating factor for ISW observation by SAR.

Summer was the season with the highest number of ISWs, more than three times the frequency of ISWs occurrence for autumn and winter. Summer stratification seems to be the main factor for such high concentration of ISWs in this season. Previous studies show that vertical stratification is present in the region from the outer shelf up to the inshore limit of the midshelf during the summer season; during the winter the midshelf practically disappears and stratification is concentrated in the outer shelf [30]. The high standard deviations (Figure 5) indicate that these waves depend on specific and changing environmental conditions for their generation as well as for observation by SAR. The presence of ISWs in all seasons confirms that vertical stratification is present even in winter, at least in the outer shelf and slope

TABLE 1: ISWs main characteristics in the SECS obtained from SAR imagery.

	Mean	Std. Deviation	Maximum	Minimun
Wind speed over ISWs (ms^{-1})	4.4	2.1	11.43	0.66
Local depth of occurrence (m)	111	51	509	14
Direction of propagation (degrees)	319	—	—	—
ISW propagation speed (ms^{-1})	0.64	0.16	1.04	0.32
Wave crest length (km)	43.5	28	152.5	10.5
Wavelength 1st soliton (km)	1.04	0.44	2.36	0.3
Wavelength 2nd soliton (km)	0.9	0.42	2.11	0.22
Wavelength 3rd soliton (km)	0.85	0.37	1.7	0.32
Wavelength of Internal Tide (km)	28.6	6.94	46.57	14.4

regions, a fact confirmed by N^2 climatology (Figure 8). The very small number of ISW packets observed in the inner shelf was expected considering that this is a well-mixed region all year round (Figure 8) and therefore not supporting the presence of internal waves.

The high number of observed ISW packets in the SECS indicates that they are generated every tidal cycle all year round. After formation, they propagate toward the coast and eventually dissipate by breaking, inducing turbulence and mixing. We raise here a hypothesis—yet to be tested—that a continuous ISWs breaking in the region could be acting together with wind mixing to maintain the inner shelf well mixed throughout the year.

The ISWs were well distributed in the north/south direction along the SECS all year round, except north of 22°S where the observation of ISWs was always low (Figures 6 and 7). We conjecture that this is probably a combined result of weak stratification, narrow continental shelf, stronger winds, and the Brazil Current intrusion over the outer shelf, reflecting ITs back to deep ocean. The outer continental shelf is where most of the ISW packets were observed (Figure 9); 74% of the summer and 84% of the winter ISWs were observed in this region. This seems to be linked to the presence of a permanent thermocline there. The mid continental shelf supports ISWs during the summer and spring; during the winter it is very narrow or almost absent. Shoaling is also a major mechanism contributing to the decrease of ISWs shoreward or their absence in the inner shelf.

The number of observed ISW packets did not change significantly between spring and neap tides despite a 60% variation of tidal currents. Similar cases have been reported elsewhere. This behavior has been explained by either an interaction of remotely generated shoaling baroclinic tides with local barotropic tides or by the fact that the limiting magnitude of the IT has already been reached during neap tides. Our data set is insufficient to make any statement in this regard for the region; numerical modeling and *in situ* data acquisition are needed before any conclusion can be drawn.

The highest values of the M_2 barotropic tidal forcing F are concentrated at the shelf break between 200 and 500 m isobaths (Figure 11). Almost all of the ISW packets showed onshore propagation (Figure 10), with sequential packets typically separated by distances on the order of the M_2 IT wavelengths (~10–40 km). These characteristics strongly indicate that the observed solitons are formed by nonlinear disintegration of shoreward propagating interfacial ITs formed at shelf break. Analyzing the slopes of M_2 IT ray characteristics for the region ($\alpha = (\omega^2 - f^2)/(N^2(z) - \omega^2)$) against topography gradient (γ), [27] show that slope is near to the critical condition for IT generation ($\alpha \sim \gamma$) at approximately 800 m depth. IT energy ray characteristics were modeled propagating both seaward and shoreward (Figure 12 of [27]). Therefore, sporadic "local" ISW generation at the shelf by IT energy rays exciting oscillations in the thermocline after reflection at bottom or surface should not be excluded if stratification conditions are adequate [25].

The mean phase speed of ISWs calculated from sequential packets was $0.64 \, ms^{-1}$ (Figure 13), being just slightly higher during summer and spring as compared to autumn and winter. Theoretical solutions show that higher speeds should be associated with stronger vertical stratification [66]. Average wavelength was about 1.04 km for the first soliton, decreasing to 0.85 km on the third as a result of nonlinearity. The average along-crest length was 43.5 km, with a maximum of 152.5 km.

Finally, we may say that despite not being a region of strong internal tides and having the presence of a western boundary current—Brazil Current is a permanent feature at the outer shelf—the analysis of ASAR data set revelead that ISWs are a ubiquitous feature in this region and, therefore, may play a relevant dynamic role in the SECS and possibly affecting biological production, sediment dispersion, and transport.

Acknowledgments

The authors would like to thank Petrobrás/Cenpes in the name of Dr. Cristina Bentz and ANP (National Brazilian Petroleum Agency) for supporting the installation and maintenance of the satellite reception station used for the acquisition and processing of the SAR images used in this paper. The second author thanks INPE for providing its facilities and CNPq (Process no. 130517/2010-0) for his financial support during his graduate program activities. The authors appreciate the helpful comments of two anonymous reviewers which contributed to improve this paper.

References

[1] J. R. Apel and F. I. Gonzalez, "Nonlinear features of internal waves off Baja California as observed from the Seasat imaging radar," *Journal of Geophysical Research*, vol. 88, no. 7, pp. 4459–4466, 1983.

[2] J. C. B. da Silva, A. L. New, M. A. Srokosz, and T. J. Smyth, "On the observability of internal tidal waves in remotely-sensed ocean colour data," *Geophysical Research Letters*, vol. 29, no. 12, pp. 1–10, 2002.

[3] C. R. Jackson, "Internal wave detection using Moderate Resolution Imaging Spectroradiometer (MODIS)," *Journal of Geophysical Research*, vol. 112, no. C11012, 2007.

[4] J. C. B. da Silva, A. L. New, and A. Azevedo, "On the role of SAR for observing "local generation" of internal solitary waves off the Iberian Peninsula," *Canadian Journal of Remote Sensing*, vol. 33, no. 5, pp. 388–403, 2007.

[5] F. T. Ulaby, R. K. Moore, and A. K. Fung, *Microwave Remote Sensing: Active and Passive*, vol. 3, Artech House, Norwood, Mass, USA, 1986.

[6] W. Alpers, "Theory of radar imaging of internal waves," *Nature*, vol. 314, no. 6008, pp. 245–247, 1985.

[7] J. C. B. da Silva, S. A. Ermakov, I. S. Robinson, D. R. G. Jeans, and S. V. Kijashko, "Role of surface films in ERS SAR signatures of internal waves on the shelf 1. Short-period internal waves," *Journal of Geophysical Research C*, vol. 103, no. 4, pp. 8009–8031, 1998.

[8] J. C. B. da Silva, S. A. Ermakov, and I. S. Robinson, "Role of surface films in ERS SAR signatures of internal waves on the shelf 3. Mode transitions," *Journal of Geophysical Research C*, vol. 105, no. 10, pp. 24089–24104, 2000.

[9] V. Kudryavtsev, D. Akimov, J. Johannessen, and B. Chapron, "On radar imaging of current features: 1. Model and comparison with observations," *Journal of Geophysical Research C*, vol. 110, no. 7, pp. 1–27, 2005.

[10] M. A. Evans, S. MacIntyre, and G. W. Kling, "Internal wave effects on photosynthesis: experiments, theory, and modeling," *Limnology and Oceanography*, vol. 53, no. 1, pp. 339–353, 2008.

[11] R. D. Pingree and G. T. Mardell, "Slope turbulence, internal waves and phytoplankton growth at the Celtic Sea shelf-break," *Philosophical Transactions of the Royal Society of London A*, vol. 302, pp. 663–682, 1981.

[12] H. Sandstrom and J. A. Elliott, "Internal tide and solitons on the Scotian Shelf: a nutrient pump at work," *Journal of Geophysical Research*, vol. 89, no. 4, pp. 6415–6426, 1984.

[13] P. M. Holligan, R. D. Pingree, and G. T. Mardell, "Oceanic solitons, nutrient pulses and phytoplankton growth," *Nature*, vol. 314, no. 6009, pp. 348–350, 1985.

[14] J. N. Moum, D. M. Farmer, W. D. Smyth, L. Armi, and S. Vagle, "Structure and generation of turbulence at interfaces strained by internal solitary waves propagating shoreward over the continental shelf," *Journal of Physical Oceanography*, vol. 33, no. 10, pp. 2093–2112, 2003.

[15] D. Bogucki, T. Dickey, and L. G. Redekopp, "Sediment resuspension and mixing by resonantly generated internal solitary waves," *Journal of Physical Oceanography*, vol. 27, no. 7, pp. 1181–1196, 1997.

[16] L. S. Quaresma, J. Vitorino, A. Oliveira, and J. da Silva, "Evidence of sediment resuspension by nonlinear internal waves on the western Portuguese mid-shelf," *Marine Geology*, vol. 246, no. 2-4, pp. 123–143, 2007.

[17] A. D. Heathershaw, "Some observations of internal wave current fluctuations at the shelf-edge and their implications for sediment transport," *Continental Shelf Research*, vol. 4, no. 4, pp. 485–493, 1985.

[18] J. Zhou and X. Zhang, "Resonant interaction of sound wave with internal solitons in the coastal zone," *Journal of the Acoustical Society of America*, vol. 90, no. 4 I, pp. 2042–2054, 1991.

[19] J. R. Apel, J. R. Holbrook, A. K. Liu, and J. J. Tsai, "The Sulu sea internal soliton experiment," *Journal of Physical Oceanography*, vol. 15, no. 12, pp. 1625–1651, 1985.

[20] A. L. New and R. D. Pingree, "An intercomparison of internal solitary waves in the Bay of Biscay and resulting from Korteweg-de Vries-type theory," *Progress in Oceanography*, vol. 45, no. 1, pp. 1–38, 2000.

[21] C. R. Jackson, J. C. B. da Silva, and G. Jeans, "The generation of nonlinear internal waves," *Oceanography*, vol. 25, no. 2, pp. 108–123, 2012.

[22] T. Gerkema, "A unified model for the generation and fission of internal tides in a rotating ocean," *Journal of Marine Research*, vol. 54, no. 3, pp. 421–450, 1996.

[23] A. L. New and R. D. Pingree, "Large-amplitude internal soliton packets in the central Bay of Biscay," *Deep Sea Research Part A*, vol. 37, no. 3, pp. 513–524, 1990.

[24] A. L. New and R. D. Pingree, "Local generation of internal soliton packets in the central bay of Biscay," *Deep Sea Research Part A*, vol. 39, no. 9, pp. 1521–1534, 1992.

[25] T. Gerkema, "Internal and interfacial tides: beam scattering and local generation of solitary waves," *Journal of Marine Research*, vol. 59, no. 2, pp. 227–255, 2001.

[26] A. F. Pereira and B. M. Castro, "Internal tides in the southwestern Atlantic off Brazil: observations and numerical modeling," *Journal of Physical Oceanography*, vol. 37, no. 6, pp. 1512–1526, 2007.

[27] A. F. Pereira, B. M. Castro, L. Calado, and I. C. A. da Silveira, "Numerical simulation of M_2 internal tides in the South Brazil Bight and their interaction with the Brazil Current," *Journal of Geophysical Research C*, vol. 112, no. 4, Article ID C04009, 2007.

[28] C. R. Jackson, *An Atlas of Internal Solitary-Like Waves and Their Properties*, Alexandria, Va, USA, 2nd edition, 2004.

[29] B. M. Castro Filho and L. B. Miranda, "Physical oceanography of the Western Atlantic Continental Shelf located between 4°N and 34°S coastal segment," in *The Sea*, K. H. Robinson and K. H. Brink, Eds., vol. 11, pp. 209–251, Wiley, Berlin, Germany, 1998.

[30] B. M. Castro, J. A. Lorenzzetti, I. C. A. Silveira, and L. B. Miranda, "Thermohaline structure and circulation in the region between Cape Sao Tomé and Chui," in *The Oceanographic Environment of the contInental Shelf and Slope in the Southeast-South Region off Brazil*, C. L. Rossi-Wongtschowski and L. S.-P. Madureira, Eds., pp. 11–120, EDUSP, São Paulo, Brazil, 2006.

[31] S. R. Signorini, "On the circulation and the volume transport of the Brazil Current between the Cape of São Tomé and Guanabara Bay," *Deep-Sea Research*, vol. 25, no. 5, pp. 481–490, 1978.

[32] E. J. D. Campos, Y. Ikeda, B. M. Castro, S. A. Gaeta, J. A. Lorenzzetti, and M. R. Stevenson, "Experiment studies circulation in the Western south atlantic," *Eos*, vol. 77, no. 27, pp. 253–259, 1996.

[33] A. R. Piola, O. O. Möller Jr., R. A. Guerrero, and E. J. D. Campos, "Variability of the subtropical shelf front off eastern South America: winter 2003 and summer 2004," *Continental Shelf Research*, vol. 28, no. 13, pp. 1639–1648, 2008.

[34] A. R. de Mesquita and J. Harari, "On the harmonic constants of tides and tidal currents of the South-eastern brazilian shelf," *Continental Shelf Research*, vol. 23, no. 11–13, pp. 1227–1237, 2003.

[35] EUMETSAT, 2012, http://www.eumetsat.int/Home/Main/DataAccess/EUMETSATDataCentre/index.htm?l=en.

[36] SCOW, 2012, http://cioss.coas.oregonstate.edu/scow/.

[37] ETOPO, 2012, http://dss.ucar.edu/datasets/ds759.3/.

[38] G. D. Egbert and S. Y. Erofeeva, "Efficient inverse modeling of barotropic ocean tides," *Journal of Atmospheric and Oceanic Technology*, vol. 19, no. 2, pp. 183–204, 2002.

[39] National Oceanographic Data Centre, 2012, http://www.nodc.noaa.gov/OC5/WOA09/pr_woa09.html.

[40] P. G. Baines, "On internal tide generation models," *Deep Sea Research Part A*, vol. 29, no. 3, pp. 307–338, 1982.

[41] S. T. Dokken, R. Olsen, T. Wahl, and M. V. Tantillo, "Identification and characterization of internal waves in SAR images along the coast of Norway," *Geophysical Research Letters*, vol. 28, no. 14, pp. 2803–2806, 2001.

[42] M. A. Donelan and W. J. Pierson Jr., "Radar scattering and equilibrium ranges in wind-generated waves with application to scatterometry," *Journal of Geophysical Research*, vol. 92, no. 5, pp. 4971–5029, 1987.

[43] P. Brandt, R. Romeiser, and A. Rubino, "On the determination of characteristics of the interior ocean dynamics from radar signatures of internal solitary waves," *Journal of Geophysical Research C*, vol. 104, no. 12, pp. 30039–30045, 1999.

[44] Y. Ouyang, J. Chong, Y. Wu, and M. Zhu, "Simulation studies of internal waves in SAR images under different SAR and wind field conditions," *IEEE Transactions on Geoscience and Remote Sensing*, vol. 49, no. 5, pp. 1734–1743, 2011.

[45] C. G. Justus, W. R. Hargraves, A. Mikhail, and D. Graber, "Methods for estimating wind speed frequency distributions," *Journal of Applied Meteorology*, vol. 17, no. 3, pp. 350–353, 1978.

[46] W. Huang, J. Johannessen, W. Alpers, J. Yang, and X. Gan, "Spatial and temporal variations of internal wave sea surface signatures in the northern south China sea studied by spaceborne SAR imagery," in *Proceedings of the 2nd SeaSAR Symposium*, Frascati, Italy, January 2008.

[47] J. L. Stech and J. A. Lorenzzetti, "The response of the south Brazil bight to the passage of wintertime cold fronts," *Journal of Geophysical Research*, vol. 97, no. 6, pp. 9507–9520, 1992.

[48] S. M. Kelly and J. D. Nash, "Internal-tide generation and destruction by shoaling internal tides," *Geophysical Research Letters*, vol. 37, no. 23, Article ID L23611, 2010.

[49] D. R. G. Jeans and T. J. Sherwin, "The evolution and energetics of large amplitude nonlinear internal waves on the Portuguese shelf," *Journal of Marine Research*, vol. 59, no. 3, pp. 327–353, 2001.

[50] T. J. Sherwin, V. I. Vlasenko, N. Stashchuk, D. R. G. Jeans, and B. Jones, "Along-slope generation as an explanation for some unusually large internal tides," *Deep-Sea Research Part I*, vol. 49, no. 10, pp. 1787–1799, 2002.

[51] O. O. Möller Jr., A. R. Piola, A. C. Freitas, and E. J. D. Campos, "The effects of river discharge and seasonal winds on the shelf off southeastern South America," *Continental Shelf Research*, vol. 28, no. 13, pp. 1607–1624, 2008.

[52] S. V. Babu and A. D. Rao, "Mixing in the surface layers in association with internal waves during winter in the northwestern Bay of Bengal," *Natural Hazards*, vol. 57, no. 3, pp. 551–562, 2011.

[53] M. A. Merrifield and P. E. Holloway, "Model estimates of M_2 internal tide energetics at the Hawaiian Ridge," *Journal of Geophysical Research C*, vol. 107, no. 8, pp. 5–1, 2002.

[54] A. Azevedo, S. Correia, J. C. B. da Silva, and A. L. New, "Hot-spots of internal wave activity off iberia revealed by multisensor remote sensing satellite observations—spotiwave," in *Proceedings of the 2nd Workshop on Coastal and Marine Applications of SAR*, pp. 125–132, Svalbard, Norway, September 2003.

[55] A. Azevedo, J. C. B. da Silva, and A. L. New, "On the generation and propagation of internal solitary waves in the southern Bay of Biscay," *Deep-Sea Research Part I*, vol. 53, no. 6, pp. 927–941, 2006.

[56] J. C. B. da Silva, A. L. New, and J. M. Magalhaes, "Internal solitary waves in the Mozambique Channel: observations and interpretation," *Journal of Geophysical Research-Oceans*, vol. 114, no. C05001, 12 pages, 2009.

[57] J. A. Lorenzzetti, J. L. Stech, W. L. Mello Filho, and A. T. Assireu, "Satellite observation of Brazil Current inshore thermal front in the SW South Atlantic: space/time variability and sea surface temperatures," *Continental Shelf Research*, vol. 29, no. 17, pp. 2061–2068, 2009.

[58] T. Gerkema and J. T. F. Zimmerman, "Generation of nonlinear internal tides and solitary waves," *Journal of Physical Oceanography*, vol. 25, pp. 1081–1094, 1995.

[59] J. Pedlosky, *Waves in the Ocean and Atmosphere. Introduction to Wave Dynamics*, Springer, Berlin, Germany, 2003.

[60] L. L. Fu and B. Holt, "Internal waves in the Gulf of California: observations from a spaceborne radar," *Journal of Geophysical Research*, vol. 89, no. 2, pp. 2053–2060, 1984.

[61] J. Small, Z. Hallock, G. Pavey, and J. Scott, "Observations of large amplitude internal waves at the Malin Shelf edge during SESAME 1995," *Continental Shelf Research*, vol. 19, no. 11, pp. 1389–1436, 1999.

[62] M. Teixeira, A. Warn-Varnas, J. Apel, and J. Hawkins, "Analytical and observational studies of internal solitary waves in the Yellow Sea," *Journal of Coastal Research*, vol. 22, no. 6, pp. 1403–1416, 2006.

[63] J. R. Apel, "Oceanic internal waves and solitons," in *Synthetic Aperture Radar Marine User's Manual*, C. R. Jackson and J. R. Apel, Eds., pp. 189–207, National Oceanic and Atmospheric Administration, Silver Spring, Md, USA, 2004.

[64] A. K. Liu, Y. Steve Chang, M. K. Hsu, and N. K. Liang, "Evolution of nonlinear internal waves in the East and South China Seas," *Journal of Geophysical Research C*, vol. 103, no. 3334, pp. 7995–8008, 1998.

[65] J. C. B. da Silva, A. L. New, and J. M. Magalhaes, "On the structure and propagation of internal solitary waves generated at the Mascarene Plateau in the Indian Ocean," *Deep-Sea Research Part I*, vol. 58, no. 3, pp. 229–240, 2011.

[66] A. R. Osborne and T. L. Burch, "Internal solitons in the Andaman Sea," *Science*, vol. 208, no. 4443, pp. 451–460, 1980.

7

Occurrence and Distribution of Polycyclic Aromatic Hydrocarbons in Water and Sediment Collected along the Harbour Line, Mumbai, India

V. Dhananjayan,[1] S. Muralidharan,[2] and Vinny R. Peter[2]

[1] Industrial Hygiene and Toxicology Division, Regional Occupational Health Centre (Southern), ICMR, Kannamangala PO, Bangalore 562 110, India
[2] Division of Ecotoxicology, Sálim Ali Centre for Ornithology and Natural History, Coimbatore 641 108, India

Correspondence should be addressed to V. Dhananjayan, dhananjayan_v@yahoo.com

Academic Editor: Xosé A. Álvarez-Salgado

This study investigated the occurrence of polycyclic aromatic hydrocarbons (PAHs) in water and sediment samples collected along the harbour line, Mumbai, India. The $\sum$PAHs quantified in water and sediment samples were ranged from 8.66 ng/L to 46.74 ng/L and from 2608 ng/g to 134134 ng/g dry wt., respectively. Significantly high concentration of $\sum$PAHs was found in water samples of Sewri and sediment samples of Mahul ($P < 0.05$). PAH concentrations detected in the present study were several folds higher than the existing sediment quality criteria suggested by various statutory agencies. The PAH composition patterns in water and sediments suggest the dominance of high molecular weight compounds and indicate important pyrolytic and petrogenic sources. The occurrence of PAHs in the marine environment has attracted the attention of the scientific community as these compounds are frequently detected in seawater and sediments at increasing levels and can have adverse health effects on marine organisms and humans. PAH concentrations detected at Sewri-Mahul site were sufficiently high to pose a risk to marine organisms if they are exposed continuously to this concentration. Hence, continuous monitoring of the ecosystem is highly warranted.

1. Introduction

All over the world over there have been imminent problems of pollution in many of the coastal regions resulting in significant damage to marine ecosystems. Polycyclic aromatic hydrocarbons (PAHs) are a group of over 100 different chemicals that are formed during the incomplete burning of coal, oil and gas, garbage, and other organic substances [1, 2]. These contaminants generate considerable interest because some of them are highly carcinogenic in laboratory animals and have been implicated in breast, lung, and colon cancers in humans [3–5]. Accordingly, they are included in the US EPA and the EU priority pollutants list. PAHs can reach surface waters and sediment in different ways, including atmospheric deposition, urban run-off, municipal and industrial effluents, and oil spillage or leakage [6, 7]. Owing to their low aqueous solubility and strong hydrophobic nature, these contaminants tend to associate with particulate material in the aquatic environment, with the underlying sediments as their ultimate sink [8].

Recent efforts by the EPA have been aimed at establishing sediment quality criteria in an effort to further reduce human exposure to PAH, especially via ingestion of shellfish. Sediments may be a significant source of PAH to the overlying water column particularly in areas where historical PAH input to the sediments has been high. Because some of the PAHs in sediments are not readily available to partition (i.e., PAH associated with soot) and are therefore not bioavailable [9], measuring the PAH in the overlying water may provide a better indication of exposure of marine organisms to PAH than discrete sediment sampling. Also, the distribution of PAH in the sediments may be quite heterogeneous and small numbers of samples may not provide a representative distribution, while overlying water measurements can be made much more extensively in a short period of time [10].

Occurrence and Distribution of Polycyclic Aromatic Hydrocarbons in Water and Sediment Collected
along the Harbour Line, Mumbai, India

55

It has also been emphasized that open seas play a role as a final sink for persistent toxic contaminants and accumulate on marine organisms [11]. The coastal zone represents that part of the land affected by its proximity to the sea and the part of the sea that is affected by its proximity to the land. Papers concerning PAHs in seawater and sediment in India are scarce, due to their very low concentrations in that matrix, in comparison with the concentrations in wastes, sewage, and contaminated soils. Unprecedented increase of human activities and other sources in and around has imposed considerable stress on the surrounding marine environment, and Mumbai harbor is no exception. Mumbai is one of the fastest growing metropolitan cities. Its population is projected to increase from 18.3 million, as per 2001 census, to 22.4 million in 2011. The quality of environment is deteriorating due to release of contaminants from industries and municipal wastes. It was also estimated that 2485 MLD wastes are released in to the marine environment from both industries and domestic sewage [12]. The major objective of the present study was to determine the distribution of PAHs in water and sediment, to assess the toxicological implications of these contaminants, and to generate baseline information on the pollution status along harbour line, Mumbai, India.

2. Materials and Methods

2.1. Study Area. Mumbai is one of the major cities in India which is located along the western coast of the country. City with a human population density of 25,000 persons/km^{-2} generates $2.2 \times 10^6 \, m^3 \, d^{-1}$ of domestic sewage out of which about $2 \times 10^6 \, m^3 \, d^{-1}$ enters marine waters including creeks and bays, largely untreated [13]. It has great diversification of industries in metropolitan region. About 8% of industries in the country are located around Mumbai in the upstream. A variety of industries, including refineries and petrochemical complexes, from this area are releasing their effluents largely untreated into the sea. There are number of ports wherein the ship and cargo handling activities contribute to marine pollution. Sewri-Mahul and Nhava mud flats about 1000 ha have been identified as an important bird area (IBA) [14]. Sewri-Mahul mudflats (19°01'00"N, 72°52'60"E) (Figure 1) which extent over an area of 10 km long and 3 km wide are dominated by mangroves all along the coast. The Sewri Bay is situated just off the wide mouth of Thane Creek along the northern periphery of Mumbai's eastern harbour.

2.2. Sample Collection. Seawater samples (1 L volume) for determination of PAH were collected during 2008 along the harbour line (Sewri, Mahul, Nhava), Mumbai, India (Figure 1). A total of 27 (9 sample pooled into 3 samples from each location) water samples were collected at 1 m depth by submerging a pre-cleaned glass amber bottle by hand. Water was filtered under vacuum through a Whatman GF/F filter (previously muffled at 300°C overnight). The samples were stored in PTFE bags and frozen at −20°C until processing. Twenty seven sediment (9 samples pooled in to 3 from each location) were collected during the

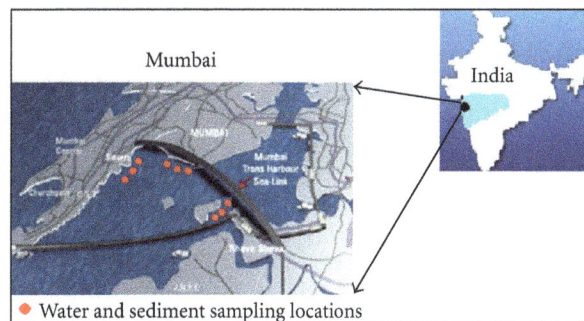

FIGURE 1: Study area showing water and sediment sample collection locations at Mumbai harbor line, Maharashtra, India.

research period. Pre-cleaned PVC core sampler was used for sampling. Sediment surfaces were sliced to 1 cm-thick slices immediately after sampling. The sediment samples were sealed in polythene covers and transported to the laboratory at SACON, Coimbatore, in ice box and stored at −20°C in the deep freezer until analysis.

2.3. Sample Processing. Water samples were extracted with dichloromethane (glass distilled grade; $3 \times 100 \, mL$). The first aliquot was used to rinse the sampling bottle and the remaining was used for solvent extraction. The two solvent extracts were combined over anhydrous sodium sulphate (also previously muffled at 300°C overnight). The collected extract was transferred to a pre-cleaned "hypovial", capped with a PTFE septum, and stored at −20°C prior to analysis. The extract was allowed to return to room temperature, and then reduced to 2 mL by means of a rotary evaporator with a water-bath operated at <30°C so as to minimize loss of volatile components. The reduced extract was then transferred to a 10 mL test-tube and reduced to near dryness under a stream of nitrogen, before being made up to 1 mL volume by addition of Acetonitrile. Sample extracts were then filtered into an auto sampler vial.

Freeze dried sediment was air dried, ground with pestle and mortar, and sieved through 0.5 mm sieve prior to further processing. Ten gram of dry sediment mixed with sodium sulfate and copper granules were extracted twice in 100 mL cyclohexane for 30 min. The extracts were concentrated to 0.5 mL under a gentle stream of nitrogen. The concentrated extracts were fractionated by a silica gel column (4 mm i.d. 90 mm). The column was then eluted first with 3.5 mL of hexane and the solution discarded. Further elution was by benzene (5 mL) to obtain PAHs [16]. All the extracts were concentrated by gentle N2 blowdown to about 100 μL.

2.4. Chemical Analysis. All the samples were quantified for PAHs using HPLC (Agilent 1100) consisting of programmable fluorescence detector at excited and emission wavelength of 260 nm and 500 nm, respectively. A 20 μL aliquot of the extract was injected through an auto sampler into C18 column (Zorbax $4.6 \times 250 \, mm$) of 5 μm particle size. The temperature of the column was maintained at 20°C. Water/Acetonitrile (ACN) was used as mobile phase with a flow of 1 mL/min. The initial content of ACN will be 50% and then increased into 60% (0–3 min) and 95% (3–14 min). This level will be held constant for 24 minutes until the end of the analysis. Recoveries of the compounds from fortified samples (50 ng/g) ranged from 94 to 103% and the results were not corrected for per cent recovery and the results were expressed in wet weight basis. Analyses were run in batches of 10 samples plus four quality controls (QCs) including one reagent blank, one matrix blank, one QC check sample, and one random sample in duplicate. The minimum detection limit for all the compounds analyzed was 10 ng/g.

Statistics. All the data were log transformed to get normal distribution. One way analysis of variance (ANOVA) was performed to assess the variation locations. Means were compared using the Bonferroni multiple comparison test. The significant level was $P < 0.05$. All the calculations were done using statistical software, SPSS student version 10.

3. Results and Discussion

This paper presents results from analyses of PAHs in water and sediment samples collected along the harbor line, Mumbai, India. Concentration range of individuals and total PAHs are given in Tables 2 and 3.

3.1. PAH Concentrations in Water. The mean total concentrations of 15 PAHs ($\sum$PAHs) in water ranged from 8.66 μg/L Nhava to 46.74 μg/L at Sewri (Table 1). The highest concentration (104.07 μg/L) was observed at Sewri, where oil intrusion was clearly visible. Among the three locations, samples collected from Nhava detected lower concentration than other two places (Figure 2). Similarly, high concentration (66.74 μg/L) was also found at Mahul, which is close to sewage outlet from Mumbai. Sewri-Mahul mudflats have a surface area of about 4 sq km. The Mahul rivulet, highly polluted with wastes from Vadala, drains directly into this area. The amount of PAHs detected there is obviously related to urban runoffs, sewage discharges, and intense shipping and oil refinery activities. Ship breaking and oil seepage from industrial activities near Sewri area lead to high concentrations of total PAHs. Additionally, boats and ships transports in these regions were discharging black smokes throughout their movement, hence there are many nonpoint sources in the bay, contributing to the wide variations of PAH concentrations detected.

In terms of individual PAH composition in water, almost all of the compounds (11 out of 15) analyzed were detected at all the sampling places. Many of the samples from Sewri and Mahul were present at concentrations in excess of μg/L, suggesting that the water in the area was heavily contaminated by PAHs. Such a wide range of PAHs at different concentrations indicates that there are potentially many different sources of PAHs in the area, possibly including combustion followed by oil residues, sewage outfalls, and industrial wastewater. From the distribution of PAHs in water alone, it is difficult to differentiate these different sources of input, nevertheless,

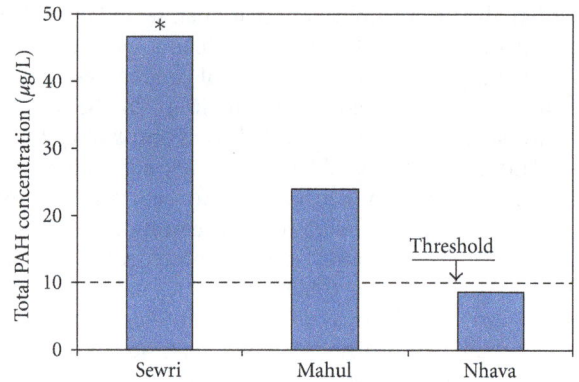

FIGURE 2: Concentration and comparison of PAHs in water sample collected from different locations along harbor line, Mumbai, India. *ANOVA, $P < 0.05$.

the data can act as an indication of the potential impacts of such high levels of PAHs on the local ecosystems.

The total PAH concentrations found in water along the harbor line, Mumbai, are several orders of magnitude higher than those detected in various other studies around the world (Table 3). Similarly, comparable high levels of PAHs were also found in seawater samples of Daya Bay, China [18], and northern Spanish [19]. Although PAHs do not show extremely high acute toxicity to aquatic organisms, the lower molecular mass compounds tend to exhibit higher lethal toxicity than the larger PAHs [33]. In addition, lethal concentration (LC50) down to less than 10 μg/L has been reported for various organisms including mysid [34, 35]. The PAH concentrations detected in water samples of Sewri and Mahul clearly show (Figure 2) that the concentrations are greater than 10 μg/L, as a result, acute toxicity may have been caused to certain exposed organisms [33]. It should be noted that different numbers of target analytes (PAH congeners) and instrument were employed in individuals studies.

3.2. PAH Concentrations in Sediment. The $\sum$PAH concentrations in sediment ranged from 17 ng/g at Sewri to 134134 ng/g at Mahul with mean concentrations of 2608–51606 ng/g dry wt. (Table 2). Although the higher load of PAHs was recorded in water samples of Sewri, sediment from Mahul recorded the highest concentrations of PAHs. The highest concentration was observed at Mahul, which is close to sewage intrusion to seashore and also receive water flow from Sewri, which is closer to Mahul. The total PAH concentration detected in sediment samples of Mahul is 20 times higher than the levels detected in Sewri and 13 times higher than those detected in sediment samples of Nhava (Figure 3).

In terms of individual PAH composition in sediment, the most compounds analysed except naphthalene were detected at all the sediment samples. Many of the PAH compounds were present at low concentration range. As far as the composition pattern of PAHs in sediments is concerned, it is mostly dominated by four-ring PAHs as shown in Table 2.

Occurrence and Distribution of Polycyclic Aromatic Hydrocarbons in Water and Sediment Collected
along the Harbour Line, Mumbai, India

57

TABLE 1: Concentrations of $\sum$PAHs (μg/L) in water collected along the harbour line, Mumbai, India.

| PAHs | Place of collection | | | | | |
| | SEWRI ($n = 3$)* | | MAHUL ($n = 3$)* | | NHAVA ($n = 3$)* | |
	Range	Mean	Range	Mean	Range	Mean
Naphthalene	<1	<1	<1	<1	<1	<1
Acenaphthene	<1–2.98	1.33	<1–2.13	1.39	<1	<1
Fluorene	<1–2.55	1.52	<1	<1	<1	<1
Phenanthrene	<1	<1	<1–3.14	1.08	<1	<1
Anthracene	<1	<1	<1	<1	<1	<1
Fluoranthene	<1–5.11	1.7	<1–1.84	1.17	<1–3.05	1.02
Pyrene	<1	<1	<1	<1	<1	<1
Benzo(a)anthracene	<1–2.93	1.78	<1–3.48	1.18	<1–3.16	1.35
Chrysene	<1	<1	<1	<1	<1	<1
Benzo(b)fluoranthene	<1–23.2	8.83	<1–17.3	5.61	<1–8.05	2.58
Benzo(k)fluoranthene	<1–2.06	1.29	<1–2.75	1.25	<1	<1
Benzo(a)pyrene	<1–1.94	1.32	<1–4.67	1.57	<1–2.03	1.34
Dibenzo(a,h)anthracene	<1–28.2	10.9	<1–12.9	4.38	<1–1.97	1.04
Benzo(g,h,i)perylene	<1–15.5	7.97	<1–5.53	2.52	<1	<1
Indeno(1,2,3-cd)pyrene	<1–19.6	10.1	<1–13	3.93	<1–2.23	1.33
Total PAHs	<1–104.07	46.74	<1–66.74	24.08	<1–20.49	8.66

<1: values below detectable concentration, *pooled from 9 water samples.

TABLE 2: Concentrations of $\sum$PAHs (ng/g, dry wt.) in sediment collected along the harbour line, Mumbai, India.

| PAHs | Place of collection | | | | | |
| | SEWRI ($n = 3$)* | | MAHUL ($n = 3$)* | | NHAVA ($n = 3$)* | |
	Range	Mean	Range	Mean	Range	Mean
Naphthalene	<1	<1	<1	<1	<1	<1
Acenaphthene	<1–245	110	<1–5265	1821	<1	<1
Fluorene	<1	<1	<1–28	9	<1	16
Phenanthrene	<1–354	120	<1–51	31	9–24	410
Anthracene	<1–331	110	<1–6.2	2	<1–1228	658
Fluoranthene	<1	<1	<1–70496	24857	527–955	17
Pyrene	<1–12	8	<1–419	271	8–23	89
Benzo(a)anthracene	<1–6	2	<1–107	54	37–143	152
Chrysene	<1–22	8	<1–344	170	13–424	193
Benzo(b)fluoranthene	<1–2796	932	<1	<1	110–470	293
Benzo(k)fluoranthene	<1–472	204	<1–40914	17486	<1–480	9
Benzo(a)pyrene	<1–265	90	<1–364	201	<1–28	739
Dibenzo(a,h)anthracene	<1–359	135	104–1450	999	316–1417	210
Benzo(g,h,i)perylene	17.3–920	337	92–13696	5153	<1–491	1251
Indeno(1,2,3-cd)pyrene	<1–1498	552	168–994	552	<1–3395	<1
Total PAHs	17–7280	2608	364–134134	51606	1020–9078	4037

<1: values below detectable concentration, *pooled from 9 sediment samples.

Pereira et al. [36] also showed that four-ring PAHs dominated PAH distributions in sediments from San Francisco Bay. Pyrolysis/combustion of fossil materials yield such PAH assemblages, which are subsequently introduced into the marine environment by coastal and river runoff [37, 38].

Industrial and domestic wastes are often another important local source. The changes in the composition pattern of PAHs between sampling sites may occur due to variations in additional input sources. The coastal region of Mumbai receives industrial discharges up to 230 million liters per

TABLE 3: Comparison of total PAH concentrations (μg/L) in sub-surface water from various marine sites around the world.

Location	Year of Sampling	N	Range	References
Chesapeake Bay, USA	1995	17	0.02–0.0657	Gustafson and Dickhut [9]
Baltic Sea	1995	14	0.300–0.594	Maldonado et al. [15]
Western Xiamen Bay, China	1998	—	0.106–0.945	Zhou et al. [16]
Alexandria coast, Egypt	—	—	0.013–0.120	El Nemr and Abd-Allah [17]
Daya Bay, China	1999	16	4.228–29.32	Zhou and Maskaoui [18]
Northern Spanish	2002	25	0.19–28.8	González et al. [19]
Saronikos Gulf (Greece)	—	17	0.425–0.459	Valavanidis et al. [20]
Gerlache Inlet sea (Antarctica)	—	—	0.005–0.009	Stortini et al. [21]
Deep Bay, South China	2004	15	0.0247–0.069	Qiu et al. [22]
Western Taiwan Strait, China	2009	—	0.0123–0.058	Wu et al. [23]
Mumbai Harbour Line, India	2008	15	8.66–46.74	Present study

N: number of PAH compounds analysed in each study.

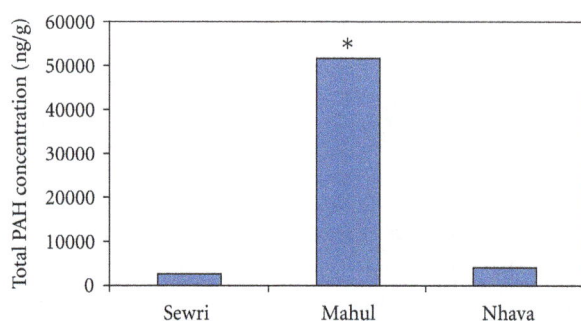

FIGURE 3: Concentration and comparison of PAHs in sediment sample collected from different locations along harbor line, Mumbai, India. *ANOVA, P < 0.05.

day (MLD) and domestic wastes of around 2200 MLD of which 1800 MLD are untreated [12]. As a result, the high concentrations of PAHs in sediment could be caused by the large amount of soil runoff and sewage discharged from this area into seawater. The results seem to suggest that PAHs in the area are derived from both the combustion of fossil fuels and petrogenic inputs.

The levels of sediment contamination by PAHs in along the harbor line, Mumbai, India is fivefold lower than the levels reported in sediment of Kitimat Harbour, Canada [25]. The total concentrations of 15PAHs in sediment of present study are several folds higher than that the levels found in Victoria Harbour, Hong Kong [24], Baltic Sea [26], Masan Bay, Korea [27], and other places (Table 4). It is also worth noting that different numbers of parent PAH compounds were analysed in different studies, and that the compounds measured may be different; so the comparison of total PAH levels from different studies have to be treated with caution. Additionally, high concentrations of total and carcinogenic PAHs were also reported in fish samples collected from same location [39].

In order to assess whether sediments in Mumbai harbor will cause toxic effects, the PAH levels in sediments were compared against effects-based guideline values such as the effects range-low (ER-L), effects range-median (ERM), and apparent effects threshold values developed by the US National Oceanic and Atmospheric Administration [40]. Among the three study locations, the maximum mean total PAH concentrations (51606 ng/g) were found in sediment samples collected from Mahul followed by Nhava (4037 ng/g). These levels were significantly higher than the ER-L value (4000 ng/g). The total PAH concentrations found in sediment samples in the present study locations exceeded the maximum values of 2000–4000 ng/g in sediment quality guidelines proposed by Ontario Ministry of the Environment [41].

4. Conclusion

This study provides important data set on PAH levels in the water and sediments along harbor line, Mumbai, India. The levels of PAHs in water and sediment were several folds higher than the other study reports from various countries. The PAH distribution profile indicated potential source dependence, as the levels were generally higher in the vicinity of known inputs such as oil terminals and ports. There are implications for the quality of seafood from many aquaculture areas destined for human consumption. The findings point to the urgent need to establish a monitoring programme for persistent organic pollutants such as PAHs, not only in water and sediment but also in the organisms themselves to ensure that any excess in concentrations over environmental quality standards is rapidly reported and necessary actions are taken. Their presence in marine sediments combined with other potentially toxic compounds can result in negative effects, which have yet to be investigated to any great extent.

Occurrence and Distribution of Polycyclic Aromatic Hydrocarbons in Water and Sediment Collected
along the Harbour Line, Mumbai, India

59

TABLE 4: Comparision of total PAH concentrations (ng/g, dry wt.) in sediment from various marine sites around the world.

Location	Year of Sampling	N	Range	References
Victoria Harbour, Hong Kong, China	1992	8	700–26100	Hong et al. [24]
Kitimat Harbour, Canada	—	15	310–528000	Simpson et al. [25]
Baltic Sea	1996	18	3.96–22100	Baumard et al. [26]
Masan Bay, Korea	1998	16	41.5–1100	Khim et al. [27]
Western Xiamen Bay, China	1998	16	247–480	Zhou et al. [16]
East China Sea	—	—	17–157	Bouloubassi et al. [28]
Bohai Sea and Yellow Sea	—	10	20.4–5734	Ma et al. [29]
Northern Adriatic Sea	1996	22	30–600	Notar et al. [30]
Daya Bay, China	1999	16	115–1134	Zhou and Maskaoui [18]
Deep Bay, South China	2004	15	353.8–128.1	Qiu et al. [22]
Southwestern Barents Sea	2006	22	58.8–326	Boitsov et al. [31]
Cienfuegos Bay, Cuba	—	—	180–5500	Tolosa et al. [32]
Mumbai Harbour Line, India	2008	15	17–134134	Present study

Acknowledgments

The authors sincerely thank Maharashtra State Road Development Corporation (MSRDC), India, for financial assistance. They are grateful to Drs. V. S. Vijayan, S. N. Prasad, Lalitha Vijayan and R. Jayakumar, SACON, for their support. They appreciate S. Patturajan and Muragesan for their assistance in all their laboratory works.

References

[1] M. M. Mumtaz, J. D. George, K. W. Gold, W. Cibulas, and C. T. De Rosa, "ATSDR evaluation of health effects of chemicals. IV. Polycyclic aromatic hydrocarbons (PAHs): understanding a complex problem," *Toxicology and Industrial Health*, vol. 12, no. 6, pp. 742–971, 1996.

[2] T. E. McGrath, J. B. Wooten, C. W. Geoffrey, and M. R. Hajaligol, "Formation of polycyclic aromatic hydrocarbons from tobacco: the link between low temperature residual solid (char) and PAH formation," *Food and Chemical Toxicology*, vol. 45, no. 6, pp. 1039–1050, 2007.

[3] M. Pufulete, J. Battershill, A. Boobis, and R. Fielder, "Approaches to carcinogenic risk assessment for polycyclic aromatic hydrocarbons: a UK perspective," *Regulatory Toxicology and Pharmacology*, vol. 40, no. 1, pp. 54–66, 2004.

[4] A. Ramesh, S. A. Walker, D. B. Hood, M. D. Guillén, K. Schneider, and E. H. Weyand, "Bioavailability and risk assessment of orally ingested polycyclic aromatic hydrocarbons," *International Journal of Toxicology*, vol. 23, no. 5, pp. 301–333, 2004.

[5] K. B. Okona-Mensah, J. Battershill, A. Boobis, and R. Fielder, "An approach to investigating the importance of high potency polycyclic aromatic hydrocarbons (PAHs) in the induction of lung cancer by air pollution," *Food and Chemical Toxicology*, vol. 43, no. 7, pp. 1103–1116, 2005.

[6] A. Gogou, I. Bouloubassi, and E. G. Stephanou, "Marine organic geochemistry of the Eastern Mediterranean: 1. Aliphatic and polyaromatic hydrocarbons in Cretan Sea surficial sediments," *Marine Chemistry*, vol. 68, no. 4, pp. 265–282, 2000.

[7] S. G. Wakeham, "Aliphatic and polycyclic aromatic hydrocarbons in Black Sea sediments," *Marine Chemistry*, vol. 53, no. 3-4, pp. 187–205, 1996.

[8] P. F. Landrum and J. A. Robbins, "Bioavailability of sediment associated contaminants to benthic invertebrates," in *Sediments: Chemistry and Toxicity on in-Place Pollutants*, R. Baudo, J. P. Giesy, and H. Muntau, Eds., pp. 237–263, Lewis Publishers, 1990.

[9] K. E. Gustafson and R. M. Dickhut, "Distribution of polycyclic aromatic hydrocarbons in Southern Chesapeake Bay surface water: evaluation of three methods for determining freely dissolved water concentrations," *Environmental Toxicology and Chemistry*, vol. 16, pp. 452–461, 1997.

[10] K. A. Maruya, R. W. Risebrough, and A. J. Horne, "Partitioning of polynuclear aromatic hydrocarbons between sediments from San Francisco Bay and their porewaters," *Environmental Science and Technology*, vol. 30, no. 10, pp. 2942–2947, 1996.

[11] S. Tanabe, M. S. Prudente, S. Kan-Atireklap, and A. Subramanian, "Mussel watch: marine pollution monitoring of butyltins and organochlorines in coastal waters of Thailand, Philippines and India," *Ocean and Coastal Management*, vol. 43, no. 8-9, pp. 819–839, 2000.

[12] M. D. Zingde and K. Govindan, "Health status of coastal waters of Mumbai and regions around," in *Environmental Problems of Coastal Areas in India*, V. K. Sharma, Ed., pp. 119–132, Book Well Publishers, New Delhi, India, 2000.

[13] M. D. Zingde, *Indian National Science Academy, New Delhi*, 1999.

[14] M. Z. Islam and A. R. Rahmani, *Indian Bird Conservation Network*, Bombay Natural History Society and Bird life International (UK), 2004.

[15] C. Maldonado, J. M. Bayona, and L. Bodineau, "Sources, distribution, and water column processes of aliphatic and polycyclic aromatic hydrocarbons in the northwestern Black Sea water," *Environmental Science and Technology*, vol. 33, no. 16, pp. 2693–2702, 1999.

[16] J. L. Zhou, H. Hong, Z. Zhang, K. Maskaoui, and W. Chen, "Multi-phase distribution of organic micropollutants in Xiamen Harbour, China," *Water Research*, vol. 34, no. 7, pp. 2132–2150, 2000.

[17] A. El Nemr and A. M. A. Abd-Allah, "Contamination of polycyclic aromatic hydrocarbons (PAHs) in microlayer and

subsurface waters along Alexandria coast, Egypt," *Chemosphere*, vol. 52, no. 10, pp. 1711–1716, 2003.

[18] J. L. Zhou and K. Maskaoui, "Distribution of polycyclic aromatic hydrocarbons in water and surface sediments from Daya Bay, China," *Environmental Pollution*, vol. 121, no. 2, pp. 269–281, 2003.

[19] J. J. González, L. Viñas, M. A. Franco et al., "Spatial and temporal distribution of dissolved/dispersed aromatic hydrocarbons in seawater in the area affected by the Prestige oil spill," *Marine Pollution Bulletin*, vol. 53, no. 5-7, pp. 250–259, 2006.

[20] A. Valavanidis, T. Vlachogianni, S. Triantafillaki, M. Dassenakis, F. Androutsos, and M. Scoullos, "Polycyclic aromatic hydrocarbons in surface seawater and in indigenous mussels (Mytilus galloprovincialis) from coastal areas of the Saronikos Gulf (Greece)," *Estuarine, Coastal and Shelf Science*, vol. 79, no. 4, pp. 733–739, 2008.

[21] A. M. Stortini, T. Martellini, M. Del Bubba, L. Lepri, G. Capodaglio, and A. Cincinelli, "n-Alkanes, PAHs and surfactants in the sea surface microlayer and sea water samples of the Gerlache Inlet sea (Antarctica)," *Microchemical Journal*, vol. 92, no. 1, pp. 37–43, 2009.

[22] Y. W. Qiu, G. Zhang, G. Q. Liu, L. L. Guo, X. D. Li, and O. Wai, "Polycyclic aromatic hydrocarbons (PAHs) in the water column and sediment core of Deep Bay, South China," *Estuarine, Coastal and Shelf Science*, vol. 83, no. 1, pp. 60–66, 2009.

[23] Y. L. Wu, X. H. Wang, Y. Y. Li, and H. S. Hong, "Occurrence of polycyclic aromatic hydrocarbons (PAHs) in seawater from the Western Taiwan Strait, China," *Marine Pollution Bulletin*, vol. 63, no. 5–12, pp. 459–463, 2011.

[24] H. Hong, L. Xu, L. Zhang, J. C. Chen, Y. S. Wong, and T. S. M. Wan, "Environmental fate and chemistry of organic pollutants in the sediment of Xiamen and Victoria harbours," *Marine Pollution Bulletin*, vol. 31, no. 4–12, pp. 229–236, 1995.

[25] C. D. Simpson, A. A. Mosi, W. R. Cullen, and K. J. Reimer, "Composition and distribution of polycyclic aromatic hydrocarbon contamination in surficial marine sediments from Kitimat Harbor, Canada," *Science of the Total Environment*, vol. 181, no. 3, pp. 265–278, 1996.

[26] P. Baumard, H. Budzinski, P. Garrigues, H. Dizer, and P. D. Hansen, "Polycyclic aromatic hydrocarbons in recent sediments and mussels (Mytilus edulis) from the Western Baltic Sea: occurrence, bioavailability and seasonal variations," *Marine Environmental Research*, vol. 47, no. 1, pp. 17–47, 1999.

[27] J. S. Khim, K. Kannan, D. L. Villeneuve, C. H. Koh, and J. P. Giesy, "Characterization and distribution of trace organic contaminants in sediment from Masan Bay, Korea. 1. Instrumental analysis," *Environmental Science and Technology*, vol. 33, no. 23, pp. 4199–4205, 1999.

[28] I. Bouloubassi, J. Fillaux, and A. Saliot, "Hydrocarbons in surface sediments from there Changjiang (Yangtze River) Estuary, East China Sea," *Marine Pollution Bulletin*, vol. 42, no. 12, pp. 1335–1346, 2001.

[29] M. Ma, Z. Feng, C. Guan, Y. Ma, H. Xu, and H. Li, "DDT, PAH and PCB in sediments from the intertidal zone of the Bohai Sea and the Yellow Sea," *Marine Pollution Bulletin*, vol. 42, no. 2, pp. 132–136, 2001.

[30] M. Notar, L. S. Hermina, and F. Jadran, "Composition, distribution and sources of polycyclic aromatic hydrocarbons in sediments of the Gulf of Trieste, Northern Adriatic Sea," *Marine Pollution Bulletin*, vol. 42, no. 1, pp. 36–44, 2001.

[31] S. Boitsov, H. K. B. Jensen, and J. Klungsøyr, "Natural background and anthropogenic inputs of polycyclic aromatic hydrocarbons (PAH) in sediments of South-Western Barents Sea," *Marine Environmental Research*, vol. 68, no. 5, pp. 236–245, 2009.

[32] I. Tolosa, M. Mesa-Albernas, and C. M. Alonso-Hernandez, "Inputs and sources of hydrocarbons in sediments from Cienfuegos bay, Cuba," *Marine Pollution Bulletin*, vol. 58, no. 11, pp. 1624–1634, 2009.

[33] R. J. Law, V. J. Dawes, R. J. Woodhead, and P. Matthiessen, "Polycyclic aromatic hydrocarbons (PAH) in seawater around England and Wales," *Marine Pollution Bulletin*, vol. 34, no. 5, pp. 306–322, 1997.

[34] S. E. Jørgensen, S. N. Nielsen, and L. A. Jørgensen, *Handbook of Ecological Parameters and Ecotoxicology.*, Elsevier, Amsterdam, The Netherlands, 1991.

[35] M. G. Barron, T. Podrabsky, S. Ogle, and R. W. Ricker, "Are aromatic hydrocarbons the primary determinant of petroleum toxicity to aquatic organisms?" *Aquatic Toxicology*, vol. 46, no. 3-4, pp. 253–268, 1999.

[36] W. E. Pereira, F. D. Hostettler, and J. B. Rapp, "Distributions and fate of chlorinated pesticides, biomarkers and polycyclic aromatic hydrocarbons in sediments along a contamination gradient from a point-source in San Francisco Bay, California," *Marine Environmental Research*, vol. 41, no. 3, pp. 299–314, 1996.

[37] R. P. Eganhouse, B. R. T. Simoneit, and I. R. Kaplan, "Extractable organic matter in urban stormwater runoff. 2. Molecular characterization," *Environmental Science and Technology*, vol. 15, no. 3, pp. 315–326, 1981.

[38] E. J. Hoffman, G. L. Mills, J. S. Latimer, and J. G. Quinn, "Urban runoff as a source of polycyclic aromatic hydrocarbons to coastal waters," *Environmental Science and Technology*, vol. 18, no. 8, pp. 580–587, 1984.

[39] V. Dhananjayan and S. Muralidharan, "Polycyclic aromatic hydrocarbons in various species of fishes from mumbai harbour, India, and their dietary intake concentration to human," *International Journal of Oceanography*, vol. 2012, Article ID 645178, 6 pages, 2012.

[40] G. B. Kim, K. A. Maruya, R. F. Lee, J. H. Lee, C. H. Koh, and S. Tanabe, "Distribution and sources of polycyclic aromatic hydrocarbons in sediments from Kyeonggi Bay, Korea," *Marine Pollution Bulletin*, vol. 38, no. 1, pp. 7–15, 1999.

[41] E. R. Long, D. D. Macdonald, S. L. Smith, and F. D. Calder, "Incidence of adverse biological effects within ranges of chemical concentrations in marine and estuarine sediments," *Environmental Management*, vol. 19, no. 1, pp. 81–97, 1995.

8

Changes in the Loop Current's Eddy Shedding in the Period 2001–2010

Fred M. Vukovich

FMV Atmospheric and Marine Consultants, 8033 Hawkshead Rd, Wake Forest, NC 27587, USA

Correspondence should be addressed to Fred M. Vukovich, fmvamc@centurylink.net

Academic Editor: Grant Bigg

A major change in the Loop Current's eddy shedding was found in the decade 2001–2010. Sixteen (16) rings separated from the Loop Current in that decade, whereas in two previous decades, 11 rings separated in each decade. More than half the rings (i.e., 56%) that separated from the Loop Current in the decade 2001–2010 had separation periods ≤8 months. In the period prior to 2001, only 26% of the rings had separation periods ≤8 months. Furthermore, the dataset average period for ring separation for the period prior to 2001, an average over a 29-year period, was about 11 months, and the dataset average Loop Current's westward tilt angle—a factor that indicates whether the Loop Current will soon shed an eddy or not—was about 16°. After the year 2000, the dataset average period for ring separation, an average over a 39-year period, decreased by about 1 month and was about 10 months. The average ring-separation period in the decade 2001–2010 was about 9 months. The dataset average of the Loop Current's westward tilt angle increased by about 5° in the period 1998–2008 and was about 20° in 2010. Potential causes for these changes are discussed.

1. Introduction

Previous studies of the cycle of warm-core ring (WCR) separation from the Loop Current have reported average separation periods of about 11-12 months [1–6]. Studies have found that the frequency distribution is bimodal with modes at 8-9 months and 13-14 months [7, 8] and with modes at 6 and 11 months [5, 6]. Vukovich [2, 4, 6] also noted that the eddy separation periods were highly variable, ranging from 6 to 19 months. Sturges and Evans [1] have suggested that the maximum period between ring separations maybe as large as 30 months, but periods that large have not been observed as yet. The minimum eddy-shedding period previously observed was 8 months [1, 9]. Vukovich [6], using a 32-year data set, found that the minimum eddy-shedding period was 6 months.

Vukovich [6] introduced the concept of the Loop Current orientation. The Loop Current orientation was defined as the westward tilt angle of the Loop Current. The westward tilt of the Loop Current is believed to be the result of a ring that develops within the Loop Current, but has not as yet separated from the Loop Current. The ring drifts westward,

bringing the Loop Current with it, so to speak, which causes the Loop Current's westward tilt angle to increase. This processes continues until the ring separates from the Loop Current. Some of the initial westward tilting of the Loop Current may be due to the presents of the West Florida Shelf, but the larger angles are most likely due to the westward drift of the rings imbedded in the Loop Current. Using a 28-year data set (1976–2003), Vukovich [6] noted that the average westward tilt angle of the Loop Current was about 17°, which, it was determined from observations, indicated that the Loop Current, on average, was not about to shed a WCR anytime in the very near future. The standard deviation was about ±14°. The mode of the dataset was 0°; that is, a north-south orientation of the Loop Current was the most observed orientation. It was also determined from observations that when the Loop Current's westward tilt angle was >30°, a WCR would separate from the Loop Current within about 30 to 60 days. There are cases when separation occurred in periods >60 days after the westward tilt angle reached 30°. In those cases, very large westward tilt angles were developed. A westward tilt angle of approximately 63° was the maximum angle observed. Angles greater than 55° were only observed

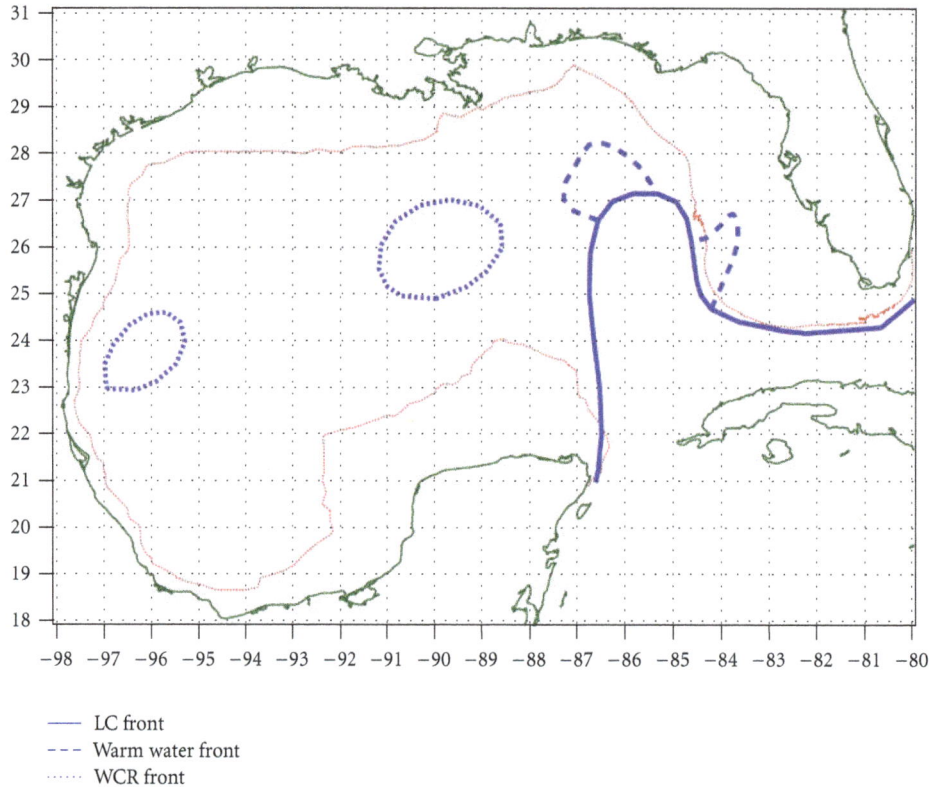

FIGURE 1: Example of the analysis of fronts in the GOM created using remote sensing data and in-situ data.

twice over the 28-year period, and angles between 50° and 54° were also only observed twice.

The data presented in the literature to date would suggest that over the last 30 or so years, the Loop Current eddy-shedding cycle has been relatively consistent with an average period for eddy shedding of about 11–12 months and the Loop Current, on average, was not about to shed a WCR anytime soon (i.e., rings would separate with periods ≥8 months). However, observations obtained in the last decade (2001–2010) indicated that changes in the Loop Current's eddy-shedding cycle have taken place. This paper presents data outlining those changes and discusses the implication of those changes relative to the forces that may affect the Loop Current's behavior.

2. Data and Procedure

The principal data resources used for this study included sea-surface temperature (SST) data from the Television and Infrared Observation Satellite-M (TIROS-M), the Heat Capacity Mapping Mission (HCCM) satellite, SEASAT, the Geostationary Operational Environmental Satellite (GOES), and numerous National Oceanic and Atmospheric Administration (NOAA) satellites; ocean color data from the Coastal Zone Color Scanner (CZCS), the Sea-viewing Wide field of View Sensor (SeaWiFS), and the Moderate Resolution Imaging Spectroradiometer (MODIS); altimeter data from the Ocean Topography Experiment (TOPEX/Poseidon), JASON, and the European Remote Sensing (ERS) satellite; various

analyzed in-situ data from ships of opportunity and from Mineral Management Service (MMS) field programs in the Gulf of Mexico (GOM); information on the Loop Current and rings in the GOM from the programs managed by various oil and gas companies. Monthly frontal analyses were created using an integral of the available remote sensing data, any available information on ocean features, and any available analyzed in situ data for the month in question when sufficient data were available. These fronts were associated with the Loop Current, rings, and other features in the GOM, and the analyses provided "characteristic" positions of the fronts/features in the GOM for the given month. An example of a frontal analysis is presented in Figure 1. The Loop Current front in these analyses is the approximate average front for the month unless a WCR separated from the Loop Current during that month. In that case, it is the position of front after separation has occurred. These frontal analyses were used to determined eddy-shedding periods and to develop time series of the Loop Current's westward tilt angle.

Satellite remote sensing data played a major role in the development of these frontal analyses. In the periods 1972 through 1978 and 1986 through 1991, only satellite SST data were available to develop the ocean front analysis. Most of the SST data were obtained from the NOAA/AVHRR, though GOES, TIROS, and HCMM data were also applied if they were available. As a result, the frontal analyses could only be determined for about 7 months in a year, depending on when the mixed layer developed in the spring/summer and eroded in the fall. However, in the summer and in parts of the spring and fall when SST data were not useful, significant use was made of "ship-of-opportunity" data to fill gaps in the frontal analyses (i.e., these data were used to obtain the month in which separation of a WCR from the Loop Current occurred) when these data were available. In the period 1979 through 1985, CZCS data were used to supplement the SST data when clear-sky data could be obtained and were principally used to detect ocean features in the warm season when the mixed layer was fully developed. After 1991 when altimeter data were available, the frontal analyses could be developed for most features for all twelve months in the year, in most cases.

The eddy shedding of the Loop Current was examined using a time-series analysis of eddy-shedding periods. The eddy-separation periods were determined by documenting the month and year of the separation of each major ring for the period 1972–2010. A major ring was defined as a large ring (i.e., rings with diameters of about 300 km or more at the time they separated from the Loop Current), which persisted for at least five months and moved into the western GOM (WGOM). The month and year of separation was that time when the major eddy completely separated from the Loop Current. In some cases, a WCR separated from the Loop Current only to be reabsorbed by the Loop Current soon thereafter. In other cases, a WCR separated from the Loop Current twice only to be reabsorbed by the Loop Current in each case. The month and year of separation of the ring from the Loop Current for this study was defined as the month and year when complete separation occurred and the ring moved into the WGOM. Discrepancies in the time of ring separation between this study and other such studies [5] are most likely due to how complete ring separation was defined and differences in features seen in the SST, ocean color, and altimeter data. The time series of eddy-shedding periods were used to create histograms of the eddy-shedding periods for different time periods and to determine various statistics for the eddy-shedding periods (e.g., average, standard deviation, mode, etc.) for different time periods. Table 4 provides the data used to calculate the statistics.

Changes in the Loop Current's orientation or westward tilt angle were also examined since these changes are also associated with the process of ring separation. The Loop Current usually tilts greatest to the west at the time a WCR is about to separate. Over 30 years of observation have shown that the westward tilt angle of the Loop Current is usually >30° at the time that a ring separates (i.e., the Loop Current is oriented approximately northwest-southeast), though, as previously noted, much larger orientation angles have been observed. In some cases, elongated westward extension of the Loop Current has been observed and the high-speed currents associated with the Loop Current were observed to influence the central GOM and parts of the WGOM [6]. Observations have also shown that after the ring separates, the Loop Current usually reestablishes itself, having its northern boundary at around 25° N, and, at that time, is usually oriented north-south (i.e., the orientation angle is 0°). As the Loop Current penetrates again into the eastern GOM (EGOM), a ring develops in the Loop Current and a westward tilt of the Loop Current begins to emerge. For this study, the Loop Current orientation was defined as the angle

made by the intersection of a line drawn parallel to the eastern and western frontal boundaries of the Loop Current that is drawn through the approximate center of the Loop Current and a longitude line (Figure 2). The Loop Current's westward tilt angle was determined for each month when monthly frontal analyses were available in the period 1976–2010. The time series data for the westward tilt angle were used to determine various statistics (i.e., the average, the mode, the standard deviation, the maximum, and the minimum orientation angle). Table 5 provides the data used to calculate the statistics.

3. Loop Current Eddy-Shedding Period

A major change in the average period at which rings separated from the Loop Current occurred in the decade from 2001 through 2010. Figure 3 presents the year-to-year change in the dataset average eddy-shedding period from 1990 to 2010. The value of the average eddy-shedding period for 1990 in the figure is the 19-year average from 1972 through 1990, that for 1991 is the 20-year average from 1972 through 1991, and so on. It can be seen that the average eddy-shedding period of the Loop Current decreased in the decade 2001 through 2010 from a high value of 11.3 months in 2003 to a value of 10.1 months in 2009. Prior to 2003, there was a small-amplitude oscillation in the dataset average eddy-shedding period. It varied from 11.3 months for 1994 to a low value of 10.6 months for 1996 and then went back to 11.3 months by 2003.

The data presented in Figure 3 starts with a 19-year average for 1990 (i.e., the average from 1972 to 1990). It may have been possible to start with a 15-year data set average for 1986, which was previously reported by Vukovich [2] and which provided an average eddy-shedding period of 10.9 months, and still obtain a reasonable average value. However, averages over periods less than 15 years are extremely susceptible to the year-to-year variability of the eddy-shedding periods and become very noisy. The trends that exist in the average eddy shedding after 1990 are trends in long-term averages and were precipitated by changes, and in some cases, major changes, in eddy-shedding periods.

The histogram of the Loop Current's eddy-shedding periods for the entire period 1972 through 2010 (Figure 4) shows a bimodel distribution with a range in the periods of 4 months to 19 months. The modes are at 6 months and 11 months, and the average eddy-shedding period over the 39-year period was 10.2 months and the standard deviation was about ±4 months (Table 1). Approximately 64% of the eddy-shedding periods were >8 months with about 36% being ≤8 months.

The distinction between the 29-year period before 2001 and the 10-year period after 2000 is illustrated when the data used to create Figure 4 are used to create histograms for the eddy-shedding periods that occurred before 2001 and those found after 2000. The histogram of the Loop Current's eddy-shedding periods for the period from 1972 through 2000 (Figure 5) shows a single mode at 6 months and a range in the periods of 5 months to 19 months. The average eddy-shedding period over the 29-year period was 10.9 months,

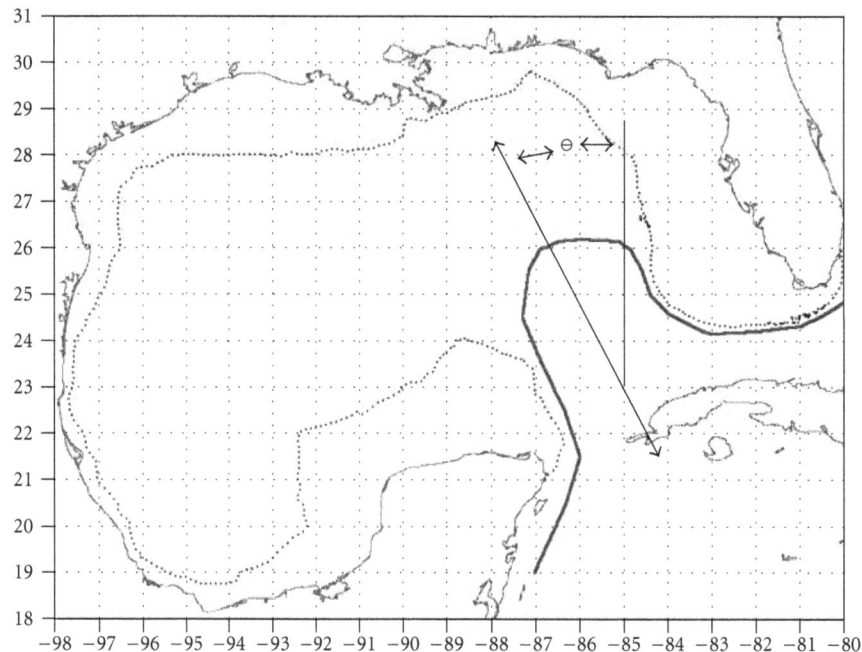

FIGURE 2: Procedure used determined the Loop Current's orientation angle.

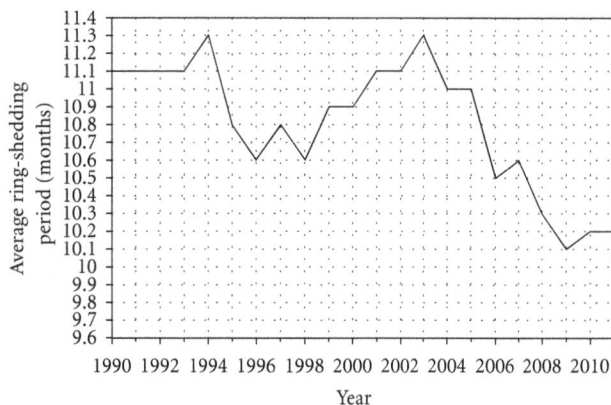

FIGURE 3: The average eddy-shedding period of the Loop Current for the period 1990–2010. The average value for 1990 is the 19-year average from 1972 through 1990, that for 1991 is the 20-year average from 1972 through 1991, and so on.

TABLE 1: Statistics on the Loop Current's eddy-shedding periods for various periods of time.

Statistic	Period of data: 1972–2010	Period of data: 1972–2000	Period of data: 2001–2010
Average separation period (month)	10.2	10.9	8.8
Standard deviation for separation period (month)	±4	±4	±5
Mode for separation period (month)	6 and 11	6	4
Percent of time that the period ≤8 months	36%	26%	56%

and the standard deviation was again about ±4 months (Table 1). Approximately 74% of the eddy-shedding periods are >8 months with about 26% being ≤8 months.

The histogram of the Loop Current's eddy-shedding periods for the period from 2001 through 2010 (Figure 6) also shows a single mode, which is found at 4 months in this case, and the range in the periods is 4 months to 18 months. The average eddy-shedding period over the 10-year period was approximately 8.8 months, and the standard deviation was about ±5 months (Table 1). It also should be noted that in the decades 1981–1990 and 1991–2000, the average separation periods were 11.8 and 10.5 months, respectively. Approximately 44% of the eddy-shedding periods are >8

months with about 56% being ≤8 months. In previous decades (i.e., 1981–1990 and 1991–2000), the eddy-shedding periods ≤8 months were 18% and 36%, respectively. In the decade 2001–2010, 16 major WCRs separated from the Loop Current. In previous decades (i.e., 1981–1990 and 1991–2000), 11 major WCRs separated from the Loop Current in each decade, and in the 9-year period 1972–1980, 10 major WCRs separated from the Loop Current. If only one ring had separated in 1971, the decade 1971–1980 would also have had 11 major WCRs separated from the Loop Current.

The period 2001–2010 was characterized with 5 years in which more than one eddy separated in each year. In two years during that period, two rings separated from the Loop Current at the same time. The first of these events took place in 2001, and a similar event took place in 2002. Prior to ring separation, the Loop had a considerable northwest-southeast

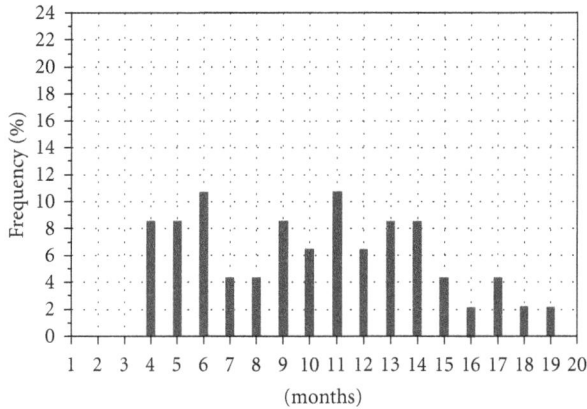

FIGURE 4: Histogram of the Loop Current's eddy-shedding periods for the period 1972–2010. The total number of events is 47.

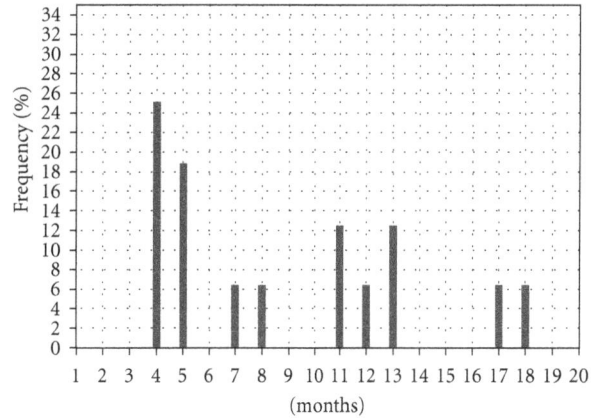

FIGURE 6: Histogram of the Loop Current's eddy-shedding periods for the period 2001–2010. The total number of events is 16.

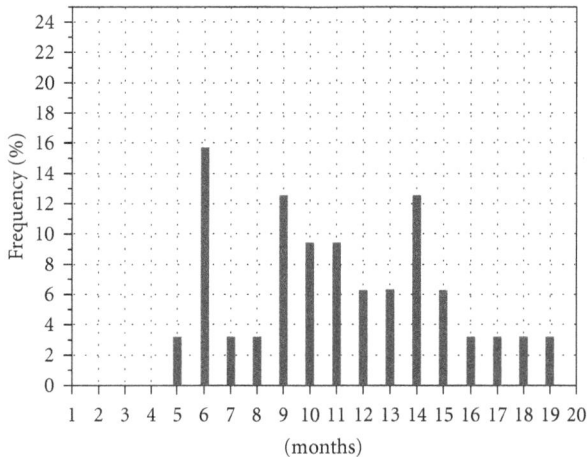

FIGURE 5: Histogram of the Loop Current's eddy-shedding periods for the period 1972–2000. The total number of events is 31.

extension (Figures 7(a) and 8(a)) in both cases. In 2002, the Loop Current was found as far west as 93° W and it affected the central GOM (Figure 8(a)) and parts of the WGOM as well as the EGOM. In 23 March 2001 image (Figure 7(a)), the Loop Current extended to 91° W; however, before the rings separated in April, the Loop Current extended to 93° W in that case also. When the Loop Current broke down, two rings were evident in the GOM. The two rings in 2002 (Figure 8(b)) were major rings, having an initial diameter of about 300 km. They drifted westward toward the western wall of the GOM and lasted 5 months or more before they dissipated. In 2001, two rings also separated from the Loop Current at the same time (Figure 7(b)). The ring west of the Loop Current was a major ring. The small ring near the northern boundary of the Loop Current was a minor ring. It had a diameter of about 170 km, drifted northward into the Desoto Canyon region, and dissipated in about 3 months. An examination of our satellite remote sensing archive, which dates from 1972 to the present, indicated that no previous event similar to those in 2001 and 2002 could be found. Examination of the literature containing analyses of the Loop

Current using ship survey data, which were too numerous to cite, as far back as 1932, provided no analyses of a similar event. The analyses in the literature do not possess sufficient continuity to state conclusively, however, such an event was not found in the period 1932–1972. The data prior to 1932 are very sparse.

Another interesting event occurred in 2006. In that year, three major rings separated from the Loop Current, but in this case, at different times in that year. The first major ring separated in February 2006 (Figure 9(a)) and had a separation period of 5 months (i.e., the previous ring separation occurred in 2005). The other two major rings had periods of 4 months. The first of these separated from the Loop Current in June 2006 (Figure 9(b)) and the other in October 2006 (Figure 9(c)). No similar event was found in either the archives of remote sensing imagery or in the analyses of the Loop Current using ship survey data in the literature dating back to 1932. Besides the 2006 triple separation of major rings event and the event in 2002 when two major rings simultaneously separated from the Loop Current, the period 2001–2010 was also characterized with three other years (i.e., 2004, 2008, and 2009) when two major rings separated from the Loop Current at different times during each of those years. The satellite remote sensing archive indicated there have previously been four instances when a major ring separated from the Loop Current twice during a particular year at different times. These occurred in 1975, 1986, 1995, and 1996. The consecutive years having two major rings separated from the Loop Current (i.e., in the years 1995 and 1996) were in part responsible for the oscillation in the data set average eddy-shedding period in the period 1994–2003 (Figure 3).

Though there was a major change in the number of and the statistics for the periods at which major rings separated from the Loop Current in the decade 2001–2010 compared to the previous time periods, there was no change in the basic statistics on the month in a year when the rings separated from the Loop Current. Table 2 shows that in the case of the entire data set for the period 1972–2010 as

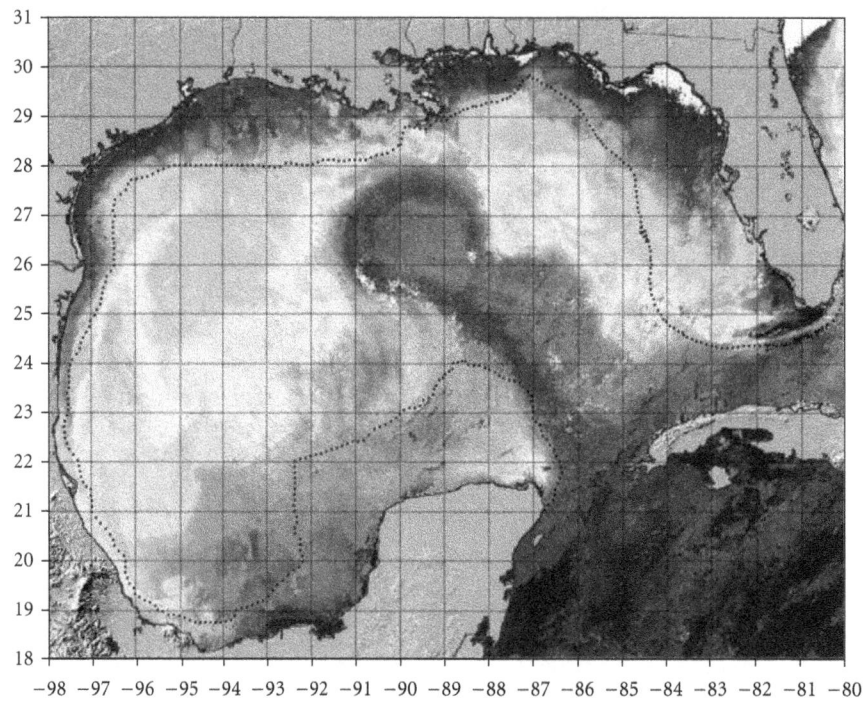

(a)

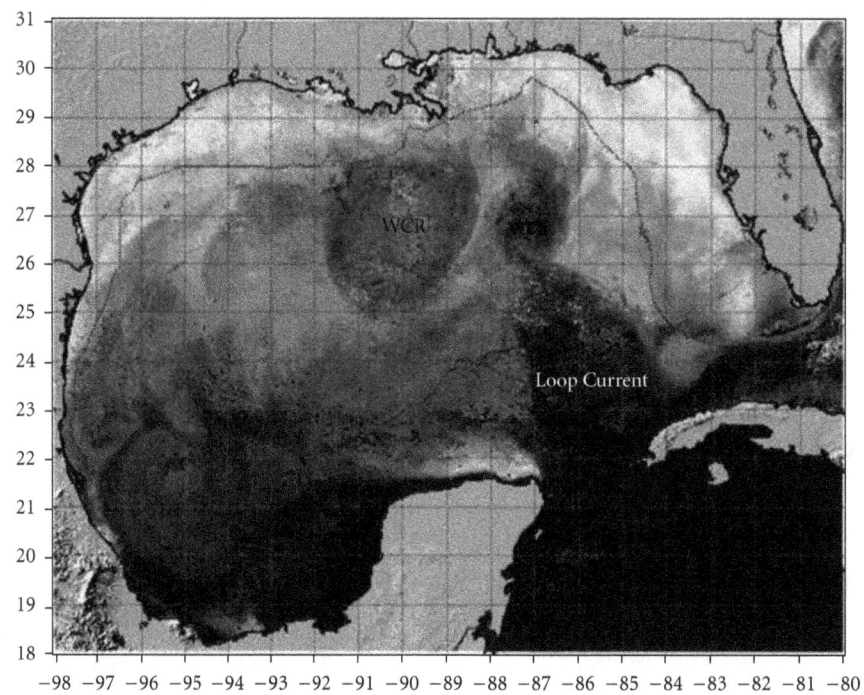

(b)

FIGURE 7: (a) Sea-surface temperature image from a NOAA satellite for 23 March 2001, showing a Loop Current with a significant northwest-southeast extension. (b) Sea-surface temperature image from a NOAA satellite for 23 April 2001, showing two rings had separated from the Loop Current.

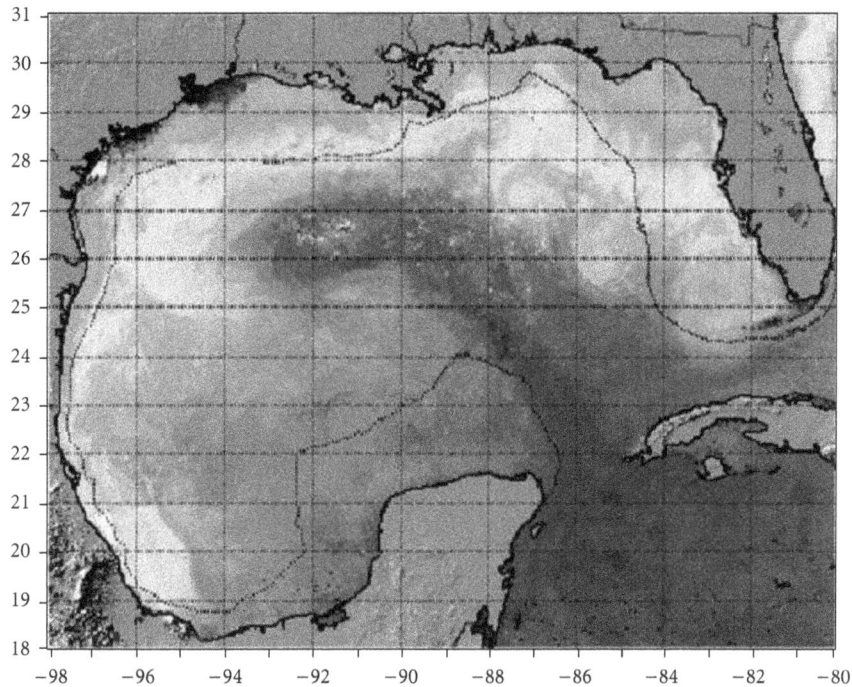

(a)

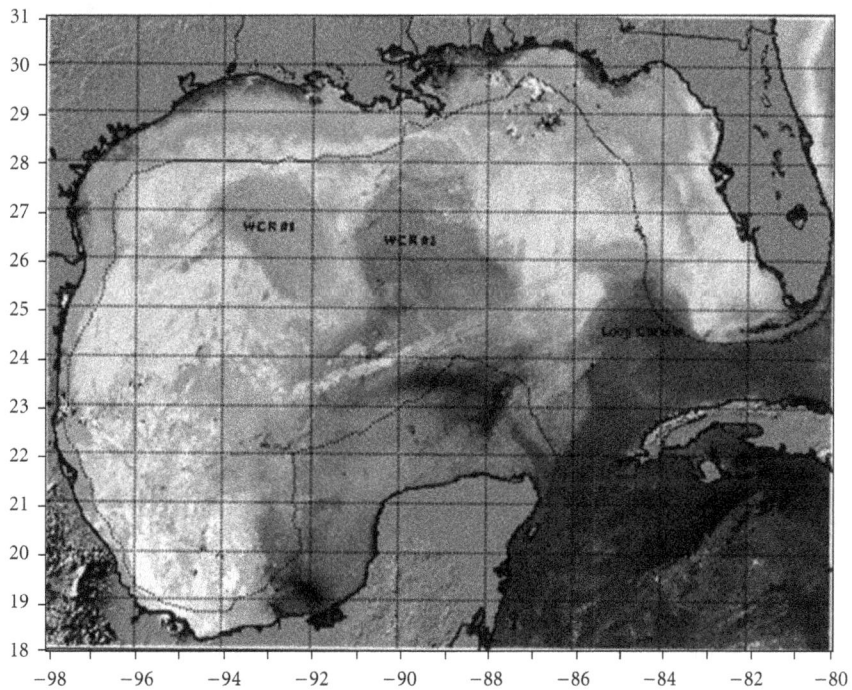

(b)

FIGURE 8: (a) Sea-surface temperature image from a NOAA satellite for 31 January 2002. (b) Sea-surface temperature image from a NOAA satellite for 15 March 2002, showing two rings had separated from the Loop Current.

well as the partitioned data sets for the time periods 1972–2000 and 2010-2010, the month, on average, when eddy-shedding took place was the sixth month (i.e., June) in each case, the standard deviation was ±3 months in each case, and the month of the highest number of major ring separations was the third month (i.e., March) in each case. These results are identical to that previously noted for a 32-year data set [6].

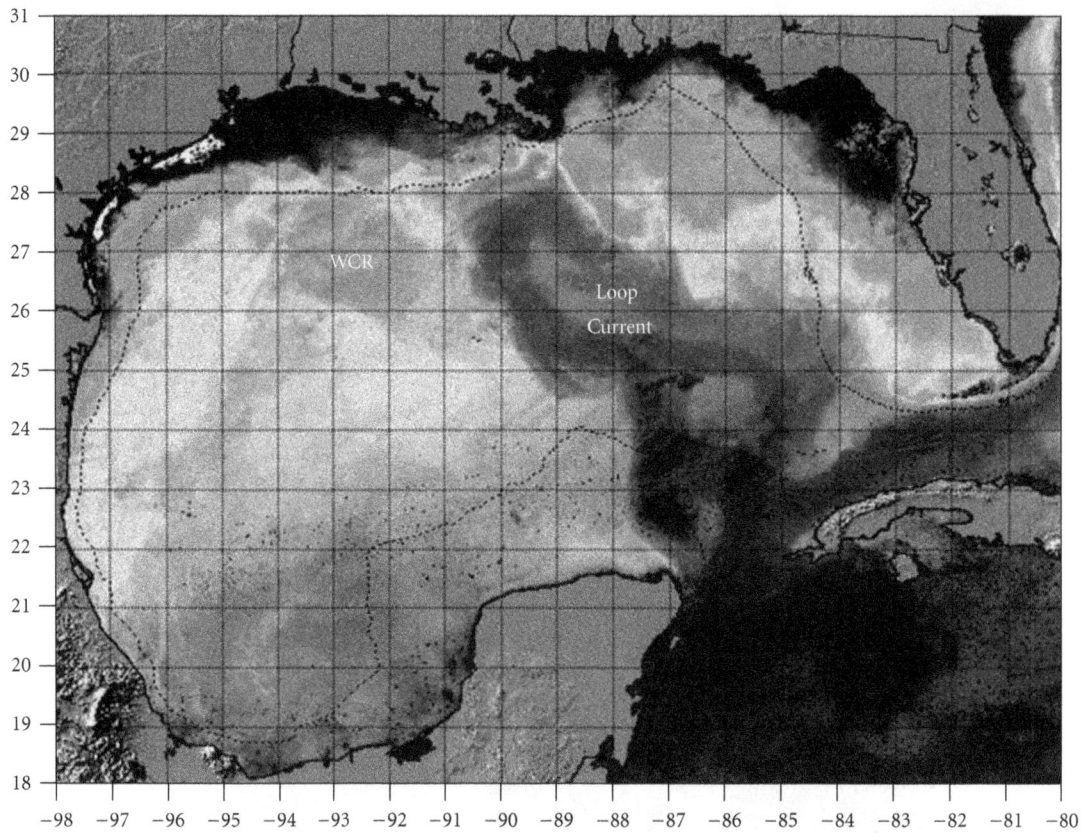

(a)

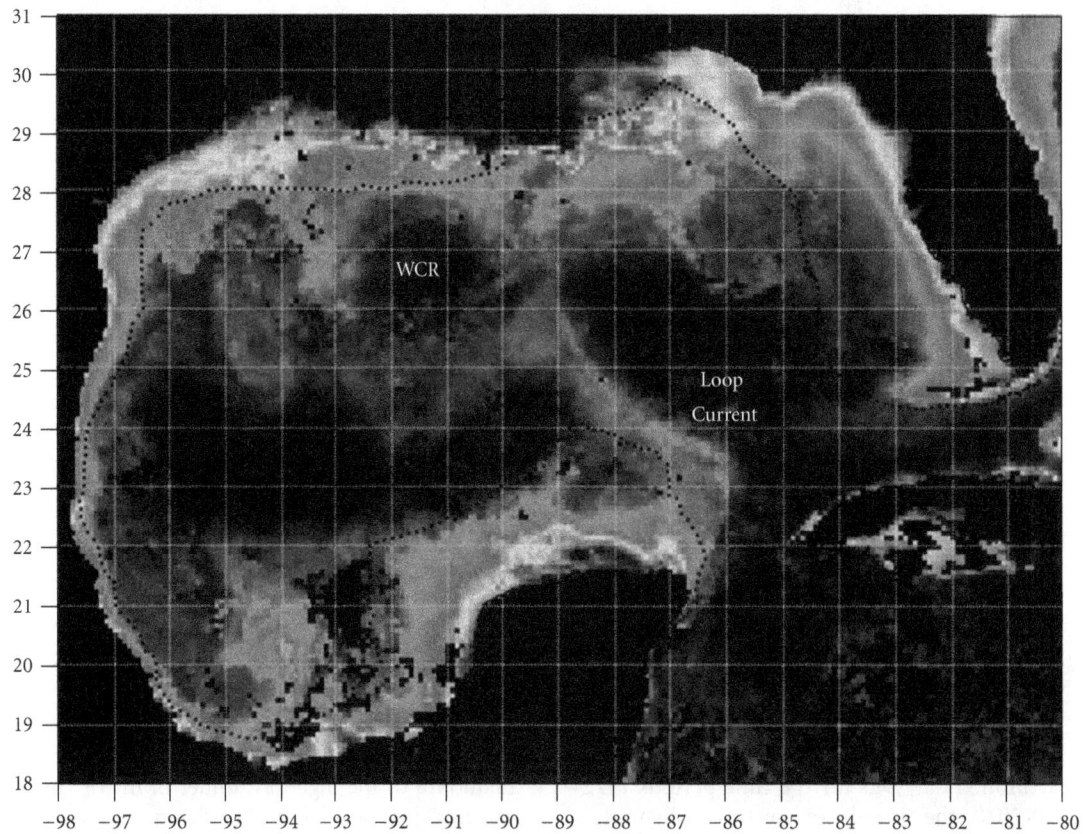

(b)

FIGURE 9: Continued.

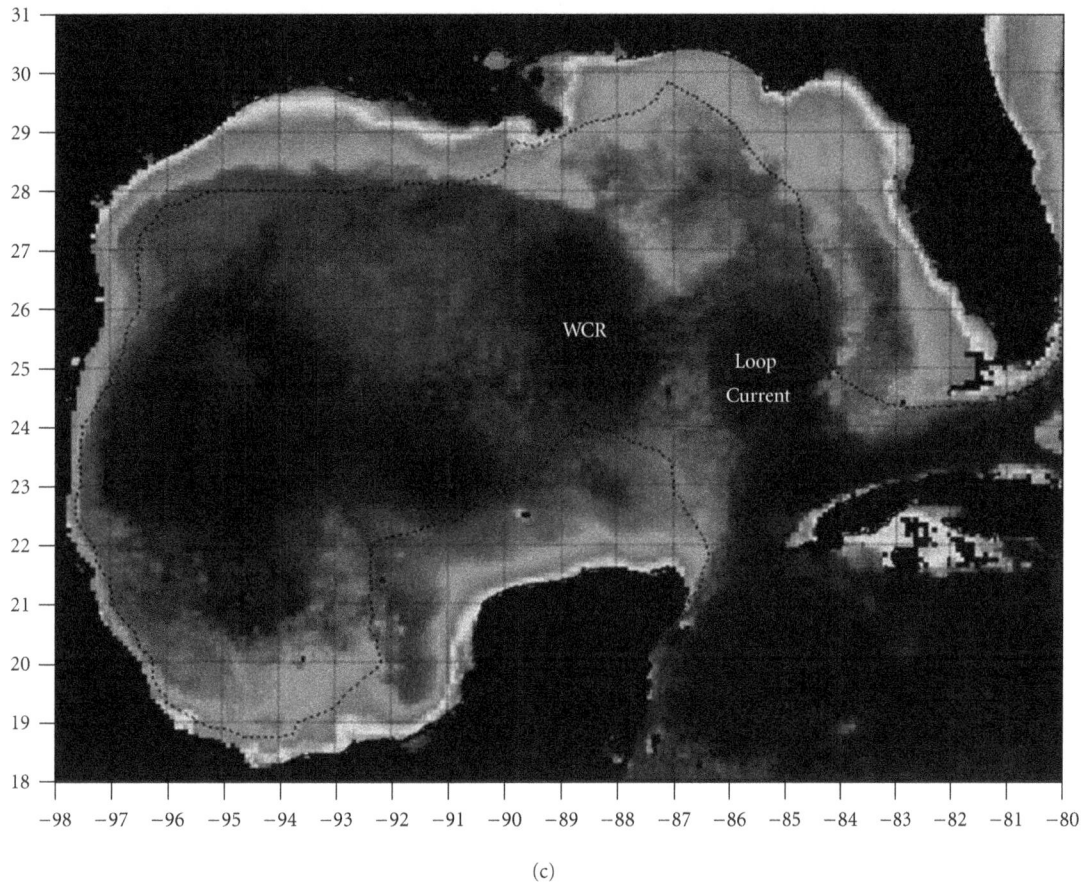

(c)

FIGURE 9: (a) Sea-surface temperature image from a NOAA satellite for 27 February 2006. (b) Average ocean color image from the MODIS satellite centered about 17 June 2006. (c) Average ocean color image from the MODIS satellite centered about 15 October 2006.

TABLE 2: Statistics on major ring separation as a function of the month in a year.

Statistic	Period of data: 1972–2010	Period of data: 1972–2000	Period of data: 2001-2010
Average separation month (month)	6	6	6
Standard deviation for separation month (month)	±3	±3	±3
Mode for separation month (month)	3	3	3

The histogram for the Loop Current's eddy shedding as a function of the month in a year for the period 1972–2010 (Figure 10(a)) shows that most of the rings, about 67%, separated in the winter (i.e., January, February, and March) and summer (i.e., July, August, and September) together (i.e., 33.3% in each season). The least number of rings separated in the fall (i.e., October, November, and December). The histogram for the period 1972–2000 (Figure 10(b)), on the other hand, shows that about an equal number of rings, about 30%, separated in winter, spring (i.e., April, May, and

June), and summer with the smallest number of separations again in the fall. The histogram for the period 2001–2010 (Figure 10(c)), like that for the entire period, shows that most of the rings, about 75% in this case, separated in winter and summer (i.e., about 37.5% in each season) with the least number of separations in the spring and fall. No rings separated in the month 12 (December) in any case. This was previously noted by Vukovich [6].

4. Loop Current Orientation

The Loop Current's orientation is a factor associated with the ring separation, and it also had major changes in the decade 2001–2010. Figure 11 presents the variation of the dataset average westward tilt angle for the 21-year period 1990–2010. Like the average eddy-shedding periods given in Figure 3, the value of the average westward tilt angle in Figure 11 is a dataset average, and for 1990, it is the 15-year average from 1976 to 1990, that for 1991 is the 16-year average from 1976 to 1991, and so on. In the period 1990 through 2000, the average westward tilt angle varied at most by about 1°, from a low of 15.7° to a high of about 16.8°. After the year 2000, the westward tilt angle increased by about 5°, reaching a high value of about 20.3° in 2008. This indicates that the Loop Current's orientation was, on average, nearer to that when

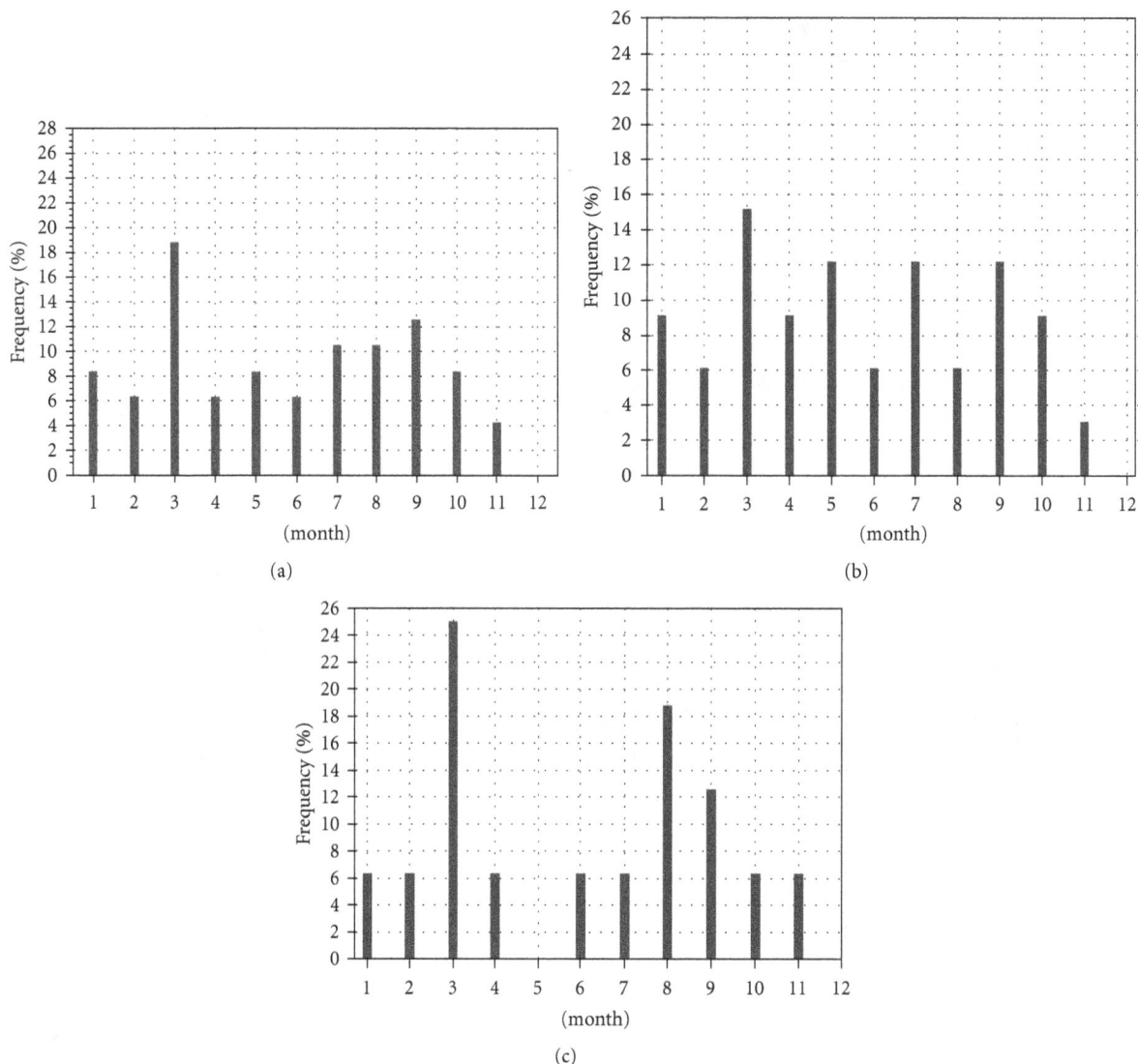

FIGURE 10: (a) Histogram of major ring separation from the Loop Current as a function of month in a year for the period 1972–2010. The total number of events is 47. (b) Histogram of major ring separation from the Loop Current as a function of month in a year for the period 1972–2000. The total number of events is 31. (c) Histogram of major ring separation from the Loop Current as a function of month in a year for the period 2001–2010. The total number of events is 16.

ring separation normally occurs (i.e., orientation angles of 30° or more) during that decade and was more inclined to have ring separation in that period than before 2001.

In order for the dataset average westward tilt angle to increase in this manner, very large monthly values of the orientation angle persisted in the 2001–2010 period. As a matter of fact in the first two years of that period (i.e., 2001 and 2002), very large westward tilt angles were documented for the Loop Current associated with the remarkable westward extension of the Loop Current (Figures 7(a) and 8(a)). The Loop Current had about northwest-southeast orientation and extended as far west as 93° W longitude. The westward tilt angle noted for the 2001 event was 53° and that for the 2002 event 63°. Examination of the satellite remote sensing data archives as far back as 1972 and the literature, which provides analyses of the Loop Current using ship survey data,

as far back as 1932, indicated that a Loop Current having both large westward extensions and large westward tilt angles had not occurred previously. The largest westward tilt angle observed prior to 2001 was 59°, but in that case, there was no large westward extension of the Loop Current.

For the period 1976 through 2000, the average orientation angle was 16.5°, the standard deviation was ±14.0°, and the mode for that dataset was 0° (Table 3). The range in the orientation angle data set was 0° to 59°. Approximately 86% of the angles were <30°, indicating that the Loop Current was most often in a mode over that period in which eddy shedding would not occur. In the decade 2001 through 2010, the average orientation angle was 26.4°, about 10° higher than that for the period 1776 through 2000 and about 9° higher than the two previous decades (i.e., the average for the decade 1981–1990 was 16.5° and that for the decade 1991–2000 was

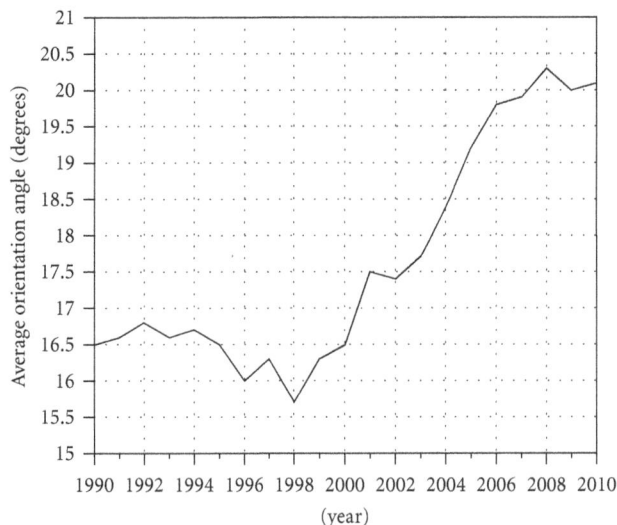

FIGURE 11: The average Loop Current westward tilt angle for the period 1990–2010. The average value for 1990 is the 15-year average from 1976 to 1990, that for 1991 is the 16-year average from 1976 to 1991, and so on.

TABLE 3: Statistics on the Loop Current's Orientation Angle for Various Periods of Time.

Statistic	All data (1976–2010)	Data before 2001 (1976–2000)	Data since 2000 (2001–2010)
Average orientation angle	20.1°	16.5°	26.4°
Standard deviation of orientation angle	15.0°	14.0°	16.9°
Mode	0°	0°	0°
Maximum angle	63°	59°	63°
Minimum angle	0°	0°	0°
Percentage of data with orientation angles <30°	75%	86%	54%
Percentage of data with orientation angles ≥30°	25%	14%	46%

16.6°). In the decade 2001 through 2010, approximately 54% of the angles were <30° or 46% were ≥30°. In the two decades prior to 2001 (i.e., the decades 1981–1990 and 1991–2000), only 14% of the orientation angles were ≥30° in each decade. In the decade 2001–2010, the Loop Current was in a state where it was just as likely to shed an eddy as not.

5. Summary and Discussion

The major findings of this study are itemized below.

(a) The dataset average period for ring separation decreased by about 1 month in the decade 2001–2010. The 39-year average (1972–2010) was about 10 months. The average over the 29-year period 1972–2000 was about 11 months.

TABLE 4: The eddy-separation period and the month in which eddy separation took place for each year. also provided is the data sources used to determine the month of separation.

Year	Separation period (months)	Separation month	Data used in determination of period
1972	•	5	SST
1973	14	7	SOP
1974	9	4	SST
1975	9	1	SST
1975	6	7	SOP
1976	13	8	SOP
1977	7	3	SST
1978	15	6	SST, SOP
1979	10	4	SST
1980	9	1	SST
1981	14	3	SST
1982	14	5	SST, OCD
1983	10	3	SST
1984	11	2	SST
1985	17	7	SST, OCD, SOP
1986	8	1	SST
1986	9	10	SST
1987	13	11	SST
1988	6	5	SST
1989	12	5	SST
1990	16	9	SOP
1991	12	9	SOP
1992	10	7	SOP
1993	11	6	SOP, SST
1994	15	9	SSH, SOP
1995	6	3	SST, SSH
1995	6	9	SSH, SOP
1996	5	2	SST, SSH
1996	11	8	SSH, SOP
1997	14	10	SST, SSH
1998	6	3	SST, SSH
1999	19	10	SST, SSH
2001	18	4	SST, OCD, SSH
2002	11	3	SST, OCD, SSH
2002	11	3	SST, OCD, SSH
2003	17	8	OCD, SSH
2004	5	1	SST, OCD, SSH
2004	8	9	OCD, SSH
2005	12	9	OCD, SSH
2006	5	2	SST, OCD, SSH
2006	4	6	SST, OCD, SSH
2006	4	10	SST, OCD, SSH
2007	13	11	SST, OCD, SSH
2008	4	3	SST, OCD, SSH

TABLE 4: Continued.

Year	Separation period (months)	Separation month	Data used in determination of period
2008	5	8	OCD, SSH
2009	7	3	SST, OCD, SSH
2009	4	7	OCD, SSH
2010	13	8	OCD, SSH

SST: Sea-Surface Temperature from Satellite Infrared Water Vapor Window Data.
OCD: Satellite Ocean Color Data from CZCS (1979–1985) or from MODIS (2000-Present).
SSH: Sea-Surface Height from Various Satellite (1991-present).
SOP: Data and/or Information Obtained from the MMS's Ship-of-Opportunity Program, MMS sponsored Field Programs, and/or Field Programs Sponsored by Various Oil and Gas Companies.

(b) Sixteen (16) WCRs separated from the Loop Current in the decade 2001–2010, whereas in two previous decades, 11 WCRs separated in each decade.

(c) The average ring-separation period in decade 2001–2010 was about 9 months, and in the decades 1981–1990 and 1991–2000, the average period for ring separation in each decade was about 11 months.

(d) More than half the rings (i.e., 56%) that separated from the Loop Current in the decade 2001–2010 had separation periods ≤8 months. In the period 1972–2000, only 26% of the rings had separation periods ≤8 months, with 18% in the period 1981–1990 and 36% in the period 1991–2000 having eddy separation periods ≤8 months.

(e) The period 2001–2010 was characterized with 5 years in which more than one major eddy separated in each year. In four of those years, two major rings separated from the Loop Current, and in one of those years, two major rings separated from the Loop Current simultaneously. In the other year, three major rings separated from the Loop current during that year at different times. Previous to 2001, so many multiple events were not observed in a decade.

(f) The dataset average of the Loop Current's westward tilt angle increased by about 5° from 1998 to 2008.

(g) The dataset average of the Loop Current's westward tilt angle was about 17° in the period 1976–2000 and about 20° in the period 1972–2010.

(h) The average westward tilt angle for the decade 2001–2010 was about 26°. In the decades 1981–1990 and 1991–2000, the average westward tilt angle was about 17° in each decade.

(i) In the period 1976–2000, the Loop Current's westward tilt angle was >30° (i.e., when the average Loop Current's westward tilt angle reaches a value >30°, ring separation normally occurs within 30–60 days) about 14% of the time, which characterizes a Loop Current in which ring separation was, on average, not imminent and/or occurred often. It was >30° about 25% in the period 1976–2010.

TABLE 5: Orientation angle as a function of month and year.

Year and month	Orientation angle
1976	
Jan	27
Feb	24
Mar	0
Apr	13
May	29
Jun	No Data
Jul	No Data
Aug	No Data
Sep	No Data
Oct	No Data
Nov	9
Dec	14
1977	
Jan	20
Feb	14
Mar	0
Apr	0
May	0
Jun	No Data
Jul	No Data
Aug	No Data
Sep	No Data
Oct	No Data
Nov	0
Dec	15
1978	
Jan	9
Feb	27
Mar	26
Apr	28
May	21
Jun	No Data
Jul	No Data
Aug	No Data
Sep	No Data
Oct	No Data
Nov	0
Dec	14
1979	
Jan	19
Feb	21
Mar	29
Apr	24
May	42
Jun	No Data
Jul	No Data
Aug	No Data
Sep	No Data

TABLE 5: Continued.

Year and month	Orientation angle
Oct	No Data
Nov	30
Dec	53
1980	
Jan	31
Feb	7
Mar	13
Apr	13
May	0
Jun	No Data
Jul	No Data
Aug	No Data
Sep	No Data
Oct	No Data
Nov	14
Dec	10
1981	
Jan	27
Feb	19
Mar	3
Apr	4
May	0
Jun	No Data
Jul	No Data
Aug	No Data
Sep	No Data
Oct	No Data
Nov	2
Dec	11
1982	
Jan	16
Feb	11
Mar	13
Apr	21
May	22
Jun	No Data
Jul	No Data
Aug	No Data
Sep	No Data
Oct	No Data
Nov	14
Dec	24
1983	
Jan	35
Feb	32
Mar	29
Apr	19
May	18
Jun	No Data

TABLE 5: Continued.

Year and month	Orientation angle
Jul	No Data
Aug	No Data
Sep	No Data
Oct	No Data
Nov	24
Dec	26
1984	
Jan	21
Feb	30
Mar	29
Apr	4
May	8
Jun	5
Jul	10
Aug	19
Sep	No Data
Oct	No Data
Nov	7
Dec	3
1985	
Jan	39
Feb	10
Mar	15
Apr	29
May	24
Jun	37
Jul	No Data
Aug	No Data
Sep	5
Oct	11
Nov	36
Dec	32
1986	
Jan	6
Feb	0
Mar	19
Apr	29
May	26
Jun	20
Jul	24
Aug	No Data
Sep	No Data
Oct	No Data
Nov	0
Dec	0
1987	
Jan	4
Feb	11
Mar	24

TABLE 5: Continued.

Year and month	Orientation angle
Apr	20
May	20
Jun	20
Jul	No Data
Aug	No Data
Sep	No Data
Oct	No Data
Nov	0
Dec	0
1988	
Jan	0
Feb	18
Mar	4
Apr	14
May	0
Jun	0
Jul	No Data
Aug	No Data
Sep	No Data
Oct	No Data
Nov	16
Dec	17
1989	
Jan	33
Feb	34
Mar	29
Apr	32
May	0
Jun	0
Jul	No Data
Aug	No Data
Sep	No Data
Oct	No Data
Nov	29
Dec	8
1990	
Jan	40
Feb	12
Mar	15
Apr	15
May	20
Jun	30
Jul	No Data
Aug	No Data
Sep	No Data
Oct	No Data
Nov	0
Dec	0
1991	

TABLE 5: Continued.

Year and month	Orientation angle
Jan	25
Feb	8
Mar	31
Apr	17
May	13
Jun	No Data
Jul	No Data
Aug	No Data
Sep	No Data
Oct	18
Nov	14
Dec	24
1992	
Jan	10
Feb	27
Mar	3
Apr	24
May	29
Jun	45
Jul	Missing
Aug	Missing
Sep	Missing
Oct	0
Nov	0
Dec	33
1993	
Jan	25
Feb	1
Mar	24
Apr	30
May	0
Jun	0
Jul	Missing
Aug	Missing
Sep	Missing
Oct	Missing
Nov	0
Dec	22
1994	
Jan	27
Feb	26
Mar	30
Apr	14
May	25
Jun	10
Jul	34
Aug	Missing
Sep	Missing
Oct	5

TABLE 5: Continued.

TABLE 5: Continued.

Year and month	Orientation angle	Year and month	Orientation angle
Nov	0	Aug	23
Dec	22	Sep	0
1995		Oct	0
Jan	21	Nov	0
Feb	23	Dec	0
Mar	3	1999	
Apr	4	Jan	4
May	2	Feb	7
Jun	24	Mar	11
Jul	Missing	Apr	42
Aug	Missing	May	21
Sep	Missing	Jun	51
Oct	11	Jul	40
Nov	6	Aug	59
Dec	16	Sep	27
1996		Oct	26
Jan	23	Nov	0
Feb	31	Dec	23
Mar	2	2000	
Apr	7	Jan	10
May	9	Feb	21
Jun	23	Mar	14
Jul	13	Apr	14
Aug	1	May	22
Sep	1	Jun	22
Oct	1	Jul	28
Nov	1	Aug	12
Dec	0	Sep	14
1997		Oct	35
Jan	17	Nov	29
Feb	22	Dec	21
Mar	18	2001	
Apr	16	Jan	39
May	17	Feb	40
Jun	Missing	Mar	53
Jul	18	Apr	44
Aug	Missing	May	28
Sep	34	Jun	9
Oct	49	Jul	18
Nov	3	Aug	44
Dec	17	Sep	45
1998		Oct	41
Jan	14	Nov	39
Feb	15	Dec	21
Mar	29	2002	
Apr	0	Jan	63
May	10	Feb	46
Jun	2	Mar	40
Jul	0	Apr	0

Year and month	Orientation angle	Year and month	Orientation angle
May	0	Feb	40
Jun	0	Mar	17
Jul	0	Apr	39
Aug	0	May	26
Sep	0	Jun	28
Oct	0	Jul	36
Nov	0	Aug	55
Dec	40	Sep	58
2003		Oct	17
Jan	22	Nov	10
Feb	7	Dec	28
Mar	18	2007	
Apr	32	Jan	0
May	48	Feb	22
Jun	39	Mar	28
Jul	30	Apr	20
Aug	36	May	24
Sep	10	Jun	14
Oct	29	Jul	25
Nov	0	Aug	33
Dec	0	Sep	40
2004		Oct	46
Jan	63	Nov	0
Feb	24	Dec	9
Mar	21	2008	
Apr	28	Jan	30
May	34	Feb	27
Jun	31	Mar	42
Jul	29	Apr	33
Aug	32	May	26
Sep	47	Jun	44
Oct	30	Jul	30
Nov	34	Aug	30
Dec	25	Sep	0
2005		Oct	28
Jan	30	Nov	28
Feb	26	Dec	40
Mar	36	2009	
Apr	37	Jan	40
May	30	Feb	44
Jun	46	Mar	0
Jul	48	Apr	26
Aug	48	May	25
Sep	40	Jun	3
Oct	48	Jul	0
Nov	36	Aug	0
Dec	36	Sep	0
2006		Oct	0
Jan	50	Nov	0

TABLE 5: Continued.

Year and month	Orientation angle
Dec	22
2010	
Jan	37
Feb	8
Mar	15
Apr	15
May	28
Jun	45
Jul	53
Aug	0
Sep	0
Oct	23
Nov	0
Dec	25

(j) In the two decades prior to 2001 (i.e., the decades 1981–1990 and 1991–2000), the Loop Current's westward tilt angle was >30° about 14% of the time in each decade. In the period 2001–2010, the Loop Current's westward tilt angle was >30° about 46% of the time, which characterizes a Loop Current in which eddy separation would occur sooner than in the years prior to 2001 and would occurr more often.

These observations suggest that a major change has occurred in the forcing function for Loop Current eddy shedding in the period 2001–2010. Oey et al. [10] provided significant information on the probable forcing functions for Loop Current ring separation. They used a numerical model to show that there are three principal forcing functions that cause WCRs to separate from the Loop Current: (1) steady transport in the Yucatan Strait; (2) Caribbean eddies (i.e., anticyclones) that "squeeze" through the Yucatan Strait; (3) wind-induced transport fluctuations through the Greater Antilles Passages. Their model shed Loop Current eddies with periods of about 9 months to 10 months with a steady transport in the Yucatan Strait. Caribbean eddies produced ring separation in their model with longer periods (i.e., on the order of about 14 to 16 months). Caribbean eddies tended to hinder the northward extension of the Loop Current into the EGOM, which led to longer periods between eddy separations. Caribbean eddies are anticyclonic eddies, which are produced in the North Brazil Current [11]. These rings grow as they travel westward. Murphy et al. [12] showed that the Caribbean eddies travel westward with an average speed of about 15 cm s^{-1} and that it took most eddies 10 months to travel from the Lesser Antilles to the Yucatan Channel.

The modeling results of Oey et al. [10] indicated that wind-induced transport fluctuations through the Greater Antilles Passages caused Loop Current eddy shedding with substantially shorter periods: periods from 3 months to 7 months. Since the decade 2001–2010 was characterized with

considerably more Loop Current ring separations with periods ≤8 months than in prior decades, this suggests that wind-induced transport fluctuations through the Greater Antilles Passages may have played a more significant role in ring separations in that decade and may have occurred with greater frequency during that decade than in prior decades.

Was there a major change in the transport through the Yucatan Channel that contributed to the evolution of the Loop Current in these cases? What role did the wind stress in the GOM play in the evolution of the Loop Current? Was the increased frequency of Loop Current eddy shedding at substantially shorter periods a result of an increased frequency of wind-induced transport fluctuations through the Greater Antilles Passages? These are questions that have no answers presently, but answers to these questions are needed. Many more years of observation will be required to determine if the changes in the Loop Current's eddy shedding observed in the last decade are a low-frequency variation or are permanent changes. Continuous monitoring the Loop Current's eddy shedding is straightforward as long as data are available from satellite altimeters, infrared radiometers, and/or ocean color sensors. However, the MODIS ocean color/IR program is coming to the end. Much of the continuous monitoring of the GOM and the global ocean from satellites will fall on the shoulders of the NPOESS program, a joint NOAA, NASA, and DOD program. It is not clear that NPOESS will provide all the satellite data needed to monitor the GOM and the global oceans properly at this time. Continuous monitoring of wind-induced transport fluctuations through the Greater Antilles Passages, Caribbean eddies, the effects of wind stress in the GOM, or some other feature yet to be determined that may be influencing the Loop Current behavior and eddy shedding, on the other hand, is not straightforward. It requires a monitoring system that does not exist. Global ocean numerical models can provide insight, and modeling of the period 2001–2010 in the GOM should be accomplished, but in the end, hard data are required for validation.

Acknowledgments

The author would like to thank Frank Monaldo of the Applied Physics Laboratory, Bob Leben of the University of Colorado, and the Ocean Color team at NASA/GSFC, which includes the SAIC support staff, for providing their processed satellite data on the web. Without those data, this study would not have been possible. The author would also like to thank the Minerals Management Service for providing large amounts of processed ship-of-opportunity data and processed data from their numerous field programs in the GOM. Lastly, he would like to thank the various oil and gas companies who provided information from their studies in the GOM.

References

[1] W. Sturges and J. C. Evans, "On the variability of the Loop Current in the Gulf of Mexico," *Journal of Marine Research*, vol. 41, no. 4, pp. 639–653, 1983.

[2] F. M. Vukovich, "Loop Current boundary variations," *Journal of Geophysical Research*, vol. 93, no. 12, pp. 15585–15591, 1988.

[3] G. A. Maul and F. M. Vukovich, "The relationship between variations in the Gulf of Mexico Loop Current and Straits of Florida volume transport," *Journal of Physical Oceanography*, vol. 23, no. 5, pp. 785–796, 1993.

[4] F. M. Vukovich, "An updated evaluation of the Loop Current's eddy-shedding frequency," *Journal of Geophysical Research*, vol. 100, no. C5, pp. 8655–8659, 1995.

[5] W. Sturges and R. Leben, "Frequency of ring separations from the Loop Current in the Gulf of Mexico: a revised estimate," *Journal of Physical Oceanography*, vol. 30, no. 7, pp. 1814–1819, 2000.

[6] F. M. Vukovich, "Climatology of ocean features in the Gulf of Mexico using satellite remote sensing data," *Journal of Physical Oceanography*, vol. 37, no. 3, pp. 689–707, 2007.

[7] W. Sturges, "The spectrum of Loop Current variability from gappy data," *Journal of Physical Oceanography*, vol. 22, no. 11, pp. 1245–1256, 1992.

[8] W. Sturges, "The frequency of ring separations from the Loop Current," *Journal of Physical Oceanography*, vol. 24, no. 7, pp. 1647–1651, 1994.

[9] D. W. Behringger, R. L. Molinari, and J. F. Festa, "The variability of anticyclonic patterns in the Gulf of Mexico," *Journal of Geophysical Research*, vol. 82, no. 34, pp. 5469–5478, 1977.

[10] L.-Y. Oey, H.-C. Lee, and W. J. Schmitz Jr., "Effects of winds and Caribbean eddies on the frequency of Loop Current eddy shedding: a numerical model study," *Journal of Geophysical Research*, vol. 108, no. 10, article 3324, 25 pages, 2003.

[11] J. A. Carton and Y. Chao, "Caribbean Sea eddies inferred from TOPEX/POSEIDON altimetry and a 1/6° Atlantic Ocean model simulation," *Journal of Geophysical Research*, vol. 104, no. 4, pp. 7743–7752, 1999.

[12] S. J. Murphy, H. E. Hurlburt, and J. J. O'Brien, "The connectivity of eddy variability in the Caribbean Sea, the Gulf of Mexico, and the Atlantic Ocean," *Journal of Geophysical Research*, vol. 104, no. 1, pp. 1431–1453, 1999.

9

Spatiotemporal Spectral Variations of AOT in India's EEZ over Arabian Sea: Validation of OCM-II

C. P. Simha,[1] P. C. S. Devara,[1] S. K. Saha,[1] K. N. Babu,[2] and A. K. Shukla[2]

[1] *Physical Meteorology and Aerology Division, Indian Institute of Tropical Meteorology, Dr. Homi Bhabha Road, Pashan, Pune 411 008, India*
[2] *DPD, Oceansat-II UP (OCM2 Validation), Space Applications Centre, ISRO, Ahmedabad 380 015, India*

Correspondence should be addressed to P. C. S. Devara, devara@tropmet.res.in

Academic Editor: Swadhin Behera

We report the results of sun-photometric measurements of Aerosol Optical Thickness (AOT) in India's Exclusive Economic Zone (EEZ) over the Arabian Sea along with synchronous Ocean Color Monitor (OCM-II) derived AOT estimates during December 12, 2009–January 10, 2010. Relatively higher values of Angstrom exponent (α) around 1.2 near coast and 0.2–0.8 in the India's EEZ, observed during the cruise period, indicate the presence of smaller particles near the coast due to anthropogenic activities; and larger particles in the India's EEZ due to advection of pollutants from Indian subcontinent via long-range transport. Results related to α and its derivative reveal four different aerosol types (urban-industrial, desert-dust, clean-marine, and mixed-type) with varying fraction during the study period. Surface radiative forcing due to aerosols is found to be 20 W/m^2 over India's EEZ. OCM-derived AOTs showed good corroboration with *in situ* measurements with a correlation coefficient of about 0.95. A reasonably good correlation was also observed between AOT and wind speed ($R = 0.6$); AOT and relative humidity ($R = 0.58$). The concurrent MODIS AOT data also agree well with those observed by the OCEANSAT (OCM-II) satellite during the campaign period.

1. Introduction

Although aerosols are minute particles, their cumulative radiative effect (both direct and indirect) on the atmosphere is tremendous [1]. They can scatter away and absorb the incident solar radiation leading to cooling of the earth's surface and simultaneous warming of the lower atmosphere up to about 6 km [2, 3]. Aerosols also pollute the atmosphere and reduce visibility [4]. Continental aerosols are mainly wind-blown mineral dust and carbonaceous and sulphate particles produced by forest fires, land-use, and industrial activities, while marine aerosols are mainly sea-salt particles are produced by wave-breaking, and sulphate particles produced by the oxidation of Dimethyl Sulphide (DMS), released by the phytoplankton [5]. As the oceans cover more than 70% of the earth's surface, they are one of the largest sources of natural aerosols. Being hygroscopic, marine aerosols are crucial in cloud formation in the marine boundary layer and are also important in the radiative coupling between the ocean and the atmosphere. While continental aerosols can be both scattering and absorbing, marine aerosols are mostly of scattering type [6], thus becoming a decisive factor in the albedo of the earth [7–10]. Aerosols are poorly characterized in climate models due to the lack of global information on their physico-chemical properties and spatial and temporal distributions [10, 11].

Most of the *in situ* data campaigns for aerosol characterization have provided significant information on spectral variability, aerosol type, particle size, and so forth; however, they could not provide information on the spatial distribution and transport process. Routine monitoring of aerosol events and their subsequent dispersal pattern are important in order to understand their role in the climatic processes. The satellite sensors provide platform for making observations covering large area as also their short-term and frequent repeatability. Ocean-color sensors for example, CZCS, SeaWiFS, MODIS, POLDER, OCEANSAT-I, OCEANSAT-II have been used to study aerosols, apart from being used to study the ocean color [3, 4]. Most of the ocean-color sensors are equipped with a few additional near infrared (NIR) bands ($\lambda > 700$ nm), which are helpful in providing vital information on atmospheric aerosols due to strong absorption by water in the NIR wavelengths. OCEANSAT-II Ocean Color Monitor (OCM-II) is one such satellite providing ocean color measurements around the sea adjoining the Indian subcontinent. Das and Mohan [4] have shown the use of OCM data for the estimation of AOT and its spatial and temporal behavior over the Indian Ocean region.

The present study focuses primarily on the temporal, spatial, and spectral variations of marine AOT over the India's EEZ (Arabian Sea) in detail. This study is performed based on the measurements conducted on-board a research vessel Sagar Kanya and OCEANSAT-II (OCM-II) sensor-derived AOTs during December 12, 2009–January 10, 2010 in the India's EEZ (Arabian Sea). During the months of December through February, north-easterly winds over the Indian subcontinent carry aerosols from the land towards the oceanic atmosphere. These months, therefore, become the most ideal period to study the effect of continental aerosols and their dispersal in the oceanic atmosphere. The satellite-derived AOT values have been compared with the *in situ* measurements, and long-range transport of air mass from the Indian subcontinent is also examined with the HYSPLIT back trajectories during the cruise period.

2. Instrumentation and Calibration

A total of four instruments were used concurrently in the present OCM-II calibration and validation experiment. They are (i) Microtops sun-photometer that measures AOT; (ii) Microtops ozonometer that measures precipitable water content (PWC) and total column ozone (TCO); (iii) Thermohygrometer/AWS (Automated Weather Station) which makes measurements of surface-level meteorological parameters; (iv) a short-wave (SW) pyranometer that determines downwelling solar radiation. Calibration is essential for these instruments to ensure reliability of its products [12]. All the above-mentioned instruments were calibrated before their deployment. The Microtops provides good quality measurements on moving platforms as explained in the sections to follow. The calibration of this instrument is done at Mauna Loa, Hawaii, a unique high-altitude location allowing access to a pure and stable atmosphere. The manufacturer, M/s Solar Light, Philadelphia, USA, undertakes this calibration campaign each year. The calibration relies on a high-performance voltage reference with the temperature coefficient $\leq 0.001\%$ per degree Celsius and long-term stability of ~0.005% per year. The full width at half maximum bandwidth at each of its wavelength channels is 2.4 ± 0.4 nm, and the accuracy of the sun-targeting angle is better than $0.1°$. Additional details of the radiometers used in the present study have been published by [13–15].

3. Site Location and Measurements Methodology

3.1. SK-266 ORV Sagar Kanya Cruise Observations. India, traditionally a seafaring country [16], has a wide India's EEZ (Exclusive Economic Zone) of about 2 million sq. km. all along the 7500 km long coastline around her. The living and nonliving resources in this zone constitute around two-thirds of the landmass of the country. Additionally, several million people, living along the coastline, are directly influenced by the oceanography of the India's EEZ, coastal environmental hazards, and related social issues. The India's EEZ also composed of many economically exploitable mineral and hydrocarbon resources. Transportation, defense, and fishing are traditional uses of the ocean. But increasing demand for fuels, metals, and construction materials has, in many instances, outstripped the production capabilities of land resources and increasingly the ocean is being exploited to meet the need. This has been particularly obvious in the case of petroleum and natural gas, for which the India's EEZ is expected to become a prime source in the next few decades. The *in situ* Aerosol Optical Thickness (AOT) and other observations were collected on-board ORV Sagar Kanya cruise (SK-266), during December 12, 2009–January 10, 2010. Figure 1 displays the cruise track covered in the India's EEZ (Arabian Sea). The ship commenced from Goa, moved towards south westwards in the ocean away from the coast and reached the India's EEZ (200 nautical miles from the sea shore, that is, $13°$ N, $68°$ E) and covered line-by-line (around 36 multiple paths during the study period), then it returned from India's EEZ ($13°$ N, $70.5°$ E) to the Goa port.

Microtops measures AOT at six wavelengths, each centered at 380, 440, 500, 675, 870, and 1020 nm, from the instantaneous solar flux measurements using its internal calibration. The on-board measurements were carried out following the standard AOT measurements protocol for ship-borne surveys. The measurement protocol was to make a triplet of observations every 5–10 minutes, while following the suggestions of the previous literature [17–19], when the solar disk was free from clouds. Also, the sun-photometer and ozonometer were switched on and off routinely to update the temperature correction [17]. The collected data were processed and grouped into series by adhering to the methodology of Smirnov et al. [19]. The arithmetic and geometric daily averages of AOT, compared to avoid sampling biases [20], agree within 0.007 or less. During the cruise, the sun-photometer and ozonometer were operated from 08:00 hrs to 16:00 hrs at 5-minute interval in the sunrise and sunset time, and in the remaining period 10-minute interval. AOT measurements were made from the ship-deck. Simultaneous surface-level meteorological observations like atmospheric pressure, wind speed/direction, temperature and relative humidity, and so forth were recorded by the AWS (Model WDL 1002; Dynalab, Pune, India). The concurrent down-welling surface SW radiation data were archived at a 1-minute interval using the gimbal-mounted SW pyranometer (Model DWR 8101; Weathertech Pvt. Ltd., Pune, India). The directional and nonlinearity errors for the pyranometer are estimated to be about 2% of the reading, whereas total daily expected uncertainty is ±3%. OCM-II satellite data was obtained corresponding to the *in situ* data collected in the Arabian Sea during December 12, 2009–January 10, 2010 covering the study area for every alternate day, because the OCM-II has once-in-two-days repeatability. OCM records reflected radiation from earth's surface in eight spectral

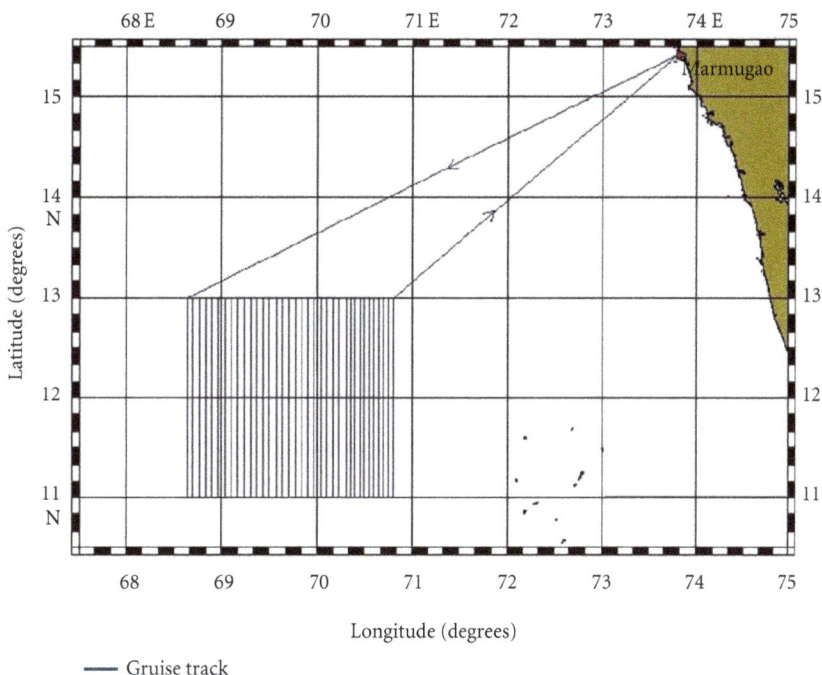

FIGURE 1: SK-266 cruise track in the India's EEZ.

bands ranging within the 412–865 nm wavelength bands. The NIR wavelength band of 865 nm has been used for the estimation of AOT over the oceanic surface.

4. Results and Discussion

4.1. Spectral AOT Variations. The spectral variations of AOT at six wavelengths for the entire cruise period are shown in Figure 2. The daily mean AOT values at 500 nm wavelength for 30 days of observations were mostly in the range from 0.14 to 0.7. Unusual AOT values occurred during the cruise period due to a sudden low-pressure system that formed near the Sri Lankan region. We have categorized the measurements into coastal and India's EEZ based on the distance from the coast. This division is arbitrary, but sufficient to illustrate the differences between the aerosol spectral properties close to and far from land. The AOT data collected from December 15, 2009–January 8, 2010 represent mainly India's EEZ environment and remaining data were considered for coastal environment. It is noted that AOT at 500 nm was about 0.23 over the India's EEZ away from the coast, similar variations have been reported by Chauhan et al. [21] and Kalapureddy and Devara [22]. When the ship approached the coast, the mean AOT at 500 nm has increased to 0.45 while the values of AOT decreased as the ship moved away from the coast in the India's EEZ, thereafter it increased again as the ship approached the coast. Figure 2 shows lower values of spectral variations in the India's EEZ in the Arabian Sea from coast as compared to that of near-coast stations, which clearly reveal distinct information on particle size distribution over the two regions. The size of marine aerosol is found larger as compared to continental aerosols.

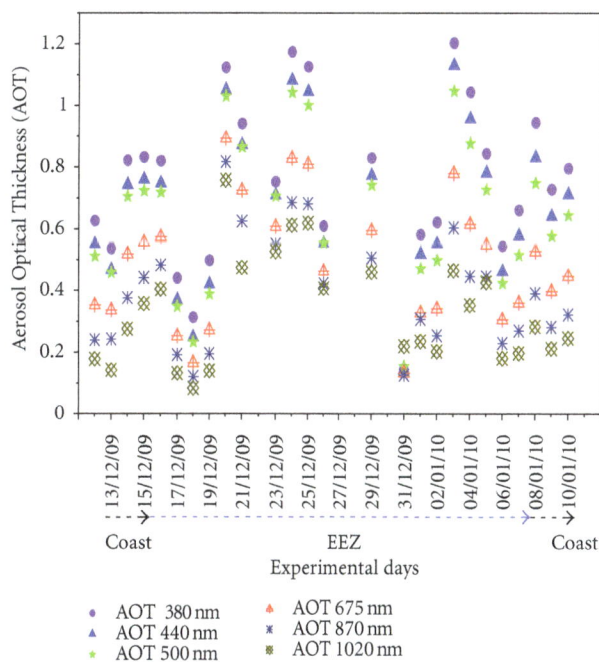

FIGURE 2: Spectral variation of the daily averages of *in situ* AOT for the cruise dates.

4.2. Ozone and PWC Variations. The daily mean columnar ozone variations are depicted in Figure 3. The TCO is found to vary from 210 to 255 DU. During the entire campaign, lower values of TCO were observed on two days, which are considered to be due to photochemical reactions with

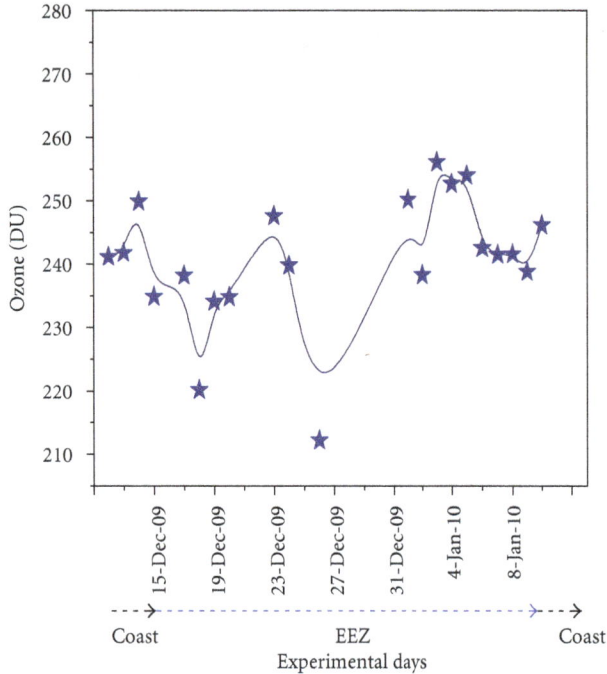

FIGURE 3: Same as Figure 2 but for ozone.

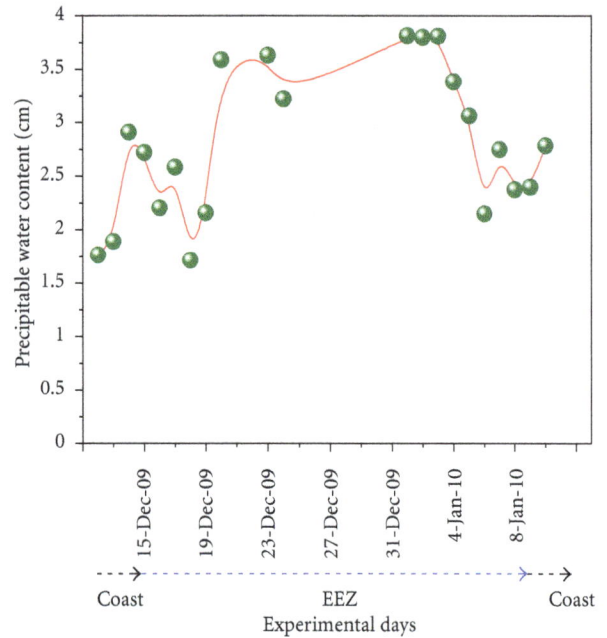

FIGURE 4: Same as Figure 2 but for precipitable water content.

NO$_x$ and CO. The PWC variations are shown plotted in Figure 4. The PWC values range from 1.5 to 3.8 cm. These values are found very low in the beginning of the campaign period due to the low pressure system on December 19, 2009, thereafter it suddenly increased to very high value ($\sim$3.0 cm) and continued to be high up to 3 January 2010.

4.3. Angstrom Parameters. Spectral variation of AOT provides information on the particle size distribution in the total atmospheric column. The inverse power-law for spectral variation of AOT is given as

$$\tau = \beta\lambda^{-\alpha}, \tag{1}$$

where, α, β, and λ represent the Angstrom exponent, turbidity factor, and wavelength used, respectively. The value of α depends on the portion of the concentration of large to small aerosol particles and is related in such a manner that higher value of α indicates dominance of smaller particles. β represents the total aerosol loading in the atmosphere and as such higher values of β give an indication of poor visibility due to higher turbidity level. The variations in Alpha and Beta during the cruise period are shown plotted in Figure 5. It is clear from the figure that mean value of α over the coastal region ($\sim$1.24) is found to be larger as compared to that over the open ocean, that is, India's EEZ ($\sim$0.2–0.8). Relatively larger values of β, found in the coastal water regions, indicate the presence of relatively higher concentration of particles as compared to the measurements over the open ocean (India's EEZ) waters. The larger values of Angstrom exponent (α) can also be attributed to the higher fine-mode aerosol loading close to the coast, arising mainly from the anthropogenic activities. The second derivative of

Alpha constitutes a measure of the rate of change of the slope with respect to wavelength. Eck et al. [23] attempted to quantify the curvature of spectral variation of AOT using the second derivative of $\ln \tau_\lambda$ versus $\ln \lambda$, that is, derivative of α with respect to $\ln \lambda$ as

$$\alpha' = \frac{d(\alpha)}{d(\ln \lambda)}. \tag{2}$$

The curvature can be an indicator of aerosol particle size, with negative curvature indicating aerosol particle size distribution dominated by fine-mode while positive curvature indicates size distribution with significant contribution by coarse-mode [23]. α' close to zero indicates constant slope AOT, indicating coarse-mode dominance, while higher values represent rapidly changing the slope. In the present study, α' was derived using observed AOT at 380, 440, 500, 675, 870, and 1020 nm wavelengths. Both the coarse- and fine-mode aerosol size distributions had been clearly noticed from Figure 6. Near the coastal environment, as discussed above, the variations in AOT, Alpha, Beta indicated fine-mode particles (negative α' value) which are considered to be due to anthropogenic activities. Once the ship entered the India's EEZ, coarse-mode particle dominance is observed (positive α' value). These features have been substantiated through long-range transport of air mass trajectories around the India's EEZ by using NOAA HYSPLIT model.

4.4. Surface Radiative Forcing. The Pyranometer-measured global SW flux in the wavelength region between 0.3 and 3.0 μm is correlated with instantaneous AOT 500 nm (corrected for the air mass factor, $1/\mu$) so as to estimate the surface radiative forcing [24]. Normalization of AOT with μ ($=\cos\theta$) is found to be necessary as the slant air column

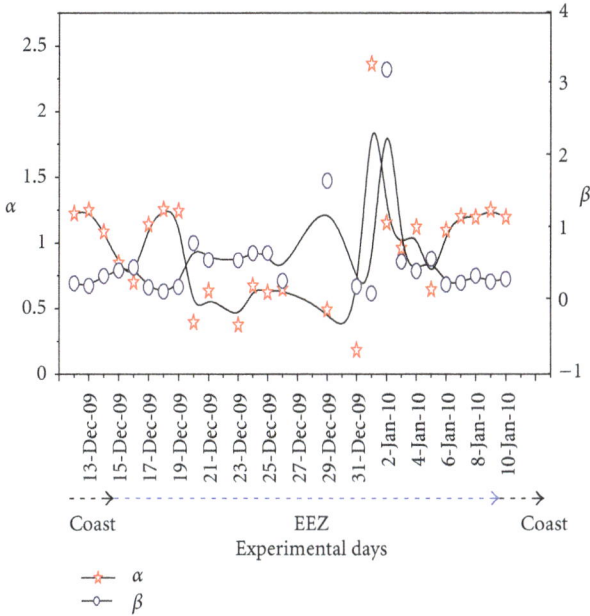

FIGURE 5: Alpha-beta variations during the cruise period.

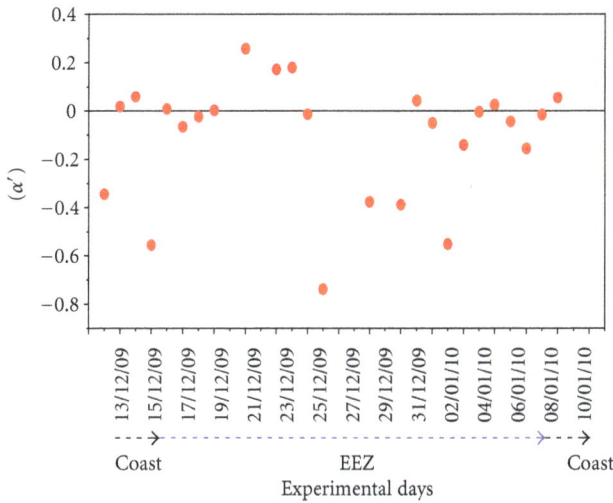

FIGURE 6: Alpha prime variations during SK-266 campaign.

length increases with increasing solar zenith angle θ. The observed flux represents the solar flux at sea surface, normal to the angle of incidence, with a cone of about 2.5° around the Sun. The AOT values for solar zenith angle greater than 60° are excluded (to avoid Earth's curvature effect) and the AOT/μ values are restricted to within 0.80. Figure 7 shows scatter plot of the measured normalized SW flux with AOT for India's EEZ. A straight line could be fitted with a negative slope of about 20.08 W/m^2 for every 0.1 increase in the AOT (from Figure 7) for India's EEZ in the Arabian Sea. The scatter between the two data sets is considered mainly to be due to (i) uncertainties in global flux spatial inhomogeneity of the atmosphere within the hemisphere and (ii) contamination of the solar flux by broken and optically thin clouds. Moreover, such a relationship is consistent with

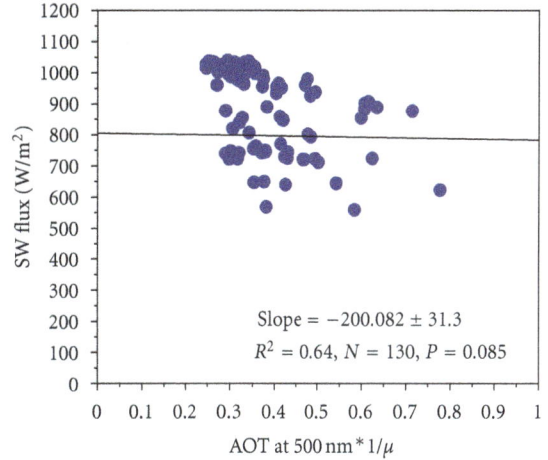

FIGURE 7: Association between the surface short-wave solar flux, and columnar AOT at 500 nm normalized for the air mass (1/μ) over the India's EEZ.

the surface solar flux values reported by earlier researchers [24–27].

4.5. Validation of OCM-II with In Situ AOT at 865 nm. The irradiance data in the HDF format from the OCM-II sensor were converted into the geophysical coordinate mode (lat-long mode) using the ENVISAT-II software [4, 21]. Thus OCM-II data were obtained at a specific wavelength in the India's EEZ and matched with the observed Microtops data. The OCM-II satellite passes in the Arabian Sea were available at 12:30 PM on alternate days. Hence the ship data were recorded from 11:00 AM to 1:30 PM and is synchronized with particular lat-long of the ship at 12:30 PM. Here we considered the OCM-II data in the $5^v \times 5^v$ grid mode to obtain AOT. The OCM-II provides AOT at 865 nm while the Microtops yields AOT at 870 nm. By using the Angstrom exponent formula, the AOTs at OCM-II and Microtops were matched at the wavelength of 865 nm. OCM-II data for cruise duration of 12 December 2009 till 10 January 2010 for pass number 12 was processed by extracting average AOT at 865 nm values from AOT product using ENVISAT-II Software. A comparative analysis of the OCM-II derived AOT at 865 nm was done with *in situ* measured AOT values using the hand-held sun-photometer. Figure 8 shows the relationship between OCM-II derived and *in situ* measured AOT values at 865 nm for match-up locations. A total of six match-up locations could be achieved during the cruise duration, because of the day-by-day satellite passage of OCM-II sensor and sudden weather disturbance near the Sri Lankan region. It is evident from the figure that OCM-II derived and *in situ* measured AOT values show a good correlation coefficient with less error ($R^2 = 0.95$ and RMSE = 10%). Most of the time OCM-II derived AOT is found to slightly over estimate the *in-situ* AOT.

4.6. Mapping of AOT over India's EEZ. The lat-long cross-section of AOT on different days over the India's EEZ during

(a)

(b)

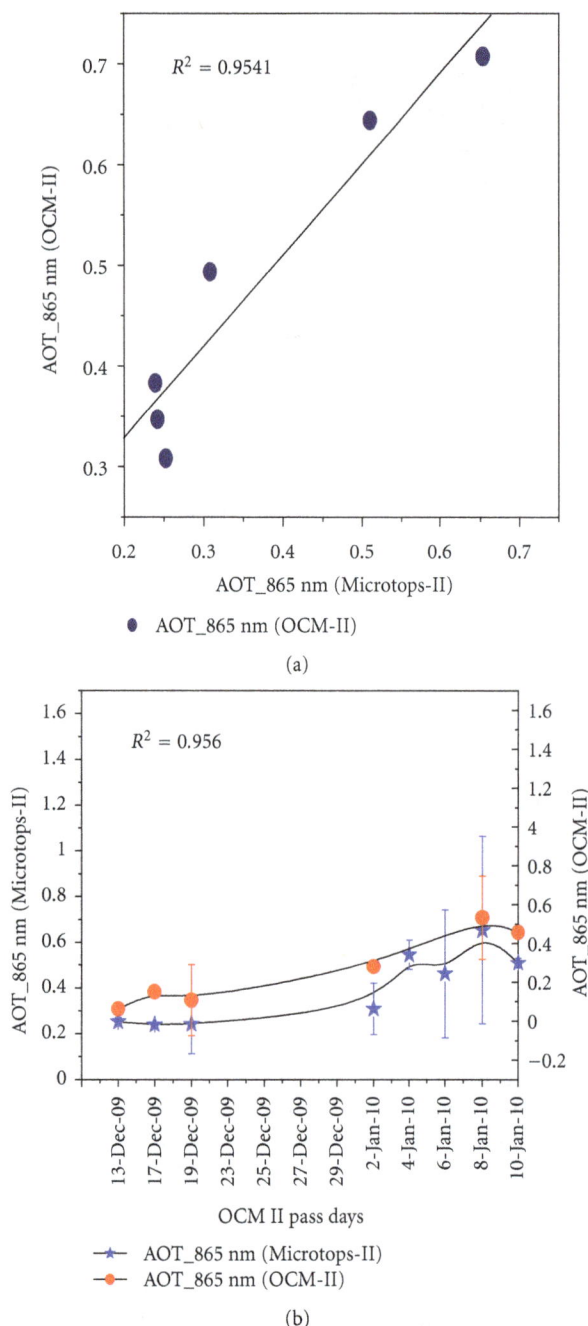

FIGURE 8: Comparison of Microtops Sun Photometer derived AOTs with OCEANSAT-II OCM-derived AOTs at 12:30 hrs (IST) during the cruise.

the cruise period is shown plotted in Figure 9. It is clear from the figure that near the coast line (14.5°N, 72°E), AOT values are high. But when the ship enters the India's EEZ (13°N, 71°E), they are very low. Due to sudden occurrence of weather disturbance over the observational site, AOT reached to very high. The lat-long cross-sections of TCO and PWC observed over the India's EEZ during the cruise period are depicted in Figures 10 and 11, respectively. It is evident that low ozone values were observed over the belt between 12°N

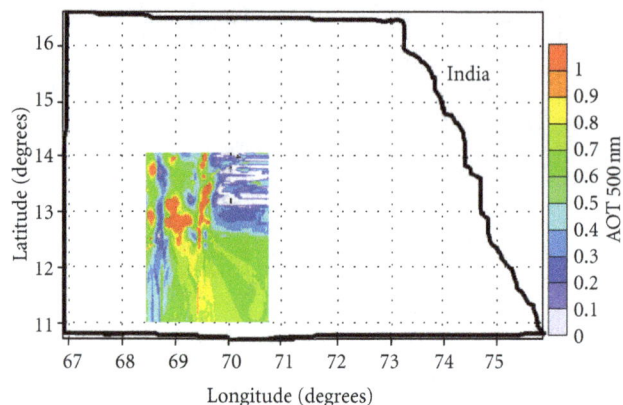

FIGURE 9: Lat-long cross-section of AOT over India's EEZ during 11 December–10 January 2010.

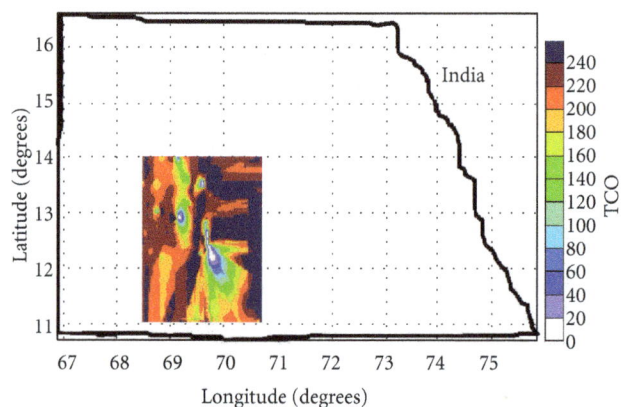

FIGURE 10: Lat-long cross-section of TCO over India's EEZ during 11 December–10 January 2010.

and 69°E, which may be due to the transport of nitrogen compounds leading to ozone depletion. The higher PWC values noticed over the belt between 13°N and 69°E are considered to be due to sudden weather disturbance formed near the Sri Lankan region.

4.7. Long-Range Transport of Aerosols. Figure 12 portrays seven-day back trajectories (at three altitudes) for the entire India's EEZ as obtained from NOAA HYSPLIT model. They clearly reveal that different air masses (originating from Indian subcontinent, Sri Lanka, Arabian Sea) influenced the AOT over the experimental site (India's EEZ). This analysis also discloses that Arabian and Indian subcontinent air masses play a significant role in influencing the India's EEZ. It further reveals significant changes, which are considered to be due to observed day-to-day variability of aerosols properties. The lower altitude air masses originate from the Arabian Sea and Indian subcontinent without traveling longer distance while those at higher altitudes travel longer distance and effect significantly the long-range transport of air masses over the Sri Lankan region.

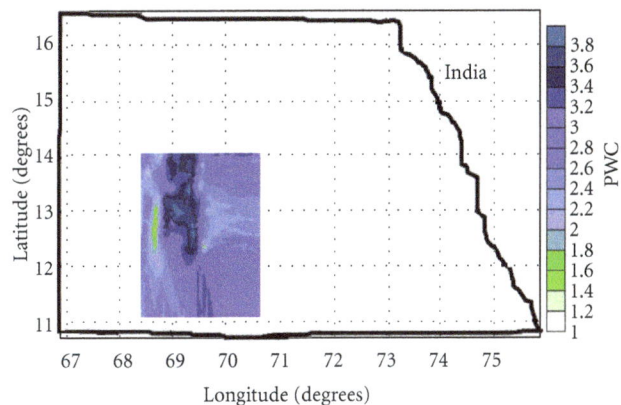

FIGURE 11: Lat-long cross-section of PWC over INDIA'S EEZ during 11 December–10 January 2010.

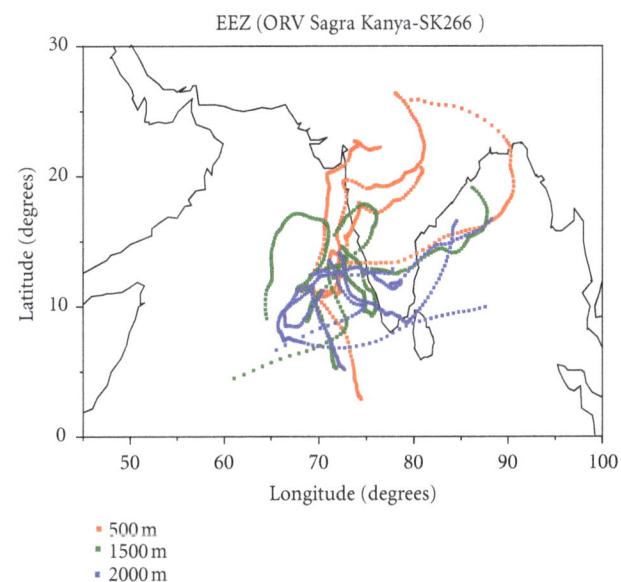

FIGURE 12: Seven-day air back trajectories for the days of aerosol measurements over the India's EEZ region at 500 m, 1500 m, and 2000 m heights. The back trajectories are plotted at hourly interval. The symbols are shown at 24-h interval.

4.8. Relationship between AOT and Surface Meteorological Parameters. Meteorological instruments on the ship's deck provided information on the prevailing wind speed, direction, deck-level temperature (T), pressure (P), and relative humidity (RH). These data were continuously recorded at 3 hour interval during the cruise period and also at every hour during the Microtops observations when the ship was stabilized in position with the help of dynamic thrusters. The measurements of T and RH are made at the upper deck-level, about 10 m above the water level. The daily mean temperature, RH, and wind speed in the India's EEZ region during the campaign period are plotted in Figure 13. The daily mean temperature was around 28-29.5°C or less during the India's EEZ which decreases to around 27°-28°C on 27–29 December 2009 and 5 January 2010. RH is found

to decrease from about 89% on 18 December 2009 and 6 January 2010 to less than 72%. Thereafter, the RH increased and is found to be in the 80–89% range on 14, 20, 24, 28 December 2009. Daily mean RH increased to more than 80% on the days associated with fog. During the winter period over northern India, the low-level winds, in general, are calm, north/northeasterly, and arrive from the polluted northern hemisphere. During this period, we observed south-westerly winds at surface-level. RH was found to vary significantly during the day on foggy and hazy days. The daily average wind speed varied from 2.5 to 11 m/s, during the first week and last week of the campaign, that is, 13 December 2009 and 2 January 2010 the wind speed is low (~2.5 m/s), in the remaining time it varied from 4 to 19 m/s. The daily average AOT at 870 nm varied along with met parameters during the cruise period. It is evident from the figure that AOT increases with higher wind speed and vice versa, and AOT also increases with RH, which confirms that the aerosols over the India's EEZ region are of hygroscopic nature and have affinity towards moisture. AOT variation with temperature is not very consistent as compared to those of RH and wind speed. Correlation coefficients of 0.6 and 0.58 were observed between AOT with wind speed and relative humidity, respectively.

4.9. Discrimination of Aerosol Types over India's EEZ. When attempting a realistic characterization of the aerosol properties, data of both AOT and Alpha have to be used [25] since they both strongly depend on wavelength. Thus scatter plots of AOT versus Alpha have been obtained to determine different aerosol types for a specific location through the determination of physically interpretable cluster regions in the diagram. Therefore, the detailed spectral information in different pairs of wavelengths can help in determining and discriminating between different aerosol types [26–28]. These AOT-Alpha patterns have been observed at several locations and for determining different aerosol types such as biomass smoke, anthropogenic aerosols, desert dust [29–33]. In all the above studies, a wide range of values for low AOT at 500 nm were obtained, thus reflecting relatively clear continental conditions with strong variability in the dominant aerosol properties. Because of the poor correlation between AOT and Alpha, values of both quantities must be considered for a realistic analysis.

Figure 14 shows the density plot of AOT 500 nm-Alpha (380–1020 nm) over the whole India's EEZ in the Arabian Sea. These contour maps were constructed in 0.1 steps for both AOT at 500 nm and Alpha (380–1020 nm) values. In the AOT 500 nm-Alpha (380–1020 nm) plot rectangle areas denoted as urban-industrial (UI), clean-maritime (CM), desert-dust (DD), and mixed type (MT) aerosols, respectively, the boundaries of these areas concerning to the selected threshold values of AOT at 500 nm and Alpha. In such studies, the selection of the threshold values can be very important. In the present study, the AOT at 500 nm ranges from 0.14 to 1.0, and the Alpha (380–1020 nm) from 0.4 to 1.7. The criterion for discriminating different aerosols, reported by Kalapureddy et al. [34], has been followed in the present study. For the same reason, the Alpha

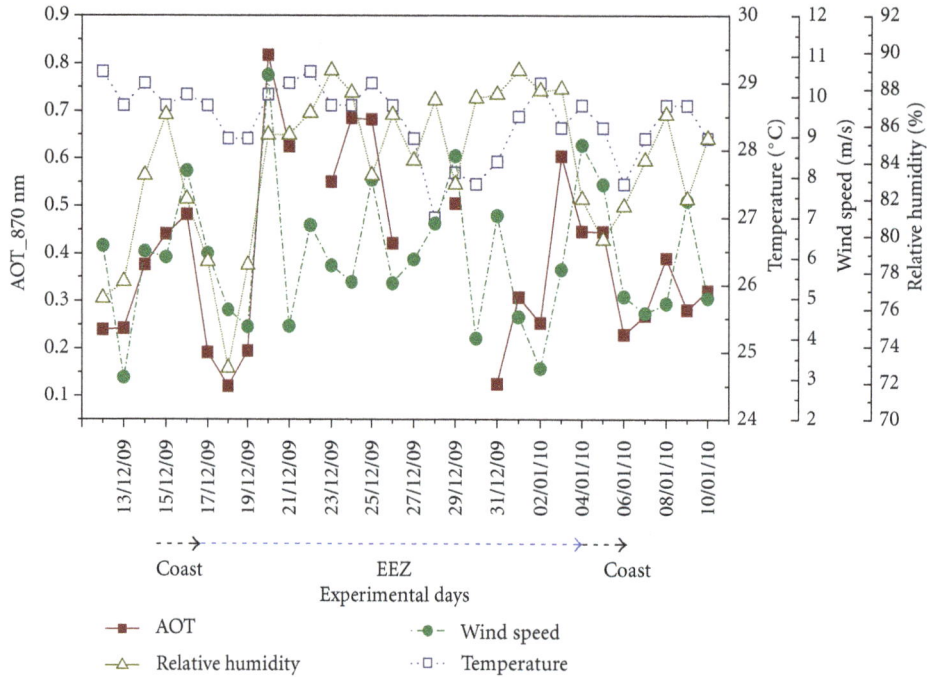

FIGURE 13: Variation of AOT with surface meteorological parameters.

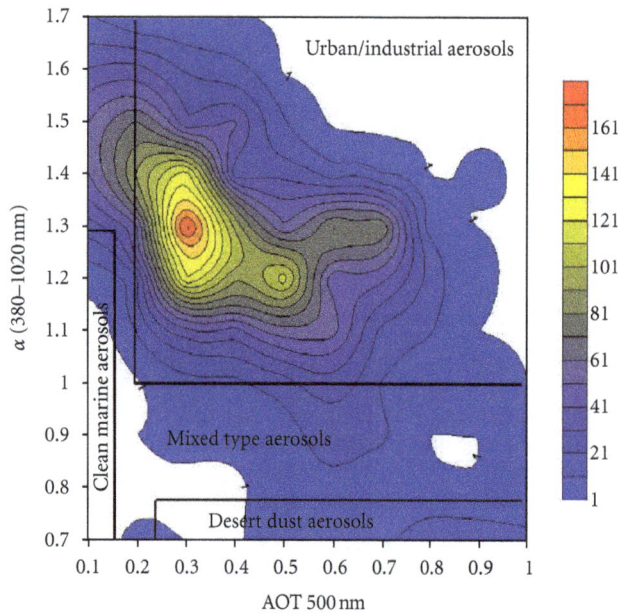

FIGURE 14: Contour density maps of the AOT 500 nm versus Alpha 380–1020 nm in the India's EEZ of Arabian Sea.

of data points is observed for low to moderate turbidity conditions, from which it is difficult to extract any useful information; only extreme values seem to be indicative of clear associations. Viewing the density plot of the whole India's EEZ, four areas of larger density are observed. However, the majority of the AOT 500 nm values lie between 0.15 and 0.6. The Alpha (380–1020 nm) values are mainly depicted in the 0.4–1.7 interval. Previous studies over the AS region reveal the near similar AOT 500 nm (0.3) and Alpha (1.0) during premonsoon (March–May) season [35–38]. Moreover, the above studies while examining the synergy of observations (physical and chemical properties of aerosols) and models also explored the aerosol type information as the fractional contribution of different aerosol in the AOT which was about 10% by mineral dust and sea-salt contribution is around 20% and anthropogenic sources contribution is above 60% over AS around premonsoon season which are nearly comparable with the present results.

The four maximum density areas represent different aerosol types. Thus the absolute maximum density area, observed for the pair of AOT at 500 nm, and Alpha (380–1020 nm) is an indicator of moderate turbidity conditions dominated by UI aerosol type. Another maximum density area is depicted for (AOT at 500 nm, Alpha (380–1020 nm) corresponds to MT aerosol, having larger fraction of coarse-mode particles under more turbid conditions. The other two secondary maximum density areas are significantly lower, but quite characteristic of different aerosol types. Thus the (AOT at 500 nm, Alpha (380–1020 nm) pair is characteristic of DD particles, while that AOT at 500 nm value of 0.5 and Alpha value of 1.1 is indicative of the transport of continental polluted plumes over AS. These different aerosol sources will

(380–1020 nm) threshold value for characterizing desert-dust aerosols was taken slightly higher (0.7 against 0.5) since the desert particles transported over oceanic regions can be mixed with the other aerosols and the larger particles are deposited near the source region; these processes lead to the increase of the Alpha values. Regarding the whole India's EEZ region!represented in Figure 14, a rather great scatter

OCM-II
AOT 865 = 0.347
Microtop
AOT 865 = 0.242

OCM-II
AOT 865 = 0.62
Microtop
AOT 865 = 0.509

(a)   (b)

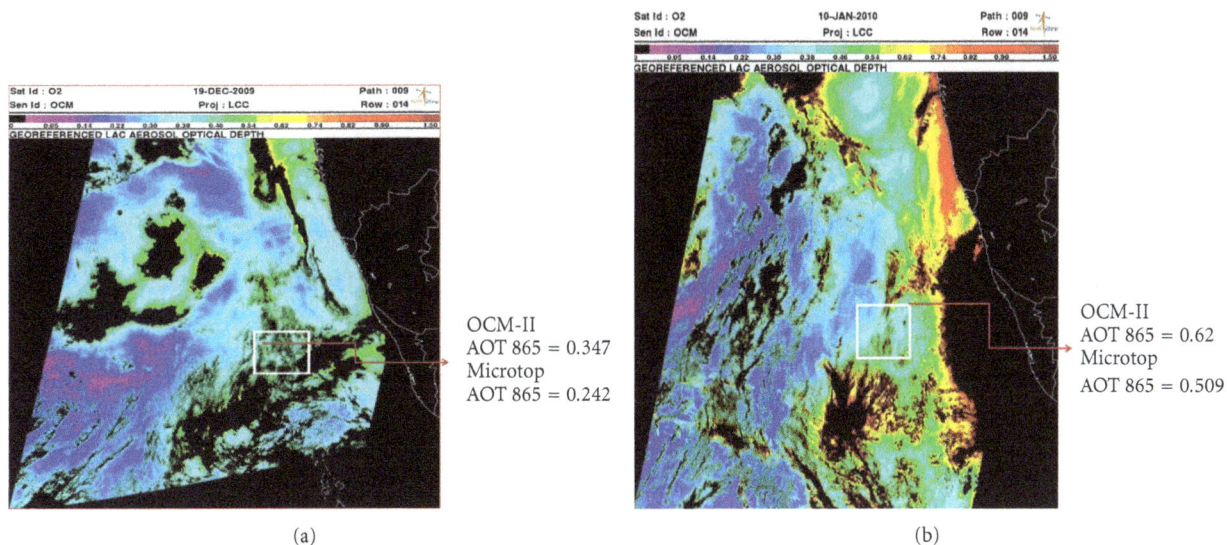

FIGURE 15: AOT-865 nm distribution as captured by OCEANSAT-2-OCM over the parts of the Arabian Sea.

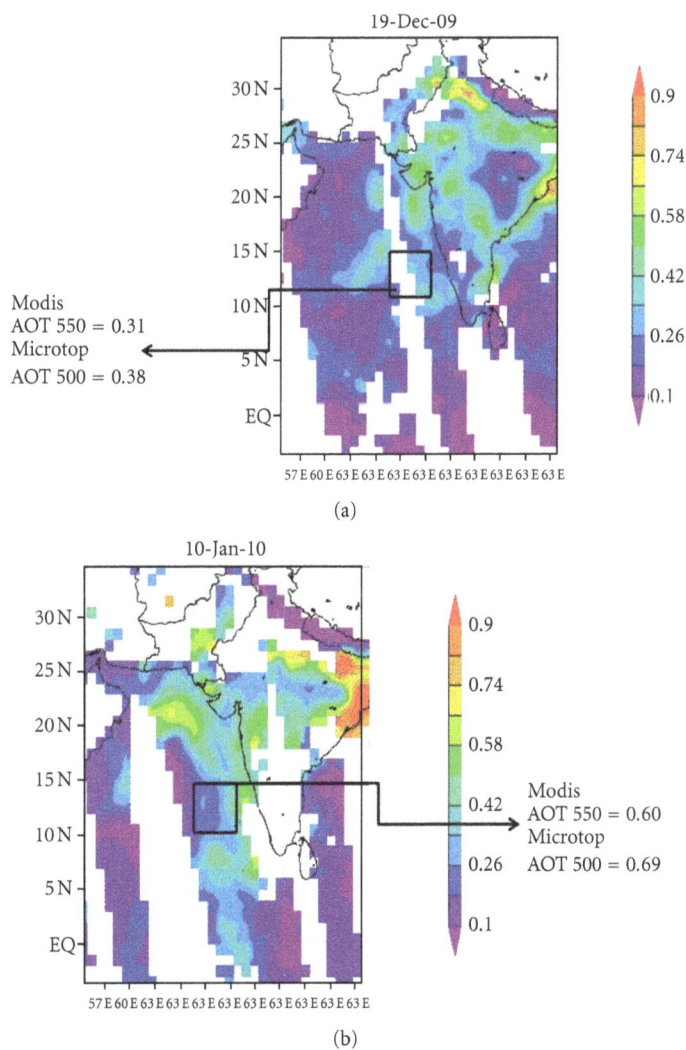

Modis
AOT 550 = 0.31
Microtop
AOT 500 = 0.38

(a)

Modis
AOT 550 = 0.60
Microtop
AOT 500 = 0.69

(b)

FIGURE 16: MODIS-Aqua AOT at 550 nm and Microtops AOT at 500 nm over the parts of the Arabian Sea.

TABLE 1: Major specifications, features, and the geophysical products of OCEANSAT-II OCM.

Parameter	Specification	
Instantaneous Geometric Field-Of- View (IGFOV) at nominal altitude (m)	360×250	
Swath (km)	1420	
Number of spectral bands	8	
Spectral range (nm)	402–885	
Spectral bands (nm)	B1: 404–424	
	B2: 431–451	
	B3: 476–496	
	B4: 500–520	
	B5: 546–566	
	B6: 610–630	
	B7: 725–755	
	B8: 845–885	
Quantization bits	12	
Along track steering	±20	
Data acquisition modes	Local area coverage (LAC) & Global area coverage (GAC)	
Geophysical parameter	Variable Range	Targeted error budget
Aerosol Optical Thickness (AOT) over oceans at 865 nm	0.0–1.0	<20%

have direct effect on the wide range of Alpha (380–1020 nm) values, from 0.4 to 1.7, especially for low AOT at 500 nm (<0.15), suggesting a range of aerosol types from pure fine-mode pollutants to predominately coarse-mode particles. For higher values of AOT at 500 nm, corresponding to dust conditions, Alpha is relatively low, a feature that compares well with that reported in the literature [39, 40]. The properties of dust may also differ because of the variability of sources or the distance from the source region [41]. Thus, the detailed spectral information given by the determination of Alpha in different spectral bands helps us in determining and discriminating the aerosol types.

4.10. AOT Distribution as Seen from OCEANSAT-II and MODIS Satellites over the India's EEZ Region. The data products from OCM-2 are available at 360 meter spatial resolution for regional studies which are also called local area coverage (LAC) products. Table 1 provides the technical details of the OCEANSAT-II OCM instrument. The global area coverage (GAC) products are available at 4 km spatial resolution for global studies. The Level 1B top of the atmosphere (TOA) radiance data from all the eight bands of OCM sensor is used along with ancillary information to generate various biogeophysical data products. The main geophysical products from OCM-2 are chlorophyll-a concentration, total suspended sediment concentration, vertical diffuse attenuation coefficient (Kd-90) at 490 nm and AOD at 865 nm. These data products from OCM-2 can also be used for studying oceanic primary production, fisheries resources,

global carbon cycle, spatial and temporal patterns of algal blooms, dynamics of coastal currents, physics of mixing and relationship between ocean physics and large scale patterns of productivity, land-ocean interaction in the coastal zone and sediment dynamics on regional as well as global scales. The availability of ocean-color data from OCM-2 opened new frontiers in remote sensing for open-sea as well as coastal regions on regional and global scales due to its relatively higher spatial resolution of around 360 meters. The obtained LAC data products during the cruise period are analyzed by using ENVISAT-II software at SAC, ISRO, Ahmedabad, India, and typical days were selected (19 December 2009 and 10 January 2010) as per the sky conditions are shown in Figures 15(a) and 15(b). Strong correlation was observed between Microtops and OCM-II data observed on these two days. The MODIS Aqua data products are also compared with the *in-situ* Microtops observations as shown in Figures 16(a) and 16(b). High correlation between AOTs deduced from MODIS and Microtops can be seen very clearly from the figure.

4.11. Aerosol Cluster during India's EEZ Campaign. Mean wind field during the SK-266 at different levels—1000 hPa and vorticity at 850 hPa are shown in Figure 17. It can be noticed that winds in the northern and the north-western Arabian Sea in all the levels are conducive for the transport of coarse mineral dust from Arabian deserts and the arid regions of west Asia. On the other hand, in the case of south-eastern Arabian Sea, even though the winds are favorable,

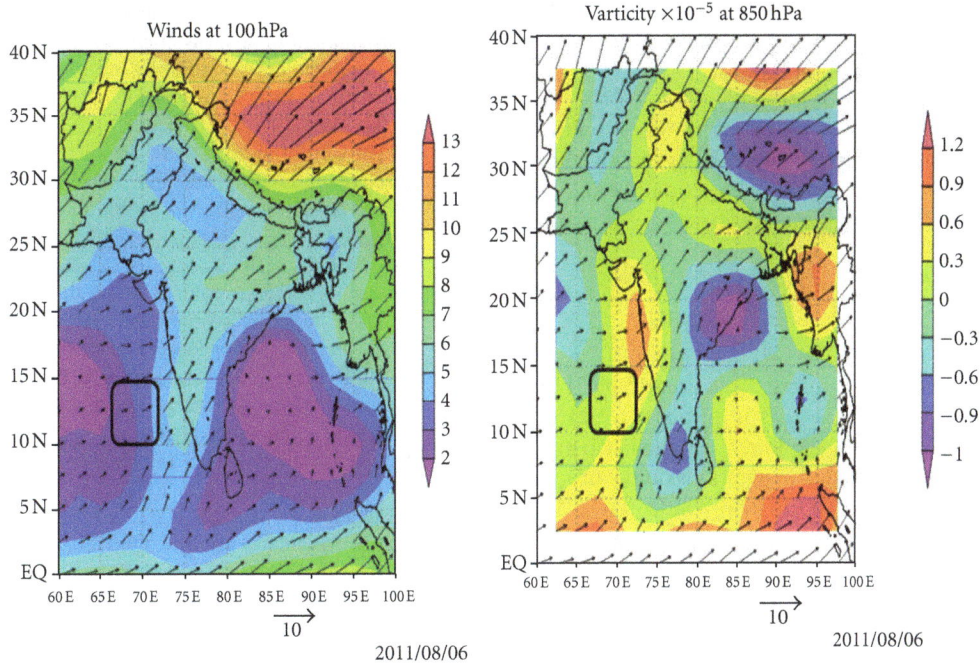

FIGURE 17: Spatial distribution of mean winds at 1000 hPa and vorticity at 850 hPa over the India's EEZ during the SK266.

dust particles transported from the AS would be lost on the way through gravitational settling since the distance involved is too large. Instead, the aerosols here will be mostly composed of particles from the urban and industrial centers along the west-coast collected by the lower altitude northerly winds [34, 42]. A scrutiny of the wind field at lower altitudes, particularly, at the surface (responsible for marine aerosol production) does not show any major day-to-day variations about the mean behavior during the period of the event. Figure 17 shows the vorticity at 850 hPa over the Arabian Sea. It was clearly seen that the anti-cyclonic circulation situated slightly west to the central peninsula was stronger on southern India and India's EEZ of the Arabian Sea. The southerly winds associated with this cyclonic flow are also very strong leading to the enhancement of wind convergence on these days of formation of the cluster of southern Indian peninsular (Figure 17). Therefore, it may be concluded that the short-term enhancement of the wind convergence at 850 hPa has given rise to the accumulation of aerosols initiating the formation of the aerosol cluster. It would be interesting to note here that such elevated aerosol cluster formations over the Arabian Sea were detected during the Indian Ocean experiment (INDOEX) conducted in 1999 [43] which were attributed to upper level transport of continental aerosols from Arabia, eastern Pakistan, and the Thar desert where the frequency of dust storm activity is large during this period of the year [44–46]. In addition to dust particles, the initial composition of the aerosol cluster will contain some contribution of fine-mode aerosols from Indian subcontinent and coastal industries brought in by the cyclonic winds during the SK-266.

5. Conclusions

The Microtops observations carried out on-board Sagar Kanya in the India's EEZ of the Arabian Sea and concurrent OCM-II satellite data have been analyzed. The salient results of the study are highlighted below.

(i) Spectral variation of the AOT during the campaign shows significant influence of anthropogenic and natural activities on the pristine AOT ($\sim$0.14) and coastal AOT ($\sim$0.7), and sudden response of weather disturbance to AOT.

(ii) Lower values of Ozone due to photochemical reactions, and sudden increase of PWC due to formation of low pressure area in the India's EEZ.

(iii) Decrease in global solar flux with every 0.1 increase in AOT, for SZA < 60°, over the India's EEZ of Arabian Sea was 20 W/m^2.

(iv) Long-range transport of air mass, consisting of north-eastern Indian land and south-eastern Sri Lankan and Arabian Sea aerosols play major role in the observed increase in fine-mode aerosol loading over the India's EEZ region.

(v) Over the observational site, UI aerosol density seems to be maximum and is followed by CM and DD type whereas MT type is found to be the least.

(vi) AOTs in the sensing region show strong correlation with wind speed and relative humidity.

(vii) High correlation was observed between OCM-II level-2 product with *in situ* ground-based Microtops and MODIS satellite observations.

(viii) Spatial distributions of mean surface-level wind at 1000 hPa and vorticity at 850 hPa have shown southwesterly winds and strong convection during the study period, resulting in aerosol cluster formations in the India's EEZ.

Acknowledgments

The study reported here was carried out jointly by IITM, Pune and SAC-ISRO, Ahmedabad. The authors are grateful to the Editor and anonymous Reviewers for their constructive and insightful comments and suggestions on the original paper. One of the authors (CPS) is highly indebted to SAC for financial support in the form of Research Fellowship. Thanks are also due to Director, IITM, Pune, India for encouragement and infrastructure support.

References

[1] IPCC, "Report to IPCC from Scientific Assessment Group (WGI)," in *Intergovernmental Panel on Climate Change (IPCC)*, Cambridge University Press, Cambridge, UK, 1995.

[2] J. A. Coakley, R. D. Cess, and F. B. Yurevich, "The effect of tropospheric aerosols on the earth's radiation budget: a parameterization for climate models," *Journal of the Atmospheric Sciences*, vol. 40, no. 1, pp. 116–138, 1983.

[3] Y. J. Kaufman, D. Tanré, H. R. Gordon et al., "Passive remote sensing of tropospheric aerosol and atmospheric correction for the aerosol effect," *Journal of Geophysical Research D*, vol. 102, no. 14, pp. 16815–16830, 1997.

[4] I. Das and M. Mohan, "Detection of marine aerosols using ocean colour sensors," *Mausam*, vol. 54, pp. 327–334, 2003.

[5] R. J. Charlson, S. E. Schwartz, J. M. Hales et al., "Climate forcing by anthropogenic aerosols," *Science*, vol. 255, no. 5043, pp. 423–430, 1992.

[6] O. Dubovik, B. Holben, T. F. Eck et al., "Variability of absorption and optical properties of key aerosol types observed in worldwide locations," *Journal of the Atmospheric Sciences*, vol. 59, no. 3, pp. 590–608, 2002.

[7] C. Tomasi, V. Vitale, A. Lupi, A. Cacciari, S. Marani, and U. Bonafé, "Marine and continental aerosol effects on the upwelling solar radiation flux in Southern Portugal during the ACE-2 experiment," *Annals of Geophysics*, vol. 46, no. 2, pp. 467–479, 2003.

[8] Intergovernmental Panel on Climate Change (IPCC), "Summary for policy makers," in *Climate Change 2007: The Physical Science Basis. Contribution of Working Group I to the Fourth Assessment Report of the Intergovernmental Panel on Climate Change*, S. Solomon, D. Qin, M. Manning et al., Eds., Cambridge University Press, Cambridge, UK, 2007.

[9] M. Rinaldi, S. Decesari, E. Finessi et al., "Primary and secondary organic marine aerosol and oceanic biological activity: recent results and new perspectives for future studies," *Advances in Meteorology*, vol. 2010, Article ID 310682, 10 pages, 2010.

[10] X. Liu, R. C. Easter, S. J. Ghan et al., "Toward a minimal representation of aerosols in climate models: description and evaluation in the Community Atmosphere Model CAM5," *Geoscientific Model Development*, vol. 5, pp. 709–739, 2012.

[11] G. W. Mann, K. S. Carslaw, D. A. Ridley et al., "Intercomparison of modal and sectional aerosol microphysics representations within the same 3-D global chemical transport model," *Atmospheric Chemistry and Physics*, vol. 12, pp. 4449–4476, 2012.

[12] P. C. S. Devara, S. K. Saha, P. E. Raj et al., "A four-year climatology of total column tropical urban aerosol, ozone and water vapour distributions over Pune, India," *Journal of Aerosol and Air Quality Research*, vol. 5, pp. 103–114, 2005.

[13] P. C. S. Devara, R. S. Maheskumar, P. E. Raj, K. K. Dani, and S. M. Sonbawne, "Some features of columnar aerosol optical depth, ozone and precipitable water content observed over land during the INDOEX-IFP99," *Meteorologische Zeitschrift*, vol. 10, no. 2, pp. 123–130, 2001.

[14] M. Morys, F. M. Mims, S. Hagerup et al., "Design, calibration, and performance of MICROTOPS II handheld ozone monitor and Sun photometer," *Journal of Geophysical Research D*, vol. 106, no. 13, pp. 14573–14582, 2001.

[15] C. Ichoku, R. Levy, Y. J. Kaufman et al., "Analysis of the performance characteristics of the five-channel Microtops II Sun photometer for measuring aerosol optical thickness and precipitable water vapor," *Journal of Geophysical Research D*, vol. 107, article 4179, 17 pages, 2002.

[16] K. S. Behera, Ed., *Maritime Heritage of India*, Aryan Books International, 1999.

[17] J. N. Porter, M. Miller, C. Pietras, and C. Motell, "Ship-based sun photometer measurements using microtops sun photometers," *Journal of Atmospheric and Oceanic Technology*, vol. 18, no. 5, pp. 765–774, 2001.

[18] K. D. Knobelspiesse, C. Pietras, and G. S. Fargion, "Sunpointing-error correction for sea deployment of the MICROTOPS II handheld sun photometer," *Journal of Atmospheric and Oceanic Technology*, vol. 20, pp. 767–771, 2003.

[19] A. Smirnov, B. N. Holben, I. Slutsker et al., "Maritime aerosol network as a component of aerosol robotic network," *Journal of Geophysical Research D*, vol. 114, no. 6, Article ID D06204, 10 pages, 2009.

[20] N. T. O'Neill, A. Ignatov, B. N. Holben, and T. F. Eck, "The lognormal distribution as a reference for reporting aerosol optical depth statistics; empirical tests using multi-year, multisite AERONET sunphotometer data," *Geophysical Research Letters*, vol. 27, no. 20, pp. 3333–3336, 2000.

[21] P. Chauhan, N. Sanwlani, and R. R. Navalgund, "Aerosol optical depth variability in the northeastern Arabian sea during winter monsoon: a study using in-situ and satellite measurements," *Indian Journal of Marine Sciences*, vol. 38, no. 4, pp. 390–396, 2009.

[22] M. C. R. Kalapureddy and P. C. S. Devara, "Pre-monsoon aerosol optical properties and spatial distribution over the Arabian Sea during 2006," *Journal of Atmospheric Research*, vol. 95, no. 2-3, pp. 186–196, 2010.

[23] T. F. Eck, B. N. Holben, J. S. Reid et al., "Wavelength dependence of the optical depth of biomass burning, urban, and desert dust aerosols," *Journal of Geophysical Research D*, vol. 104, no. 24, pp. 31333–31349, 1999.

[24] A. Jayaraman, D. Lubin, S. Ramachandran et al., "Direct observations of aerosol radiative forcing over the tropical Indian Ocean during the January-February 1996 pre-INDOEX cruise," *Journal of Geophysical Research D*, vol. 103, no. 12, pp. 13827–13836, 1998.

[25] B. N. Holben, D. Tanré, A. Smirnov et al., "An emerging ground-based aerosol climatology: aerosol optical depth from AERONET," *Journal of Geophysical Research D*, vol. 106, no. 11, pp. 12067–12097, 2001.

[26] V. E. Cachorro, R. Vergaz, and A. M. De Frutos, "A quantitative comparison of α-Å turbidity parameter retrieved in different spectral ranges based on spectroradiometer solar radiation measurements," *Atmospheric Environment*, vol. 35, no. 30, pp. 5117–5124, 2001.

[27] G. Pace, A. di Sarra, D. Meloni, S. Piacentino, and P. Chamard, "Aerosol optical properties at Lampedusa (Central Mediterranean). 1. Influence of transport and identification of different aerosol types," *Atmospheric Chemistry and Physics*, vol. 6, no. 3, pp. 697–713, 2006.

[28] D. G. Kaskaoutis, H. D. Kambezidis, N. Hatzianastassiou, P. G. Kosmopoulos, and K. V. S. Badarinath, "Aerosol climatology: on the discrimination of aerosol types over four AERONET sites," *Atmospheric Chemistry and Physics Discussions*, vol. 7, no. 3, pp. 6357–6411, 2007.

[29] T. F. Eck, B. N. Holben, O. Dubovik et al., "Column-integrated aerosol optical properties over the Maldives during the northeast monsoon for 1998–2000," *Journal of Geophysical Research D*, vol. 106, no. 22, pp. 28555–28566, 2001.

[30] T. F. Eck, B. N. Holben, D. E. Ward et al., "Characterization of the optical properties of biomass burning aerosols in Zambia during the 1997 ZIBBEE field campaign," *Journal of Geophysical Research D*, vol. 106, no. 4, pp. 3425–3448, 2001.

[31] M. Masmoudi, M. Chaabane, D. Tanré, P. Gouloup, L. Blarel, and F. Elleuch, "Spatial and temporal variability of aerosol: size distribution and optical properties," *Journal of Atmospheric Research*, vol. 66, no. 1-2, pp. 1–19, 2003.

[32] D. H. Kim, B. J. Sohn, T. Nakajima et al., "Aerosol optical properties over east Asia determined from ground-based sky radiation measurements," *Journal of Geophysical Research D*, vol. 109, no. 2, Article ID D02209, 18 pages, 2004.

[33] K. O. Ogunjobi, Z. He, K. W. Kim, and Y. J. Kim, "Aerosol optical depth during episodes of Asian dust storms and biomass burning at Kwangju, South Korea," *Atmospheric Environment*, vol. 38, no. 9, pp. 1313–1323, 2004.

[34] M. C. R. Kalapureddy, D. G. Kaskaoutis, P. E. Raj et al., "Identification of aerosol type over the Arabian Sea in the premonsoon season during the Integrated Campaign for Aerosols, Gases and Radiation Budget (ICARB)," *Journal of Geophysical Research D*, vol. 114, no. 17203, Article ID D17203, 12 pages, 2009.

[35] S. K. Satheesh, V. Ramanathan, X. Li-Jones et al., "A model for the natural and anthropogenic aerosols over the tropical Indian Ocean derived from Indian Ocean Experiment data," *Journal of Geophysical Research D*, vol. 104, no. 22, pp. 27421–27440, 1999.

[36] K. K. Moorthy, A. Saha, B. S. N. Prasad, K. Niranjan, D. Jhurry, and P. S. Pillai, "Aerosol optical depths over peninsular India and adjoining oceans during the INDOEX campaigns: spatial, temporal, and spectral characteristics," *Journal of Geophysical Research D*, vol. 106, no. 22, pp. 28539–28554, 2001.

[37] V. Ramanathan, P. J. Crutzen, J. Lelieveld et al., "Indian Ocean Experiment: an integrated analysis of the climate forcing and effects of the great Indo-Asian haze," *Journal of Geophysical Research D*, vol. 106, no. 22, pp. 28371–28398, 2001.

[38] S. S. Babu, V. S. Nair, and K. K. Moorthy, "Seasonal changes in aerosol characteristics over Arabian Sea and their consequence on aerosol short-wave radiative forcing: results from ARMEX field campaign," *Journal of Atmospheric and Solar-Terrestrial Physics*, vol. 70, no. 5, pp. 820–834, 2008.

[39] A. Smirnov, B. N. Holben, O. Dubovik et al., "Atmospheric aerosol optical properties in the Persian Gulf," *Journal of the Atmospheric Sciences*, vol. 59, no. 3, pp. 620–634, 2002.

[40] K. O. Ogunjobi, Z. He, and C. Simmer, "Spectral aerosol optical properties from AERONET Sun-photometric measurements over West Africa," *Journal of Atmospheric Research*, vol. 88, no. 2, pp. 89–107, 2008.

[41] I. N. Sokolik, O. B. Toon, and R. W. Bergstrom, "Modeling the rediative characteristics of airborne mineral aerosols at infrared wavelengths," *Journal of Geophysical Research D*, vol. 103, no. 8, pp. 8813–8826, 1998.

[42] K. K. Moorthy, S. K. Satheesh, S. S. Babu, and C. B. S. Dutt, "Integrated Campaign for aerosols gases and Radiation Budget (ICARB): an overview," *Journal of Earth System Science*, vol. 117, no. 1, pp. 243–262, 2008.

[43] J.-F. Léon, P. Chazette, J. Pelon, F. Dulac, and H. Randri-amiarisoa, "Aerosol direct radiative impact over the INDOEX area based on passive and active remote sensing," *Journal of Geophysical Research D*, vol. 107, no. 19, article 8006, 2002.

[44] E. J. Welton, K. J. Voss, P. K. Quinn et al., "Measurements of aerosol vertical profiles and optical properties during INDOEX 1999 using micropulse lidars," *Journal of Geophysical Research D*, vol. 107, no. 19, article 8019, 2002.

[45] B. Jha and T. N. Krishnamurthi, "Real-time meteorological reanalysis atlas during pre-INDOEX field phase—1998," Rep. 98-08, INDOEX Publ 20, Tallahassee, Department of Meteorology, Florida State University, 1998.

[46] K. Rajeev, V. Ramanathan, and J. Meywerk, "Regional aerosol distribution and its long-range transport over the Indian Ocean," *Journal of Geophysical Research D*, vol. 105, no. 2, pp. 2029–2043, 2000.

10

Marine Environmental Risk Assessment of Sungai Kilim, Langkawi, Malaysia: Heavy Metal Enrichment Factors in Sediments as Assessment Indexes

Jamil Tajam and Mohd Lias Kamal

Centre of Ocean Research, Conservation & Advances (ORCA), Division of Research, Industrial Linkage, Community Network & Alumni, Universiti Teknologi MARA, 02600 Arau, Perlis, Malaysia

Correspondence should be addressed to Jamil Tajam; jamiltajam@perlis.uitm.edu.my

Academic Editor: Heinrich Hühnerfuss

Concentrations of Cd, Co, Pb, and Zn in riverbed sediments from six sampling stations along the Sungai Kilim, Langkawi, Malaysia, were determined by using the Teflon Bomb Digestion. From this study, the concentrations of heavy metals in riverbed sediments were found ranging between 6.10 and 8.87 $\mu g/g$ dry weight for Co, 0.03 and 0.45 $\mu g/g$ dry weight for Cd, 59.8 and 74.9 $\mu g/g$ dry weight for Zn, and 1.06 and 11.69 $\mu g/g$ dry weight for Pb. From the observation, these areas were polluted by domestic waste, aquaculture, and tourism activities. For clarity, enrichment factor index was used to determine the level of sediment contamination in the study area. From this study, the average EF value is a bit high for Cd (2.15 ± 1.17) followed by Zn (1.12 ± 0.09), Pb (0.44 ± 0.32), and lastly, Co (0.36 ± 0.04). Based on the contamination categories, Cd was categorised as moderately enriched, while the rest of the metals studied were in deficient-to-minimally enriched by the anthropogenic sources.

1. Introduction

Heavy metals are one of the most poisonous and serious groups of pollutants due to their high toxicity, abundance, and ease of accumulation from various plants and animals. It has been accepted that heavy metals can exist in the environment deriving from a variety of natural and anthropogenic sources. The phenomena of erosion, acidification, and weathering processes have brought input of these metals into the environment in a natural way. According to Idris [1], the natural occurrence of heavy metals in aquatic environments and their movement through the hydrocycle in addition to the inputs from anthropogenic activities reflect their ubiquity and complexity. Meanwhile, human activities also contribute to the existence of these metals such as industrial processes, agricultural and aquaculture activities, domestic wastes, and emission from vehicles [2].

Nowadays, these anthropogenic heavy metals contribute to the uppermost pollution to the aquatic environment, especially in the sediment. Sediment plays a major role in determining the pollution pattern of marine ecosystem [3]. According to Singh et al. [4] and Mwamburi [5], the sediments can act as both carriers and sinks for contaminants, reflecting the history of pollution while also providing a record of catchment inputs into the aquatic ecosystem. On the other hand, sediment can play a significant role as a scavenger agent for heavy metals, and an adsorptive sink in marine environment [6, 7]. Tsugonai and Yamada [8] further explained that aquatic sediment can act as scavengers of metals in the environment due to its several, sulfides, organic matter, iron and manganese oxides, and clays. Hence, the use of sediments is advantageous to assess human impacts on the marine environment.

For these reasons, the assessment of pollutants in the estuary system of Sungai Kilim, Langkawi, was carried out to perform the current sediment data assessment in order to assess its level of pollution in this area. Generally, Sungai Kilim is situated in the north-eastern region of Pulau Langkawi. The river is located at approximately $99°52'8.01''$ E and $6°25'39.75''$ N. Sungai Kilim has become one of the popular sites to be visited for its attractive ecology and features. Due to its uniqueness, ecotourism benefits towards many locals and tourists [8]. The downstream of Sungai Kilim is characterized

Marine Environmental Risk Assessment of Sungai Kilim, Langkawi, Malaysia: Heavy Metal Enrichment
Factors in Sediments as Assessment Indexes

93

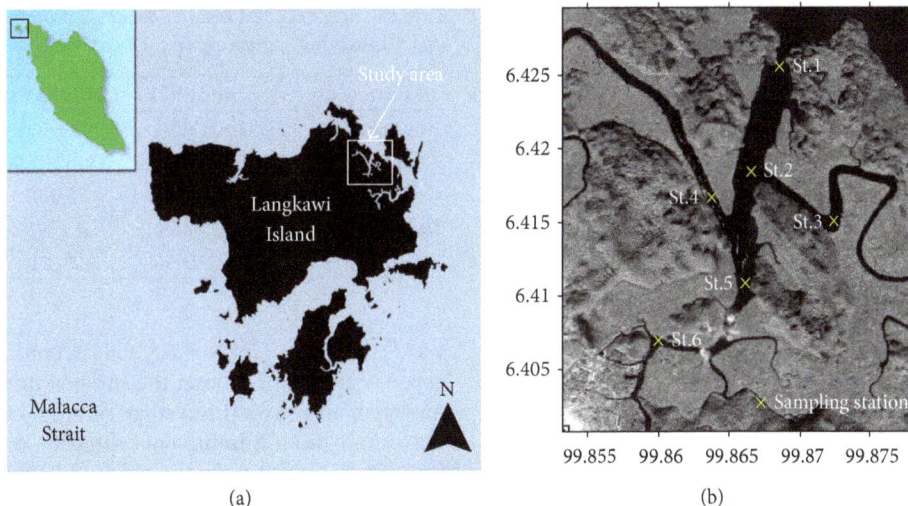

FIGURE 1: Sampling area.

by the high number of karst formations with high density of vegetative roots covering its surface. The middle stream is connected to two small rivers or tributaries. Beyond the upstream, urbanization conquered the banks where roads, restaurants, and tourist jetty are developed well for tourism activities.

Recently, quite a number of heavy metal assessment studies have been reported in this area. This assessment is particularly carried out in order to protect and conserve the environment of the Kilim River from further contamination by anthropogenic sources. Furthermore, the information gathered from this monitoring study may also provide an aid in the management of suitable policy for Langkawi Development Authority (LADA) to preserve the Sungai Kilim ecosystem as Kilim Karst Geoforest Park.

2. Methodology

2.1. Sample Collection and Preservation. Sample collection was conducted in December 2009 during the northeast monsoon season. Generally, Sungai Kilim, Langkawi (Figure 1) is very popular for a variety of ecotourism activities and due to this, major development has been done to improve the river's quality for providing the best services to tourists. These include eagle feeding activities (Station 3), floating restaurants, aquaculture (Station 4), yacht anchoring area (Station 4), and also boating services (Station 6). Therefore, six stations were established along the Sungai Kilim and marked using GPS (global positioning system) (Table 1). These locations were selected based on the fact that they might have been impacted by the nearby source of contamination. The distance between each station was approximately 1 km. Riverbed sediment samples were collected using the Van Veen Grab, where afterwards, samples were placed in plastic bags which were previously immersed in 5% nitric acid for two to three days to prevent sample contamination. The sediment samples were then preserved in the ice box at 4°C to maintain the original condition of the samples. At the laboratory, samples were dried in the oven at 105°C for 24 hours.

TABLE 1: The coordinates for each sampling station.

Station	Latitude	Longitude
1	06°25'32.46''N	099°52'06.42''N
2	06°25'05.14''N	099°51'57.91''N
3	06°24'57.78''N	099°52'20.57''N
4	06°25'00.96''N	099°51'47.58''N
5	06°24'33.54''N	099°51'56.10''N
6	06°24'24.66''N	099°51'34.92''N

For heavy-metal analysis, it was ensured that the samples had been completely dried before grinding the samples with mortar and pestle and sieved under 63 μm size. Precautions in preventing sample contamination were given priority. Samples were then stored in labelled plastic vials and kept in the drying cabinet until lab analysis.

2.2. Sample Digestion. In this study, the digestion and analytical procedures were adopted and applied from those of Kamaruzzaman [9], Jamil [10], and Trimm et al. [11] with little modifications. For this analysis, 0.05 g of the fine powder sediment (<63 μm) was weighed and put into a Teflon vessel. After that, 1.5 mL of mixed acid (2.5 HF : 3 HNO$_3$: 3 HCl) was added into the Teflon vessels using a single channel pipette, 100–1000 microlitre (μL) of the brand CappAero which was ISO 9001; 2000 certified. This digestion method is also known as the aqua regia + HF digestion method, which was also applied by Chen and Ma [12] and Deely and Fergusson [13]. Finally, the Teflon Bomb jackets were screwed tightly to prevent the appearance of silicate gel on their bodies, before placing the Teflon bombs into the oven for 6 hours at 160°C. After 6 hours, they were cooled down under room temperature, where after that, 3.0 mL of acid solution composed of ethylenediaminetetraacetic acid (EDTA) and boric acid was added. The samples were then again put into the oven at 160°C for another 6 hours. The clear solution obtained was transferred into centrifuge tubes and meshed up to 10 mL with Mili-Q water. To verify the precision of the analytical

TABLE 2: Five contaminant categories based on the EF value [25].

Enrichment factor (EF) value	Contamination degree
<2	Deficiency to minimal enrichment
2–5	Moderate enrichment
5–20	Significant enrichment
20–40	Very high enrichment
>40	Extremely high enrichment

TABLE 3: Recovery test results (concentration for Fe is in percentage (%), while other metals are in μg/g dry weight).

Heavy metals	Measured SRM	Certified value	Recovery (%)
Iron, Fe	$2.053 \pm 0.115\%$	$2.008 \pm 0.039\%$	102.24
Cadmium, Cd	0.127 ± 0.011	0.148 ± 0.007	85.81
Cobalt, Co	4.726 ± 0.028	5.000	94.52
Lead, Pb	9.552 ± 0.473	11.7 ± 1.2	81.67
Zinc, Zn	45.389 ± 0.698	48.9 ± 1.60	92.82

procedures, the sediment samples were analysed in three replicates for each sampling point and a sample blank. To confirm analytical accuracy, portions of certified reference materials (SRM1646a-estuarine sediments) from the National Institute of Standards and Technology (NIST) were analysed with each batch of samples. The concentrations of metals (Cd, Co, Fe, Pb, and Zn) in the final digested solutions were then analysed using the Inductively Coupled Plasma Mass Spectrometer (Perkin Elmer. Elan 9000).

2.3. Ecological Risk Assessment. In this study, Fe is used as the normalized metal as it is an acceptable normalization element to be used in the calculation of enrichment factor since Fe distribution was not related to other heavy metals [13]. Fe usually has a relatively high natural concentration and is therefore not expected to be substantially enriched from anthropogenic sources in estuarine sediments [14]. These facts are in accordance with Daskalakis and O'Connor [15] who proposed that Fe is associated with fine solid surfaces, where its geochemistry is similar to that of many trace metals and its natural sediment concentration tends to be uniform. Moreover, a wide range of studies have used Fe and Al normalizations as an alternative to grain-size normalization [16, 17]. The enrichment factors (EFs) for each metal were calculated based on the formula following [18–20]:

$$\text{EF} = \frac{(C_s/C_{\text{Fe}})_{\text{sample}}}{(C_s/C_{\text{Fe}})_{\text{crust}}}. \tag{1}$$

Meanwhile, the values for the earth's crustal elements were taken from Carmichael [21], Bodek et al. [22], and Ronov and Yaroshevsky [23], as published by Lide [24]. The resulting values were then referred to the contaminant categories proposed by Sutherland [25] to reveal and acknowledge the enrichment degree of the heavy metals as illustrated by Table 2.

3. Result and Discussion

For method validation, certified reference material (SRM1646a) was determined as a precision check. The percentage of recoveries ($n = 5$ for each metal) for certified and measured concentration of those metals was satisfactory, with the recoveries being 81.67–102.24%. Table 3 shows the recovery test results for SRM (1646a) analysis.

From this study, the heavy metal contents of the sediment were analysed, and the results were depicted in Table 4. The trend of mean concentration of heavy metals in Sungai Kilim

was Fe > Zn > Co > Pb > Cd. According to Table 4, the concentration of Fe along the stations increased toward the upstream area, with the average value of $4.803 \pm 0.422\%$. The maximum and minimum concentration of Fe were obtained from St.5 and St.1 with the values 5.230% and 4.086% respectively. Furthermore, the highest concentration of Zn was found at St.6 with a value of 74.923 μg/g dry weight, whereas the lowest concentration was found at St.1 with a concentration of 59.81 μg/g dry weight. As for Zn, 66.817 ± 4.856 μg/g dry weight of Zn has been found distributed along Sungai Kilim. In the meantime, Co and Pb displayed the highest value at St.2 (8.872 μg/g dry weight) and St.4 (11.69 μg/g dry weight), correspondingly. Lastly, Cd was distributed unevenly along the river with the range value of 0.026–0.452 μg/g dry weight. According to the conducted statistical analysis of 2-way ANOVA, there was a significant difference between the sampling stations ($P < 0.05$).

According to Rickard and Niagru [26] and Muller [27], the high concentration of heavy metals is usually not only influenced by the chemical processes but also by anthropogenic activities. Hence, these anthropogenic activities carried along the river plus the increase of development in that area are seen to have impacted the river as the domestic sewage, oil spill, and aquaculture, and many more are considered to be common sights. On average, Fe concentration in Sungai Kilim was lower compared to the earth's crust value (5.635%). It was found that Fe was more concentrated at St.5 rather than the downstream, which could be possibly caused by the weathering process of sedimentary rocks that enter the river. Williamson [28] had stated that typical levels for Fe in sedimentary rocks are given as limestone 0.33%, sandstone 0.98%, shale 4.7%, and banded iron formation 28%. The distribution of Fe was rather uniform along the river, decreasing towards downstream. This trend suggests that only a small amount of Fe has been drifting downstream.

From the observation, the highest Zn concentration value was discovered at St.6 (upstream) and constantly decreasing towards the downstream area. This might be due to the bat waste from the bat cave at the upstream area. According to Miko et al. [29], bat droppings contain a relatively high content of Zn; thus, this could justify the source of Zn in this area. In addition, Davis et al. [30] showed that urban runoff might be slightly smaller. This is in agreement with Roney et al. [31] who found that Zn and its compounds are present in the Earth's crust and in most rocks, certain minerals, and some carbonate sediments, and, as a result of weathering effects towards these materials, soluble compounds of zinc are formed and may be released to water [32]. On the other hand,

Marine Environmental Risk Assessment of Sungai Kilim, Langkawi, Malaysia: Heavy Metal Enrichment
Factors in Sediments as Assessment Indexes

95

TABLE 4: Heavy metal in each station (concentration for Fe is depicted in percentage (%), while other metals are in $\mu g/g$ dry weight).

Heavy metals	St.1	St.2	St.3	St.4	St.5	St.6	Average	One-way ANOVA
Cd	0.288	0.388	**0.452**	0.026	0.201	0.294	0.275 ± 0.137	$P < 0.05^{*}$
Co	7.989	**8.872**	7.686	6.104	7.761	7.756	7.695 ± 0.818	$P < 0.05^{*}$
Fe	4.086	4.939	4.939	4.402	**5.230**	5.222	4.803 ± 0.422	$P < 0.05^{*}$
Pb	6.508	1.057	1.844	**11.69**	5.571	5.079	5.292 ± 3.473	$P < 0.05^{*}$
Zn	59.81	66.918	62.245	68.308	68.697	**74.923**	66.817 ± 4.856	$P < 0.05^{*}$

*There are significant differences between the sampling stations.

TABLE 5: The comparison of heavy metal concentrations in the present study with other heavy metal studies throughout Malaysia.

No.	Area	Cd	Co	Fe	Pb	Zn	References
(1)	Sungai Kilim, Langkawi	0.27 ± 0.14	7.69 ± 0.82	4.80 ± 0.42	5.29 ± 3.47	66.82 ± 4.86	Present study
(2)	Terengganu River	0.9 ± 1.7	15.1 ± 7.4	6.3 ± 1.3	44.1 ± 14.4	65.6 ± 34.8	Jamil [10]
(3)	Kemaman estuary	—	16.00	—	—	—	Kamaruzzaman and Ong [40]
(4)	Sungai Kelantan	1.82 ± 0.02	—	3.860	20.82 ± 0.28	18.67 ± 0.56	Ahmad et al. [41]
(5)	Langkawi coastal water	—	—	—	41.87 ± 7.3	—	Kamaruzzaman et al. [42]
(6)	Sungai Semenyih	—	20.95 ± 12.97	—	44.71 ± 22.34	62 ± 36.34	Muhammad et al. [43]
(7)	Kuala Perlis	—	—	—	25.77	55.73	Yap and Pang [44]
(8)	Kuala Kedah jetty	—	—	—	26.81	53.21	Yap and Pang [44]
(9)	Kuala Muda	—	—	—	31.92	33.6	Yap and Pang [44]
(10)	Kuala Juru jetty	—	—	—	30.20	317.39	Yap and Pang [44]
(11)	Juru industrial drainage	—	—	—	65.32	484.14	Yap and Pang [44]
(12)	Earth bulk continental crust	0.15	25	5.63	14	70	Lide [24]

Monaci and Bargagli [33] have also found that the possible source of Zn may also be from motor oil, grease, phosphate fertilizers, sewage sludge, transmission fluid, undercoating, asphalt, and concrete.

Meanwhile, the average concentration of both Co and Pb in this study area was discovered to be slightly lower than the earth's crust value. Faroon et al. [34] stated that the primary anthropogenic sources of Co lie in phosphate fertilizers, which, in this study, could be due to bat waste and aquaculture activities. Naylor et al. [35] have reported that fish pellets also contain some cobalt elements. A very high concentration of Pb at St.4 revealed that the surroundings were heavily contaminated with leaded petrol stemmed from outboard boat engines from boating activities and places of anchored yacht that were very concentrated in that particular area compared to the other stations. The contamination of Pb has been the major attention of many researchers such as Lee et al. [36], Alloway [37], and Monaci and Bargagli [33], where they have confirmed that leaded fuel, automobile or motor exhaust, lubricating oil, and grease are the possible major sources of Pb.

Lastly, the Cd concentration for all sampling stations was dramatically higher than that of the earth's crust value, except for St.4. From the observation, the highest concentration indicated that the main source of heavy metal enrichment came from anthropogenic elements deposited directly or indirectly by human activities such as boat cruising along the study area. According to Miko et al. [29], oil combustion from the boating activities was the main factor that influenced the increase of Cd. On the other hand, Dong et al. [38] stated that

human activities including discharges of municipal wastewater, agriculture, mining, fossil fuels, and industrial wastewater are a major source of Cd contamination in the marine environment especially in estuaries. Cd however has a very low solubility in aqueous solution. It is readily adsorbed to suspended solids, where, after a series of natural processes, Cd particles will finally sink and accumulate in sediments [39].

The comparison of heavy metal compositions made between the study area and other places in Malaysia is available in Table 5. The table shows the value of heavy metal concentrations reported by previous researchers, namely Kamaruzzaman and Ong [40], Jamil [10], Ahmad et al. [41], Kamaruzzaman et al. [42], Muhammad et al. [43] and Yap and Pang [44]. Based on Table 5, Zn concentration in the present study area was found to be relatively lower compared to the Juru area and earth bulk continental crust but higher compared to other places. Meanwhile, the other metals obtained seemed to have lower concentrations with respect to the other places in Malaysia.

For a better estimation of anthropogenic input, an enrichment factor (EF) was calculated for each metal by dividing its ration to the normalizing element by the same ration found in the chosen baseline. Generally, throughout this study, it was found that Fe distribution was rather uniform along the river, which indicates a good stability of the metal in the river as a whole. The good stability of Fe along the river signifies and ascertains its capability to be adopted as the most suitable normalizing element for this study compared to other metals, particularly in this study area. According to Figure 2,

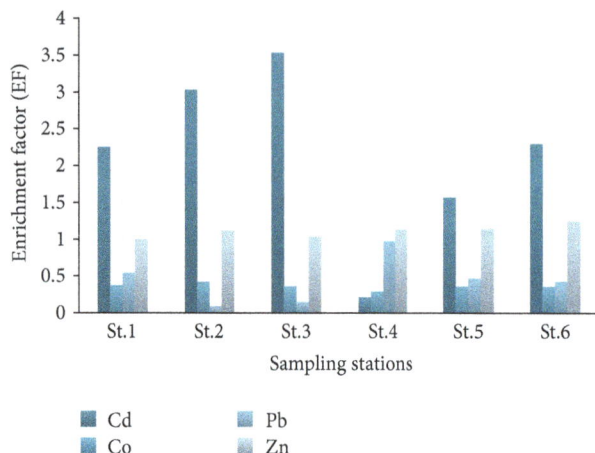

FIGURE 2: Enrichment factors for each heavy metal in the study area.

the EF value of Cd (2.15±1.17) for all stations was higher than 2.0 except for St.4 and St.5. This means that a significant portion of heavy metals was provided by anthropogenic sources mainly from the oil combustion from boating activities. However, Co (0.36±0.04), Pb (0.44±0.32), and Zn (1.12±0.09) obtained minimal enrichment (<2.0) for all stations. The data obtained will provide scientific evidence and may be referred to as the baseline data for a better understanding of our estuary ecosystem. In Malaysia, especially in Langkawi area, studies regarding heavy-metal pollution are currently quite limited. Therefore, continuous monitoring of heavy-metal pollution in this environment is definitely indispensable.

4. Conclusion

From this study, the overall trend of mean concentration of heavy metals in Sungai Kilim can be concluded by the following order: Fe > Zn > Co > Pb > Cd. However, the distribution of Cd in the sediment was found to be higher than that of the earth's crust value, while other metals (Co, Fe, Pb, and Zn) were found to be of lower values. To clarify and validate how serious the contamination level of each metal to the river is, the enrichment factor (EF) was applied. According to this approach, the river was moderately enriched by Cd. The higher concentrations of Cd in the study area were estimated to have originated from the oil combustion of boating activities along the study area.

Acknowledgments

This research was conducted under the funding of University Teknologi MARA (UiTM), through the Research Excellence Fund. The authors wish to express their gratitude to the Oceanography Laboratory team members for their invaluable assistance and hospitality throughout the sampling period.

References

[1] A. M. Idris, "Combining multivariate analysis and geochemical approaches for assessing heavy metal level in sediments from Sudanese harbors along the Red Sea coast," *Microchemical Journal*, vol. 90, no. 2, pp. 159–163, 2008.

[2] A. Demirak, F. Yilmaz, A. Levent Tuna, and N. Ozdemir, "Heavy metals in water, sediment and tissues of Leuciscus cephalus from a stream in southwestern Turkey," *Chemosphere*, vol. 63, no. 9, pp. 1451–1458, 2006.

[3] J. M. Casas, H. Rosas, M. Solé, and C. Lao, "Heavy metals and metalloids in sediments from the Llobregat basin, Spain," *Environmental Geology*, vol. 44, no. 3, pp. 325–332, 2003.

[4] K. P. Singh, D. Mohan, V. K. Singh, and A. Malik, "Studies on distribution and fractionation of heavy metals in Gomti river sediments—a tributary of the Ganges, India," *Journal of Hydrology*, vol. 312, no. 1–4, pp. 14–27, 2005.

[5] J. Mwamburi, "Variations in trace elements in bottom sediments of major rivers in Lake Victoria's basin, Kenya," *Lakes and Reservoirs: Research and Management*, vol. 8, no. 1, pp. 5–13, 2003.

[6] A. Tessier and P. G. C. Campbell, "Partitioning of trace metals in sediments: relationships with bioavailability," *Hydrobiologia*, vol. 149, no. 1, pp. 43–52, 1987.

[7] N. W. Chan, "Protecting and conserving our natural heritage: potentials, threats and challenges of Langkawi Geopark," in *Proceedings of the International Conference World Civic Forum*, Seoul, South Korea, May 2009.

[8] S. Tsugonai and M. Yamada, "226Ra in Bering sea sediment and its application as a geochronometer," *Geochemical Journal*, vol. 13, pp. 231–238, 1980.

[9] B. Y. Kamaruzzaman, *Geochemistry or the marine sediments. Its paleoceanographic significance [Ph.D. thesis]*, Hokkaido University, 1999.

[10] T. Jamil, *Physicochemical and sediment characteristics of the bottom sediment of Terengganu River, Terengganu Malaysia [M.S. thesis]*, Kolej Universiti Sains dan Teknologi Malaysia, 2006.

[11] D. L. Trimm, H. H. Beiro, and S. J. Parker, "Comparison of digestion techniques in analyses for total metals in marine sediments," *Bulletin of Environmental Contamination and Toxicology*, vol. 60, no. 3, pp. 425–432, 1998.

[12] M. Chen and L. Q. Ma, "Comparison of three aqua regia digestion methods for twenty Florida soils," *Soil Science Society of America Journal*, vol. 65, no. 2, pp. 491–499, 2001.

[13] J. M. Deely and J. E. Fergusson, "Heavy metal and organic matter concentrations and distributions in dated sediments of a small estuary adjacent to a small urban area," *The Science of the Total Environment*, vol. 153, no. 1-2, pp. 97–111, 1994.

[14] L. F. Niencheski, H. L. Windom, and R. Smith, "Distribution of particulate trace metal in Patos Lagoon estuary (Brazil)," *Marine Pollution Bulletin*, vol. 28, no. 2, pp. 96–102, 1994.

[15] K. D. Daskalakis and T. P. O'Connor, "Normalization and elemental sediment contamination in the coastal United States," *Environmental Science and Technology*, vol. 29, no. 2, pp. 470–477, 1995.

[16] V. T. Breslin and S. A. Sañudo-Wilhelmy, "High spatial resolution sampling of metals in the sediment and water column in Port Jefferson Harbor, New York," *Estuaries*, vol. 22, no. 3, pp. 669–680, 1999.

[17] H. L. Windom, S. J. Schropp, F. D. Calder et al., "Natural trace metal concentrations in estuarine and coastal marine sediments of the Southeastern United States," *Environmental Science and Technology*, vol. 23, no. 3, pp. 314–320, 1989.

[18] S. Covelli and G. Fontolan, "Application of a normalization procedure in determining regional geochemical baselines," *Environmental Geology*, vol. 30, no. 1-2, pp. 34–45, 1997.

Marine Environmental Risk Assessment of Sungai Kilim, Langkawi, Malaysia: Heavy Metal Enrichment
Factors in Sediments as Assessment Indexes

97

[19] V. Simeonov, D. L. Massart, G. Andreev, and S. Tsakovski, "Assessment of metal pollution based on multivariate statistical modeling of "hot spot" sediments from the Black Sea," *Chemosphere*, vol. 41, no. 9, pp. 1411–1417, 2000.

[20] L. Zhang, X. Ye, H. Feng et al., "Heavy metal contamination in western Xiamen Bay sediments and its vicinity, China," *Marine Pollution Bulletin*, vol. 54, no. 7, pp. 974–982, 2007.

[21] R. S. Carmichael, *CRC Practical Handbook of Physical Properties of Rocks and Minerals*, CRC Press, Boca Raton, Fla, USA, 1989.

[22] I. Bodek, W. J. Lyman, W. F. Reehl, and D. H. Rosenblatt, *Environmental Inorganic Chemistry*, Pergamon Press, New York, NY, USA, 1988.

[23] A. B. Ronov and A. A. Yaroshevsky, "Earth's crust geochemistry," in *Encyclopedia of Geochemistry and Environmental Sciences*, R. W. Fairbridge, Ed., Van Nostrand, New York, NY, USA, 1969.

[24] D. R. Lide, *CRC Handbook of Chemistry and Physics*, CRC Press, Boca Raton, Fla, USA, 85th edition, 2005.

[25] R. A. Sutherland, "Bed sediment-associated trace metals in an urban stream, Oahu, Hawaii," *Environmental Geology*, vol. 39, no. 6, pp. 611–627, 2000.

[26] D. T. Rickard and J. O. Niagru, "Aqueous environmental chemistry of lead," in *The Biochemistry of Lead in the Environment: Part A. Ecological Cycles*, pp. 219–284, Elsevier, Amsterdam, The Netherlands, 1978.

[27] F. L. L. Muller, "Colloid/solution partitioning of metal-selective organic ligands, and its relevance to Cu, Pb and Cd cycling in the firth of Clyde," *Estuarine, Coastal and Shelf Science*, vol. 46, no. 3, pp. 419–437, 1998.

[28] M. A. Williamson, "Iron," in *Encyclopedia of Geochemistry*, C. P. Marshall and R. W. Fairbridge, Eds., pp. 348–353, Kluwer Academic, Dordrecht, Germany, 1999.

[29] S. Miko, M. Kuhta, and S. Kapelj, "Environmental baseline geochemistry of sediments and percolating waters in the Modric Cave, Croatia," *Acta Carsologica*, vol. 31, no. 1, pp. 135–149, 2002.

[30] A. P. Davis, M. Shokouhian, and S. Ni, "Loading estimates of lead, copper, cadmium, and zinc in urban runoff from specific sources," *Chemosphere*, vol. 44, no. 5, pp. 997–1009, 2001.

[31] N. Roney, V. Cassandra, M. Williams, M. Osier, and S. J. Paikoff, *Toxicological Profile for Zinc*, U.S. Department Of Health And Human Services Public Health Service Agency for Toxic Substances and Disease Registry, 2005.

[32] NAS, "Inorganic solutes," in *Drinking Water and Health*, vol. 1, pp. 205–229, National Academy of Sciences; National Academy Press, Washington, DC, USA, 1977.

[33] F. Monaci and R. Bargagli, "Barium and other trace metals as indicators of vehicle emissions," *Water, Air, and Soil Pollution*, vol. 100, no. 1-2, pp. 89–98, 1997.

[34] O. M. Faroon, H. Abadin, S. Keith et al., *Toxicological Profile for Cobalt*, U.S. Department of Health and Human Services Public Health Service Agency for Toxic Substances and Disease Registry, 2004.

[35] S. J. Naylor, R. D. Moccia, and G. M. Durant, "The chemical composition of settleable solid fish waste (Manure) from commercial rainbow trout farms in Ontario, Canada," *North American Journal of Aquaculture*, vol. 61, no. 1, pp. 21–26, 1999.

[36] D. S. Lee, J. A. Garland, and A. A. Fox, "Atmospheric concentrations of trace elements in urban areas of the United Kingdom," *Atmospheric Environment*, vol. 28, no. 16, pp. 2691–2713, 1994.

[37] B. J. Alloway, *Heavy Metals in Soils*, Blackie Academic and Professional, Glasgow, UK, 2nd edition, 1995.

[38] C. D. Dong, C. F. Chen, M. S. Ko, and C. W. Chen, "Enrichment, accumulation and ecological risk evaluation of cadmium in the surface sediments of Jen-Gen River Estuary, Taiwan," *International Journal of Chemical Engineering and Applications*, vol. 3, no. 6, pp. 370–373, 2012.

[39] C. W. Chen, C. F. Chen, and C. D. Dong, "Contamination and potential ecological risk of mercury in sediments of Kaohsiung River mouth, Taiwan," *International Journal of Environmental Science and Development*, vol. 3, pp. 66–71, 2012.

[40] Y. Kamaruzzaman and M. C. Ong, "Geochemical proxy of some chemical elements in sediments of kemaman river Estuary, Terengganu, Malaysia," *Sains Malaysiana*, vol. 38, no. 5, pp. 631–636, 2009.

[41] A. K. Ahmad, I. Mushrifah, and M. Shuhaimi-Othman, "Water quality and heavy metal concentrations in sediment of Sungai Kelantan, Kelantan, Malaysia: a baseline study," *Sains Malaysiana*, vol. 38, no. 4, pp. 435–442, 2009.

[42] B. Y. Kamaruzzaman, N. T. Shuhada, B. Akbar et al., "Spatial concentrations of lead and copper in bottom sediments of Langkawi Coastal Area, Malaysia," *Research Journal of Environmental Sciences*, vol. 5, pp. 179–186, 2011.

[43] B. G. Muhammad, N. A. S. Wan, and I. Mohd, "Sebaran logam berat dalam lembangan sungai semenyih," in *Proceedings of the Regional Symposium on Environment and Natural Resources*, vol. 1, pp. 595–602, Kuala Lumpur, Malaysia, April 2002.

[44] C. K. Yap and B. H. Pang, "Assessment of Cu, Pb, and Zn contamination in sediment of north western Peninsular Malaysia by using sediment quality values and different geochemical indices," *Environmental Monitoring and Assessment*, vol. 183, no. 1-4, pp. 23–39, 2011.

11

Comparative Analysis of Sea Surface Temperature Pattern in the Eastern and Western Gulfs of Arabian Sea and the Red Sea in Recent Past Using Satellite Data

Neha Nandkeolyar,[1] Mini Raman,[2] G. Sandhya Kiran,[1] and Ajai[2]

[1] *Department of Botany, Faculty of Science, M.S. University, Baroda, Vadodara 390002, India*
[2] *Marine Optics Division, Marine and Planetary Sciences Group, Space Application Centre, Ahmedabad 380015, India*

Correspondence should be addressed to Neha Nandkeolyar; neha.nandkeolyar@gmail.com

Academic Editor: Swadhin Behera

With unprecedented rate of development in the countries surrounding the gulfs of the Arabian Sea, there has been a rapid warming of these gulfs. In this regard, using Advanced Very High Resolution Radiometer (AVHRR) data from 1985 to 2009, a climatological study of Sea Surface Temperature (SST) and its inter annual variability in the Persian Gulf (PG), Gulf of Oman (GO), Gulf of Aden (GA), Gulf of Kutch (KTCH), Gulf of Khambhat (KMBT), and Red Sea (RS) was carried out using the normalized SST anomaly index. KTCH, KMBT, and GA pursued the typical Arabian Sea basin bimodal SST pattern, whereas PG, GO, and RS followed unimodal SST curve. In the western gulfs and RS, from 1985 to 1991-1992, cooling was observed followed by rapid warming phase from 1993 onwards, whereas in the eastern gulfs, the phase of sharp rise of SST was observed from 1995 onwards. Strong influence of the El Niño and La Niña and the Indian Ocean Dipole on interannual variability of SST of gulfs was observed. Annual and seasonal increase of SST was lower in the eastern gulfs than the western gulfs. RS showed the highest annual increase of normalized SST anomaly (+0.64/decade) followed by PG (+0.4/decade).

1. Introduction

In today's era, one of the greatest challenges faced by humankind is "global warming." There has been a 0.6°C increase in global temperature in the last century, and it is projected to further increase by 1.8°C to 4°C in the 21st century [1], posing a serious threat to the socioeconomic sector worldwide. With oceans covering approximately 72% of the earth's surface, any discussion on climate change and global warming would remain incomplete without including the role of the oceans. Studies have revealed that because of the increase in greenhouse gases, the oceans are warming significantly, which is affecting the marine ecosystem [2–5], resulting in shifting of habitats and dwindling of marine biota. Most of the studies with respect to climate change have focussed on global scale changes [5–15]. However, an exhaustive analysis at the regional and local levels is needed for framing adaptive and mitigative strategies to alleviate the effect of global warming.

In the Indian context, the recent finding, that amongst the oceans, the warming of the Indian Ocean is second highest [15], is a cause of worry and calls for immediate attention. Even though climatological studies have been done in the past in the Indian Ocean encompassing the Arabian Sea and the Bay of Bengal [16–20], a comparative analysis of the changing Sea Surface Temperature (SST) pattern the gulfs of Arabian Sea (Persian Gulf, Gulf of Oman, Gulf of Aden, Gulf of Kutch, and Gulf of Khambhat) and the Red sea has not been done. The gulfs of the Arabian Sea are not just strategically important with rich sources of oil and natural gas but are also the hot spots of the marine biodiversity [21–28]. However, in the recent years, there has been an expeditious change in the marine ecosystem of the eastern and western gulfs of the Arabian Sea and the Red Sea, owing to anthropogenic interference [29–32]. Since 1990s, 40% of the coasts of the Persian Gulf have been modified [33]. The Persian Gulf and the Red

Sea areas have been reported to be warming rapidly owing to the developmental projects undertaken in the surrounding coastal countries [30–32]. The Gulf of Kutch on the Indian coast is being aggressively developed as oil importing bases because of its proximity to the Middle East countries [29]. In general, the issues of common concern in the eastern and western gulfs of the Arabian Sea and the Red Sea include various anthropogenic activities like industrialization, coastal infrastructure development projects, setting up of new ports and oil terminals, oil pollution from shipping industry, overfishing, dredging, and increase in tourism and recreational activities. [29–32, 34], resulting in habitat destruction, changes in temperature and salinity profile, and causing a significant loss of biodiversity [35].

In this regard, the present work was taken up to study the monthly, seasonal, and annual pattern of Sea Surface Temperature (SST) and to analyse its changing pattern with emphasis on interannual variability in the eastern (Gulf of Kutch and Gulf of Khambhat) and the western gulfs (Persian Gulf, Gulf of Oman, and Gulf of Aden) of the Arabian Sea and the Red sea.

2. Materials and Methods

2.1. Area of Study. The gulfs of the Arabian Sea basin (Figure 1(a)) can be grouped into the eastern and the western branches. The eastern branch includes the Gulf of Kutch and the Gulf of Khambhat on the Indian coast, whereas the western branch includes the Gulf of Oman and the Persian Gulf in the northwestern part of Arabian Sea, and the Gulf of Aden in the southwestern part of Arabian Sea, connecting it with the Red Sea.

2.1.1. Persian Gulf. The Persian Gulf (24–3°N; 48–56.5°E), (Figure 1(b)) located in an arid subtropical zone, is a semienclosed shallow, marginal sea with an average water depth of 36 m. It is spread over an area of 239,000 km² and is 990 km in length. It is connected to the Gulf of Oman through the 56 km wide Strait of Hormuz [36].

In the Persian Gulf, the seasonal differences of insolation, along with cold winds from the nearby highlands, result in extreme temperature (ranging from 16°C to 35°C) [36] and salinity (ranging from 36 to 43 parts per thousand) conditions [37]. However, despite the extreme conditions, the offshore waters of the Persian Gulf are rich in nutrients and support a variety of marine photoautotroph like the macroalgae, phytoplankton, mangroves, sea grass beds, and intertidal vegetation [25].

2.1.2. Gulf of Oman. The Gulf of Oman (22°3–26°5′N; 56.5–61°43′E) is a strait connecting the Arabian Sea with the Persian Gulf (Figure 1(c)). The Gulf of Oman is 320 km wide and 560 km long.

Some of the most powerful eddy currents of the Arabian Sea are located close to the Omani coast. The coastal areas of the Gulf of Oman support varied habitats including mangrove swamps, lagoons, and mudflats.

2.1.3. Gulf of Aden. The Gulf of Aden (GA) (10°–15°N; 43°–52°E), connecting the Red Sea with the Arabian Sea, is part of the Suez Canal shipping route between the Mediterranean Sea and the Indian Ocean (Figure 1(d)). It is about 900 km long, with average depth of 1800 m, and covers an area of about 220 × 10³ km². Its importance lies in the fact that it provides an outlet to the saline water masses of the Red Sea into the Arabian Sea [38].

2.1.4. Red Sea. The Red Sea (RS) (12°29′N–27°57′N; 34°36′E–43°30′E) is approximately 2,100 km long and 280 km wide (Figure 1(e)). The average depth is about 500 m, but at places it is more than 2,000 m deep. The Red Sea is linked to the Mediterranean Sea by the Suez Canal in the north, whereas in the south, it is connected to the Arabian Sea through the Gulf of Aden.

The waters of the Red Sea are warm and saline. The temperature ranges from 21°C–28°C in the north and 26°C to 32°C in the south. Salinities in the Red Sea range from 37% in the south to 42% in the north. A consequence of these extreme conditions is that some species within the Red Sea (like mangroves, shallow sea grasses, etc.) probably exist at the limits of their physiological tolerance [30].

2.1.5. Gulf of Kutch. The Gulf of Kutch (22°15′–23°4′N; 68°20′–70°40′E) encloses an area of 7350 km² (Figure 1(f)). Towards the western end, the Gulf is about 75 km wide and 60 m deep, while in the eastern end it is 18 km wide and less than 20 m deep. It is under the influence of strong tidal currents. High rate of evaporation, along with the release of salty water from the adjoining saltpans of Rann of Kutch, makes the eastern part of the Gulf of Kutch more saline (40%) than the western part (35%). The temperature of the Gulf waters varies between 24°C to 30°C.

The Gulf of Kutch is one of India's only coastal areas endowed with coral reefs. It provides a platform for different habitats like coral reefs, mangroves, creeks, mud flats, islands, rocky shore, sandy shore, and so forth and hence is enriched in biodiversity [29].

2.1.6. Gulf of Khambhat. Gulf of Khambhat (between 72°2′E to 72°6′E and 21° to 22°2′N) is one of the major fishing areas along western coast of India (Figure 1(f)). It is about 80 km wide at mouth and tapers to 25 km along the coast. It is about 140 km in length and is characterised by several inlets and creeks formed by the confluence of rivers. The Gulf of Khambhat is shallow with about 30 m average depth and abounds in shoals and sandbanks. It is known for its extreme tides. The tidal range at Gulf of Khambhat is known to be one of the largest along the Indian coastline [39].

2.2. Data Used. The regional effect of the global warming in the oceanic water masses needs to be assessed and quantified. However, the lack of uniform data and methodology hampers such regional analysis [1]. With the availability of National Oceanic and Atmospheric Administration's (NOAA) Advanced Very High Resolution Radiometer

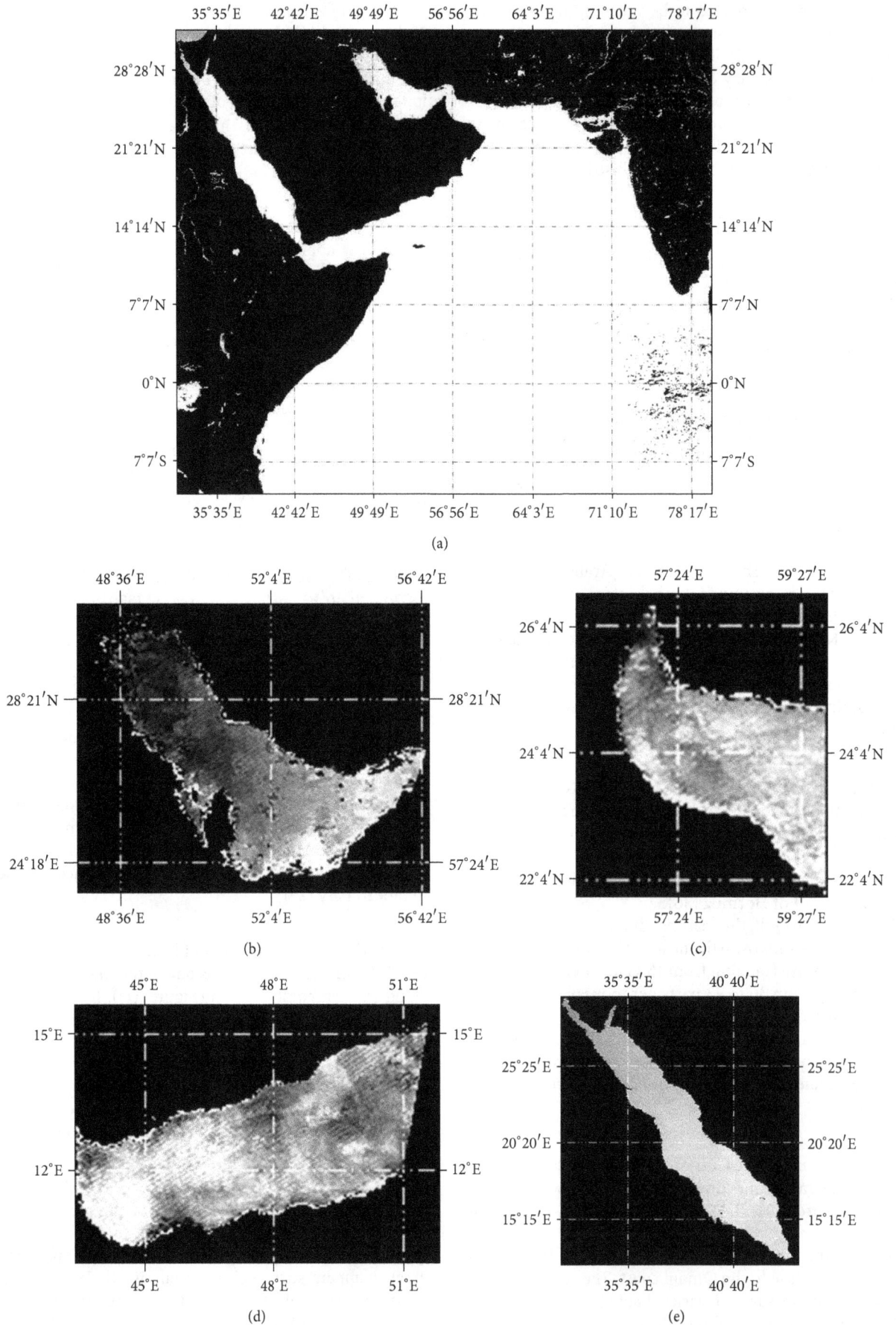

FIGURE 1: Continued.

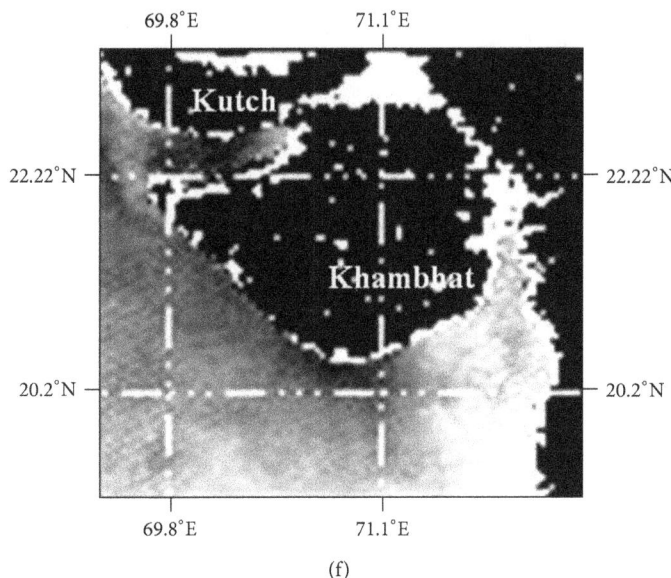

(f)

FIGURE 1: (a) Area of Study showing Arabian Sea basin. (b) Persian Gulf. (c) Gulf of Oman. (d) Gulf of Aden. (e) Red Sea. (f) Gulf of Kutch and Gulf of Khambhat.

(AVHRR) satellite data set, for 25 years of span (1985–2009), to some extent the problem of uniformity of data has been solved. Besides, the high resolution satellite derived SST data gives better results than *in situ* data in terms of its continuity and spatial coverage. Even though Reynolds SST and National Center for Environmental Prediction (NCEP) reanalyzed data sets [40] are available for longer periods, but their low spatial resolution (1° and 2.5°, resp.) and interpolation and reanalysis methods do not accurately capture the coastal areas and the upwelling regions [41].

In this study the monthly NOAA AVHRR Pathfinder (version 5.0) SST data at 4 km resolution was obtained from NASA'S Jet Propulsion Laboratory's Physical Oceanographic Centre (http://podaac.jpl.nasa.gov/), for the period 1985–2009, for climatological study of the eastern and western gulfs of the Arabian Sea and the Red Sea. SST fields were derived through the Multichannel Sea Surface Temperature (MCSST) algorithm, as given by McClain [42].

2.3. Methodology. For each of the defined areas of study, the monthly images were masked to avoid the influence of the land and clouds using the image processing software ENVI 4.1 and ERDAS 9.0. For analysis all the pixels of the study area were included. Care was taken to exclude the pixels with zero values while averaging. Following Joint Global Ocean Flux Study (JGOFS) [43], the four climatological seasons described here are

(a) northeast monsoon (December–March) (NEM),

(b) spring intermonsoon (April–May) (SIM),

(c) southwest monsoon (June–September) (SWM), and

(d) fall intermonsoon (October–November) (FIM).

The climatological mean (CM_{25}) of 25 years (1985–2009) for each month was calculated by averaging the monthly

mean ($M_{y(i)}$). The interannual variability was analyzed using the monthly normalized anomalies, computed by subtracting the monthly climatological mean (CM_{25}) from the monthly mean ($M_{y(i)}$) of each year, and normalized to the standard deviation for that month (SD_{25}) and given as

$$M_{(NA)} = \frac{M_{y(i)} - CM_{25}}{SD_{25}}.\qquad(1)$$

The seasonal and annual normalized anomalies were computed by averaging the monthly normalized anomalies over appropriate seasons and years. Coefficient of variation (CV) was used to express the magnitude of interannual variability in the annual and the seasonal SST for the eastern and the western gulfs of the Arabian Sea and the Red sea [44] and is calculated as

$$CV = \frac{\text{Standard Deviation}}{\text{Mean}} \times 100.\qquad(2)$$

To study the effect of El Niño and La Niña, comparison with ENSO (El Niño-Southern Oscillation) was carried out using the multivariate ENSO index, that is, MEI index, provided by the climate diagnostic centre (http://www.cdc.noaa.gov/).

3. Results

3.1. Comparison of the Monthly Climatological Mean SST (1985–2009). Figure 2 shows the comparative analysis of the monthly climatological mean SST (1985–2009) of the eastern and western gulfs of the Arabian Sea and the Red Sea. The SST in the Arabian Sea open ocean varies in a typical bimodal pattern with warming during spring inter monsoon (SIM) (April-May) and fall intermonsoon (FIM) (October-November) and cooling during the southwest monsoon (SWM) (June–September) and northeast monsoon (NEM) (December–March) seasons [45–47].

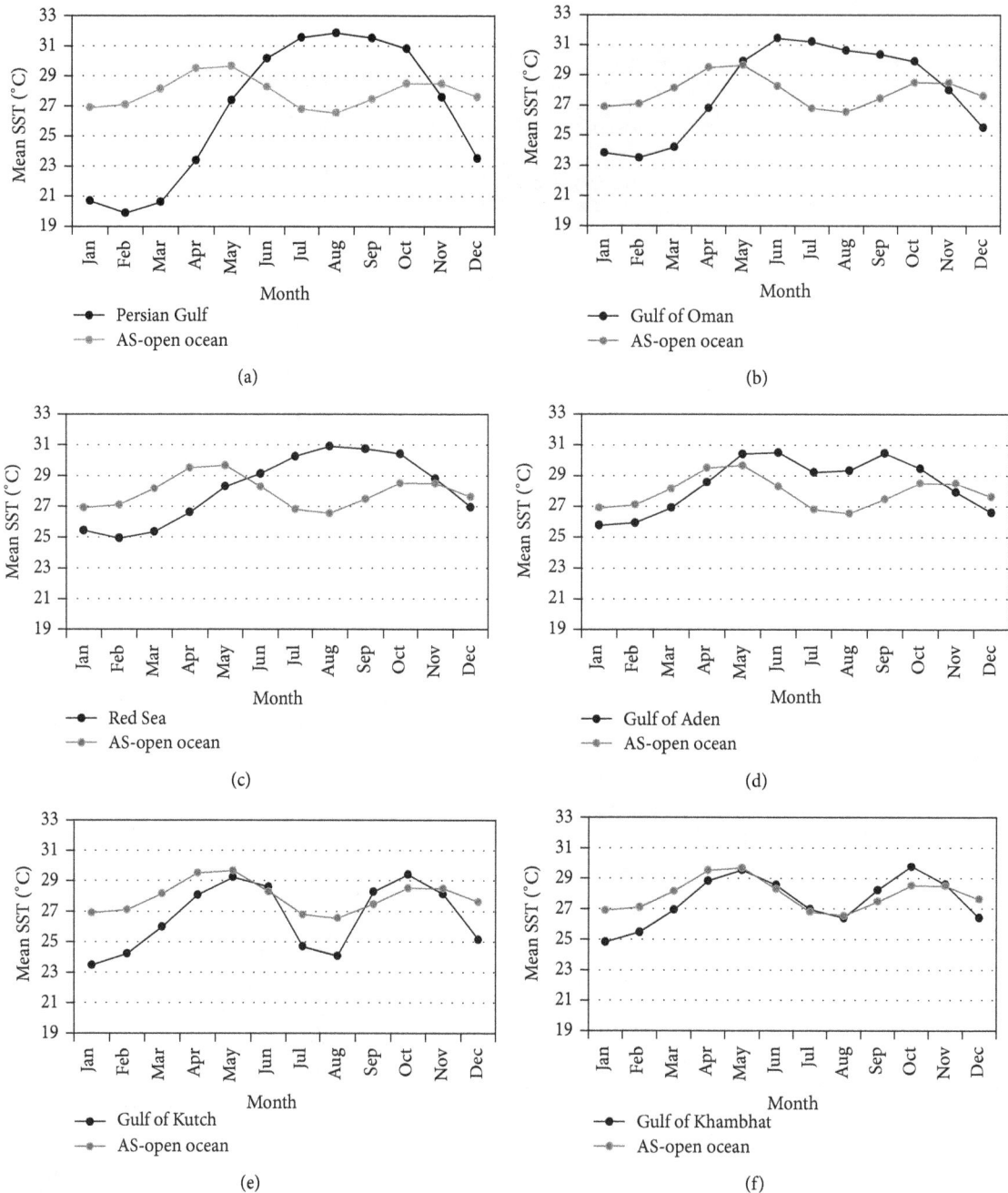

FIGURE 2: Comparison of the monthly climatological mean SST (1985–2009) of the Persian Gulf, Gulf of Oman, Red Sea, Gulf of Aden, Gulf of Kutch, and Gulf of Khambhat with Arabian Sea open ocean.

The Gulf of Aden and the eastern gulfs of the Arabian Sea, namely, Gulf of Kutch and Gulf of Khambhat followed the typical bimodal SST pattern as in the Arabian Sea. However, unlike the Arabian Sea open ocean where the minimum SST is observed during the months of July-August (SWM season), the minimum SST in these gulfs was found in the month of January (NEM season). In the bimodal pattern, the observed second peak of SST in the Gulf of Kutch and Gulf of Khambhat was in the month of October, similar to the Arabian Sea open ocean pattern, but in the Gulf of Aden, the

second peak of SST was observed in the month of September. Climatologically, from November to April, Gulf of Aden was cooler than the Arabian Sea open ocean (by an average 1°C), while from May to October, it was warmer than the Arabian Sea open ocean (by an average 2.02°C). Climatologically, both the eastern gulfs were cooler than the Arabian Sea open ocean (by an average difference of 1.32°C and 0.38°C, resp.), for all the months, except for September and October. Besides, in the months of July and August Gulf of Kutch was found to be cooler than Arabian Sea open ocean months by more than 2°C, whereas in Gulf of Khambhat no such significant difference was found.

The Persian Gulf, Gulf of Oman, and the Red Sea exhibited a distinct unimodal SST curve instead of the typical bimodal pattern as in the Arabian Sea open ocean, with minimum temperature during the NEM season and maximum temperature during the SWM season. Whereas the minimum temperature for these three water bodies was in February, the maximum temperature in the Persian Gulf and the Red sea was in August, whereas in the Gulf of Oman, the maximum temperature was in the month of June. Climatologically from November to May, the Persian Gulf, Gulf of Oman, and the Red sea were found to be cooler than the Arabian Sea open ocean (by an average 4.9°C, 2.2°C, and 1.9°C, resp.), while from June to October they were warmer than the Arabian Sea open ocean (by an average 3.7°C, 2.7°C and 2.08°C, resp.). Amongst the eastern and western gulfs of the Arabian Sea and the Red Sea, the Persian Gulf exhibited the highest range of SST (11.97°C).

3.2. Interannual Variability of SST in the Eastern and Western Gulfs of Arabian Sea and Red Sea from 1985 to 2009

3.2.1. Comparison of the Annual Normalized SST Anomaly from 1985 to 2009. A time series study of the interannual variability of SST can be seen from Figure 3. A warming trend was observed in all the gulfs of the Arabian Sea in terms of the increasing annual normalized SST anomalies from 1985 to 2009. Whereas in the eastern gulfs of the Arabian Sea, the annual increase of normalised SST anomaly from 1985 to 2009 was insignificant with the Gulf of Kutch and Gulf of Khambhat showing an increase of 0.19/decade and 0.1/decade, respectively, in the western gulfs of the Arabian Sea and the Red sea, a sharp increase in SST was observed with the maximum rate of increase of normalized SST anomaly in the Red Sea (0.64/decade) followed by the Persian Gulf (0.4/decade).

A remarkable time lag in the rise of annual SST was found between the eastern and the western gulfs of the Arabian Sea and the Red Sea. For the western gulfs of the Arabian Sea and the Red Sea, a sharp increase in the annual SST was observed from 1992-1993 to 2009. However, for the eastern gulfs, the warming was delayed till 1995, and hence instead of 1992-1993, the sharp rise of annual SST was found from 1995 onwards.

For the Persian Gulf, Gulf of Oman, and Gulf of Aden, a decrease in annual normalized anomaly (cooling) was observed from 1985 to 1991, and thereafter from 1992 to 2009, a sharp increase (warming) was found. In the Persian Gulf, the normalized anomaly decreased (by −1.61), from −0.25 in the 1985 to −1.36 in 1991 and thereafter increased (by +1.48) to +0.12 in 2009. In the Gulf of Oman normalized anomaly decreased (by −1.0) from −0.44 (1985) to −1.44 (1991) and later increased (by +1.0) to +0.45 (2009). However, in the Red Sea, the normalized anomaly decreased from 1985 to 1992, (instead of 1991), from −0.76 (1985) to −1.53 (1992), and thereafter increased to +0.64 (2009).

In the Gulf of Kutch and Gulf of Khambhat (eastern gulfs of the A rabian Sea), a significant decrease in normalized anomaly (cooling) was observed from 1985 to 1994, and thereafter from 1995 to 2009 a sharp increase in normalized anomaly (warming) was found. In the Gulf of Kutch and Gulf of Khambhat, the normalized anomaly decreased from −0.10 in 1985 to −0.92 in 1994 and subsequently increased to +0.75 in 2009. Similarly, in the Gulf of Khambhat the normalized anomaly decreased from −0.22 in 1985 to −0.75 in 1994 and later increased to +0.58 in 2009.

In the three western gulfs of the Arabian Sea, namely, Persian Gulf, Gulf of Oman, and Gulf of Aden, the maximum negative normalized SST anomaly was observed in 1991 (−1.36, −1.44, and −1.16, resp.). In the Red Sea the maximum negative normalized SST anomaly was found in 1992 (−1.5). On the other hand in both the eastern gulfs of the Arabian Sea, maximum negative normalized SST anomaly was observed in 1994 (−0.93 in Gulf of Kutch and −0.75 in Gulf of Khambhat). The maximum positive normalized SST anomaly was found to be in 1998 in Gulf of Aden (+1.54), in 1999 in Persian Gulf (+1.1), in 2001 in Red Sea (+1.1), and in 2002 in Gulf of Gulf of Oman, whereas in both the eastern gulfs of the Arabian Sea, maximum positive normalized anomaly was observed in 2007.

In the western gulfs of Arabian Sea and the Red Sea, the influence of the El Niño and the La Niña events can be seen from the crest and the troughs of the graph depicting the annual SST anomaly. For instance the La Niña years like 1985, 1986, 1989, 1996, 1999, 2000, and 2008 were found to have a high negative anomalies (cooling), and the El Niño years like 1986, 1988, 1991–1995, 1997-1998, 2002, 2006, and 2009 were observed to have a high positive normalized anomalies (warming). However, the influence of the El Niño and La Niña varied in terms of the difference in magnitude of the positive and negative anomalies. Besides, in the eastern gulfs of the Arabian Sea, only few of the El Niño (namely, 1987, 1988, 2002, and 2006) or La Niña (namely, 1989 and 2008) events influenced the warming or cooling of the gulfs.

3.2.2. Comparison of Seasonal Normalized SST Anomaly from 1985-2009. The interannual variation of SST during the four seasons, namely, northeast monsoon (NEM) (December–March), spring intermonsoon (SIM) (April-May), southwest Monsoon (SWM) (June–September), and fall inter Monsoon (FIM) (October-November), affecting the gulfs of Arabian Sea and the Red sea, was analyzed using the normalized SST anomaly indices as in Figure 4.

As seen from Figures 4(a), 4(b), 4(c), and 4(d), a similar interannual variation of SST pattern for the western gulfs of the Arabian Sea (namely, Persian Gulf, Gulf of Oman, and Gulf of Aden) and the Red Sea was observed with an increasing trend of SST in all the four seasons. However, in the eastern gulfs (Gulf of Kutch and Gulf of Khambhat), an increasing pattern of SST was found during the NEM, SIM, and FIM, but during the SWM season, cooling was observed.

During the northeast monsoon (NEM) (December–March), (Figure 4(a)), the normalized SST anomaly increased (warming) in all eastern and western gulfs of the Arabian

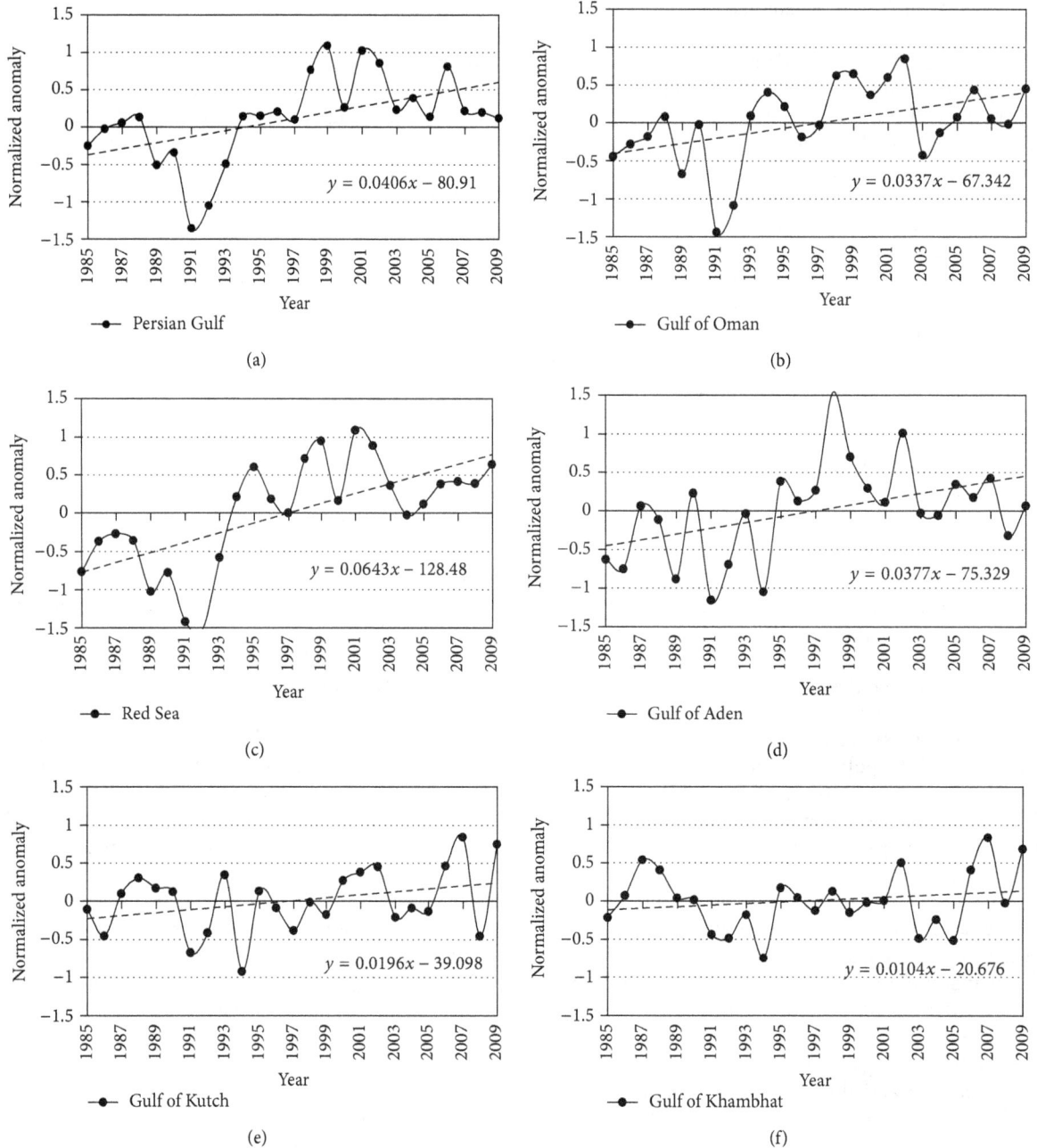

$$y = 0.0406x - 80.91$$

$$y = 0.0337x - 67.342$$

(a)   (b)

$$y = 0.0643x - 128.48$$

$$y = 0.0377x - 75.329$$

(c)   (d)

$$y = 0.0196x - 39.098$$

$$y = 0.0104x - 20.676$$

(e)   (f)

FIGURE 3: Comparison of the annual normalised SST anomaly from 1985 to 2009 of the Persian Gulf, Gulf of Oman, Red Sea, Gulf of Aden, Gulf of Kutch, and Gulf of Khambhat.

Sea and the Red Sea, but the increase was more in the western gulfs and the Red Sea as compared to the eastern gulfs, with the maximum rate of increase of SST in the Red Sea (0.55/decade). A similar interannual variability pattern during the NEM season was observed in the Persian Gulf, Gulf of Oman, and the Red Sea on one hand and Gulf of Aden, Gulf of Kutch, and Gulf of Khambhat on the other. From 1986 to 1992, in the Persian Gulf, Gulf of Oman, and the Red Sea, a decrease in SST anomaly was observed (cooling), which was followed by a phase of increasing SST anomaly (warming) from 1992 to 1999. From 1999 to 2005, the SST anomaly

decreased substantially. However, from 2005 to 2009, again a warming phase with increasing SST anomaly was observed. In the year 1998-1999, an El Niño year, a very high positive SST anomaly was observed in the western gulfs of the Arabian Sea and the Red Sea. The warming was remarkable, as in the Persian Gulf and the Red Sea, the positive anomaly reached up to +2.05 and +1.38, respectively. Even though the Gulf of Oman showed a high positive deviation of +1.5 from the climatological mean in 1999, but its maximum positive anomaly (i.e., maximum warming) was observed during 2002 (+1.92), which was again an El Niño year. Amongst the eastern gulfs

Comparative Analysis of Sea Surface Temperature Pattern in the Eastern and Western Gulfs of Arabian Sea and the Red Sea in Recent Past Using Satellite Data

105

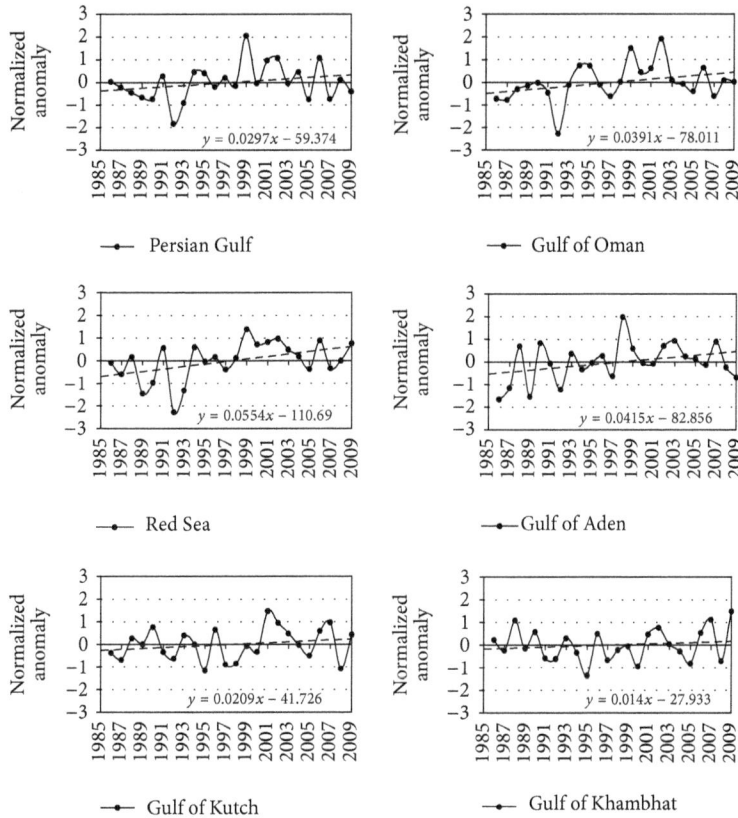

(a) NEM

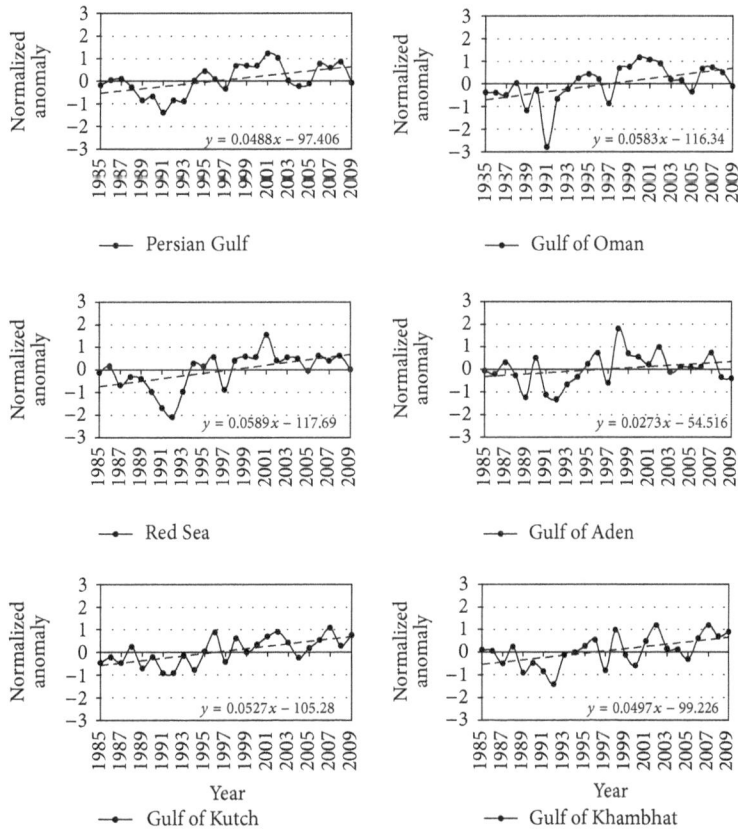

(b) SIM

Figure 4: Continued.

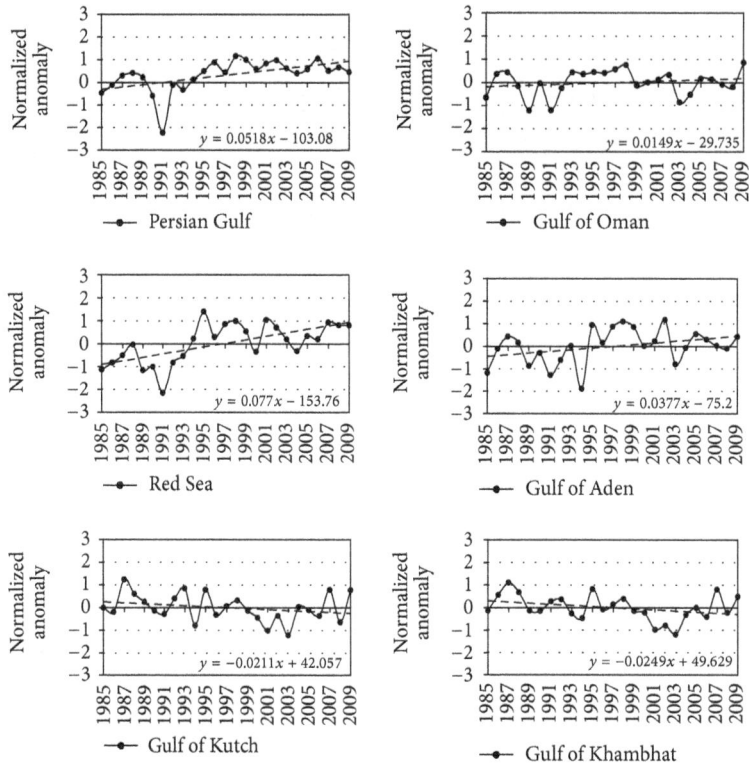

(c) SWM

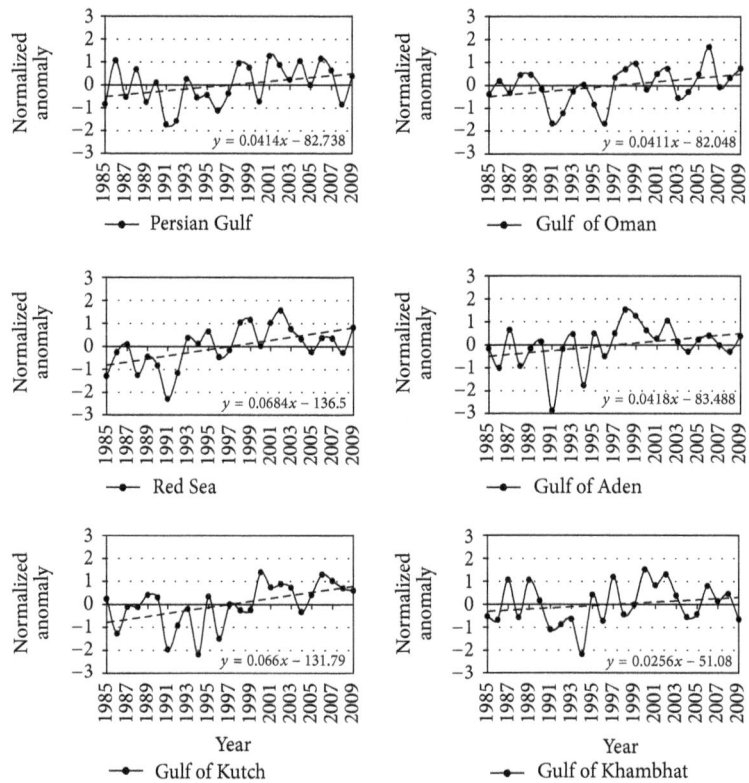

(d) FIM

FIGURE 4: Comparison of the seasonal normalised SST anomaly from 1985 to 2009 of the Persian Gulf, Gulf of Oman, Red Sea, Gulf of Aden, Gulf of Kutch, and Gulf of Khambhat. (a) Northeast monsoon (NEM) (December–March), (b) spring intermonsoon (SIM) (April-May), (c) southwest monsoon (SWM) (June–September), and (d) fall intermonsoon (FIM) (October-November).

of the Arabian Sea and the Gulf of Aden, a similar pattern of interannual variation was observed, wherein from 1986 to 1990, the SST anomaly increased remarkably, followed by a phase of decreasing SST anomaly from 1990 to 1995. However, this cooling phase got extended in the Gulf of Aden up to 1997. From 1995 to 2001-2002, the SST anomaly increased in the two eastern gulfs, whereas in the Gulf of Aden, the increase was noticed up to 2003. From 2001-2002 (2003 in case of Gulf of Aden) to 2005, the SST anomaly again decreased, resulting in a cooling phase. However, from 2005 onwards, like the rest of the western gulfs of the Arabian Sea and the Red Sea, a remarkable increase in SST anomaly was observed in the Gulf of Aden and the two eastern gulfs of the Arabian Sea. In the Gulf of Aden the maximum positive SST anomaly was in 1998 (+1.98). However, in Gulf of Kutch, the maximum positive deviation from the climatological mean was noted during 2001 (+1.47), whereas in Gulf of Khambhat it was observed to be during 2009 (+1.48). During NEM, the Persian Gulf, Gulf of Oman, and the Red Sea exhibited the maximum negative deviation of SST from the climatological mean (i.e., maximum cooling) during 1992 (−1.8, −2.28, and −2.29, resp.), whereas Gulf of Kutch, and Khambhat showed the maximum negative deviation observed during 1995 (−1.15 and −1.36, resp.).

During the spring intermonsoon (SIM) (April-May) (Figure 4(b)), the normalized SST anomaly increased (warming) in all the eastern and the western gulfs of the Arabian Sea and the Red Sea, with maximum increase in the Gulf of Oman and the Red Sea (at the rate of 0.58/decade). It was observed that from 2005 onwards, during the SIM, the normalized SST anomalies increased substantially in the Persian Gulf, Gulf of Oman, Red Sea, Gulf of Kutch, and Gulf of Khambhat. The interannual variability pattern of the eastern gulfs of the Arabian Sea and the Gulf of Aden was found to be similar, beginning with a cooling phase (decreasing SST anomaly) from 1985 to 1991-1992, followed by a warming phase (increasing SST anomaly) which continued up to 1998. Whereas in the Gulf of Aden, the next phase of cooling was noticed from 1998 till 2003, in the eastern gulfs, a time lag of a year was found, with Gulf of Kutch extending the period of cooling from 1998 to 2004 and Gulf of Khambhat extending the period from 1998 to 2005. In the eastern gulfs of the Arabian Sea, the maximum positive deviation of SST from the climatological mean (i.e., maximum warming) during SIM was in 2002. Amongst the rest of the western gulfs of the Arabian Sea and the Red Sea, the pattern of interannual variability of SST was observed to be similar during the SIM. In the Persian Gulf, Gulf of Oman, and the Red Sea, from 1985 to 1991-1992, a cooling phase (with decreasing SST anomaly) was found. It was followed by a decade of warming from 1991-1992 to 2001, when the SST anomaly increased significantly. In these gulfs and the Red Sea, from 2001 onwards till 2005, another cooling phase was observed. However, in the last 5 years, that is, from 2005 to 2009 a substantial warming was noticed with significant increase in the SST anomaly during this period. Whereas the maximum positive deviation of SST from the climatological mean in the Red Sea and the Persian Gulf was during 2001 (+1.23 and +1.55, resp.), in the Gulf of Aden, it was in 1998 (+1.8) and in the Gulf of Oman in 2000 (+1.15). The

maximum negative deviation of SST from the climatological mean (i.e., maximum cooling) in the Persian Gulf and the Gulf of Oman was in 1991 (−1.4 and −2.8, resp.), whereas in the Red Sea, Gulf of Aden, Gulf of Kutch, and Gulf of Khambhat, it was in 1992 (−2.1, −1.32, −0.92, and −1.41, resp.).

During the southwest monsoon season (SWM) (June–September) (Figure 4(c)), a cooling trend was observed in the eastern gulfs of the Arabian Sea, whereas in the western gulfs a warming trend was found in the last 25 years, with the maximum increase of SST anomaly in the Red Sea (0.77/decade). A similar pattern of SST anomaly was observed in the eastern gulfs of Arabian Sea on one hand and the western gulfs of Arabian Sea and the Red Sea on the other hand. However, a careful examination revealed that in the last 6 years (2003–2009) a substantial warming has been taking place in both the eastern and western gulfs of the Arabian Sea during the SWM season. In the eastern and western gulfs of Arabian Sea and the Red Sea, the SST anomaly increased from 1985 to 1987-1988, followed by a phase of decreasing anomaly, which continued until 1990-1991. During the extended El Niño period of 1991–1995, a sharp rise in SST anomaly was observed in the eastern gulfs of the Arabian Sea, after which SST decreased till 2003. On the other hand in the western gulfs of the Arabian Sea and the Red Sea, the SST anomaly increased substantially from 1991 till 1998 (with maximum positive anomaly in 1998). However, similar to the eastern gulfs, the SST anomalies again decreased from 1998 to 2003. But from 2003 onwards, a sharp rise of SST was observed till 2009, in both the eastern gulfs and the western gulfs of the Arabian Sea and the Red sea, with the maximum positive deviation of SST anomaly in the year 2007. The Persian Gulf showed positive SST anomalies in all the years from 1993 to 2009. Even in the Red Sea from 1993 onwards the anomalies have been positive, except for the years 2000 and 2004. The years with strong positive SST anomaly (more than +1.0) during SWM season were 1995, 1997, 1998, and 2002 in the Gulf of Aden; 1995, 1997, 1998, 2001, 2007, and 2009 in the Red Sea; 1998, 2002, and 2006 in the Persian Gulf; 1987 and 1993 in the Gulf of Kutch; and 1987 in the Gulf of Khambhat. None of the years showed strong positive SST anomalies (more than +1.0) in the Gulf of Oman. On the other hand, the years with strong negative SST anomaly (more than −1.0) during SWM season were 1985, 1991, and 1994 in the Gulf of Aden; 1985, 1989, 1990, and 1991 in the Red Sea; 1991 in the Persian Gulf; 1989, 1991, and 2003 in the Gulf of Oman; 2001 and 2003 in the Gulf of Kutch and the Gulf of Khambhat.

During the fall intermonsoon (FIM) (October-November), an increasing trend of the normalized anomalies for the eastern and the western gulfs of the Arabian Sea and the Red sea (Figure 4(d)) was observed. The Red Sea and the Gulf of Kutch had a much sharper increase in positive SST anomalies (0.68/decade and 0.66/decade, resp.) than the rest of the gulfs. For both the eastern gulfs of the Arabian Sea a similar interannual variability pattern was observed. In the Gulf of Kutch and Gulf of Khambhat, the SST anomaly increased from 1985-1986 to 1989-90 followed by a substantial cooling, which continued until 1994 (with maximum negative anomaly in 1994). From 1994 to 2000, a warming phase was followed by a cooling phase, which continued up to the year

2004 in the two eastern gulfs. However, in the last 4 years, that is, from 2004 till 2008, the SST anomaly again increased, with 2006 showing a high SST anomaly. Overall in the Gulf of Kutch and Gulf of Khambhat, a sharp rise of normalized anomalies (warming) was found from 1994 onwards, though the rise was more pronounced in the Gulf of Kutch, in which, from 1997 onwards, (except for 2004) none of the years showed negative departure from the climatological mean during this season. On the other hand, in the western gulfs of the Arabian Sea and the Red Sea, even though the pattern of interannual variability during the FIM season was similar, a significant difference in the magnitude of variability was noticed especially in the Persian Gulf and the Red Sea, owing to their enclosed locations. In the Persian Gulf, Gulf of Oman, Red sea, and Gulf of Aden the normalized SST anomalies decreased from 1985 to 1991 (cooling) and there-after increased (warming) remarkably until 1998-1999. Even-tually from 1998-1999 to 2003-2004, the SST anomalies again decreased significantly. However, similar to the eastern gulfs of the Arabian Sea, the western gulfs and the Red Sea also exhibited warming from 2003-2004 to 2009. During the FIM, in the Persian Gulf and the Gulf of Oman, apart from 1991, 1996 also showed intense cooling with negative SST anomaly reaching up to −1.67.

The influence of El Niño and La Niña on the interannual SST pattern during different seasons was observed. However, not all the warming and cooling events had similar effect on the SST anomaly. Besides, the geographical location of the particular gulf also played a crucial role in determining the impact of the ENSO.

3.2.3. Coefficient of Variation of Annual and Seasonal SST of the Eastern and Western Gulfs of Arabian Sea and Red Sea from 1985 to 2009.

The coefficient of variation (CV) gives the magnitude of the interannual variability. CV values of the annual and seasonal SST of the eastern and the western gulfs of the Arabian Sea and the Red Sea showed large variation amongst these water bodies as seen in Table 1.

The largest variation in annual SST was found in the Persian Gulf (mean CV of 18.02%), followed by the Gulf of Oman (mean CV of 11.1%), whereas Gulf of Aden was least variable (mean CV of 6.37%). The interannual variability in seasonal mean SST was highest in the Persian Gulf in all the seasons except for the SWM season. During SWM season, the maximum variability was observed to be in the Gulf of Kutch (mean CV of 10.07%) followed by Gulf of Khambhat (mean CV of 4.9%). Gulf of Aden and Gulf of Oman were least variable during the NEM and SWM seasons, whereas Gulf of Khambhat was least variable during SIM and FIM seasons.

3.3. Comparison of Normalised SST Anomaly of the Eastern and Western Gulfs of Arabian Sea and Red Sea with Multi-variate ENSO Index and Role of Indian Ocean Dipole.

The El Niño/Southern Oscillation (ENSO) is a phenomenon that integrates both atmospheric and oceanic parameters and is measured by various indices like the Southern Oscillation Index (SOI), Multivariate ENSO Index (MEI), Nino 3.4, and

TABLE 1: % Coefficient of variation for annual and seasonal SST.

Gulf/Sea	Annual	NEM	SIM	SWM	FIM
Persian Gulf	18.02%	7.88%	11.77%	2.68%	7.77%
Gulf of Oman	11.1%	3.85%	8.37%	1.93%	4.61%
Red Sea	8.02%	3.56%	4.65%	2.76%	3.83%
Gulf of Aden	6.37%	2.20%	4.68%	2.63%	3.84%
Gulf of Kutch	9.35%	5.06%	3.63%	10.07%	3.14%
Gulf of Khambhat	6.5%	4%	2.13%	4.90%	2.83%

so forth. [48–51]. El Niño and La Niña result from interaction between the ocean and the atmosphere in the tropical Pacific. The warm El Niño to cold La Niña conditions oscillate on an average of every three to five years. Of all these indices, MEI, a representative of SST, sea surface pressure, zonal and merid-ional components of the surface wind, air temperature and cloudiness of the sky, gives more structured information about ENSO phenomena and its correlation with surface temperature [52]. Hence, in this study, MEI was used to find the relationship between ENSO events and the change in the SST pattern in the gulfs of the Arabian Sea as seen in Figure 5.

Besides ENSO, the Indian Ocean Dipole (IOD) is also known to be affecting the interannual SST anomalies. One phase of the IOD (positive IOD or PIOD) causes a cooling of SST in the eastern tropical Indian Ocean, while the western tropical Indian Ocean (western Arabian Sea) tends to experi-ence a warming of sea surface temperatures. The other phase (negative IOD or NIOD) in contrast involves high SST in the eastern Indian Ocean and low in the west [49, 53, 54]. Some dipole events occur without ENSO events, and some ENSO events do not accompany the dipole events. When an ENSO event occurs, the dipole tends to be positive during El Niño and negative during La Niña. However, the amplitude of the dipole varies greatly from year to year. 1991, 1994, 1997-1998 are reported to be the positive IOD years, whereas 1992, 1996, and 1998 are reported to be negative IOD years [55].

The major El Niño events of 1986-1987, 1991-1992, 1997-1998, 2002-2003, 2006-2007, and 2009-2010 and the La Niña events of 1988-1989, 1998-1999, 2000-2001, and 2007-2008 are evident from the peaks and troughs of the MEI values, with positive MEI values indicating a warming event (El Niño) and the negative MEI value indicating a cooling event (La Niña). The influence of the ENSO on SST of the Gulfs of the Arabian Sea can be clearly seen. However, one to one connection was not found. Besides, not all the El Niño or La Niña events had similar impact in the eastern and the western gulfs of the Arabian Sea. For example, the El Niño of 1991-1992 and 1994-1995 did not influence the eastern gulfs of the Arabian Sea as compared with the one of 1997-1998. The year 1997-1998 was also a positive IOD year. Hence, the extent of warming especially in the western gulfs of the Arabian Sea and the Red Sea was high. Besides, in 1992, a high negative anomaly in the western gulfs and the Red Sea was observed, owing to the influence of the negative IOD. Moreover, the impact of an IOD was seen more in the western domains of the Arabian Sea. During the extended El Niño of 1991–1995, the normalized positive SST anomaly increased remarkably in all the gulfs except the Gulf of Kutch and

Comparative Analysis of Sea Surface Temperature Pattern in the Eastern and Western Gulfs of Arabian Sea and
the Red Sea in Recent Past Using Satellite Data

109

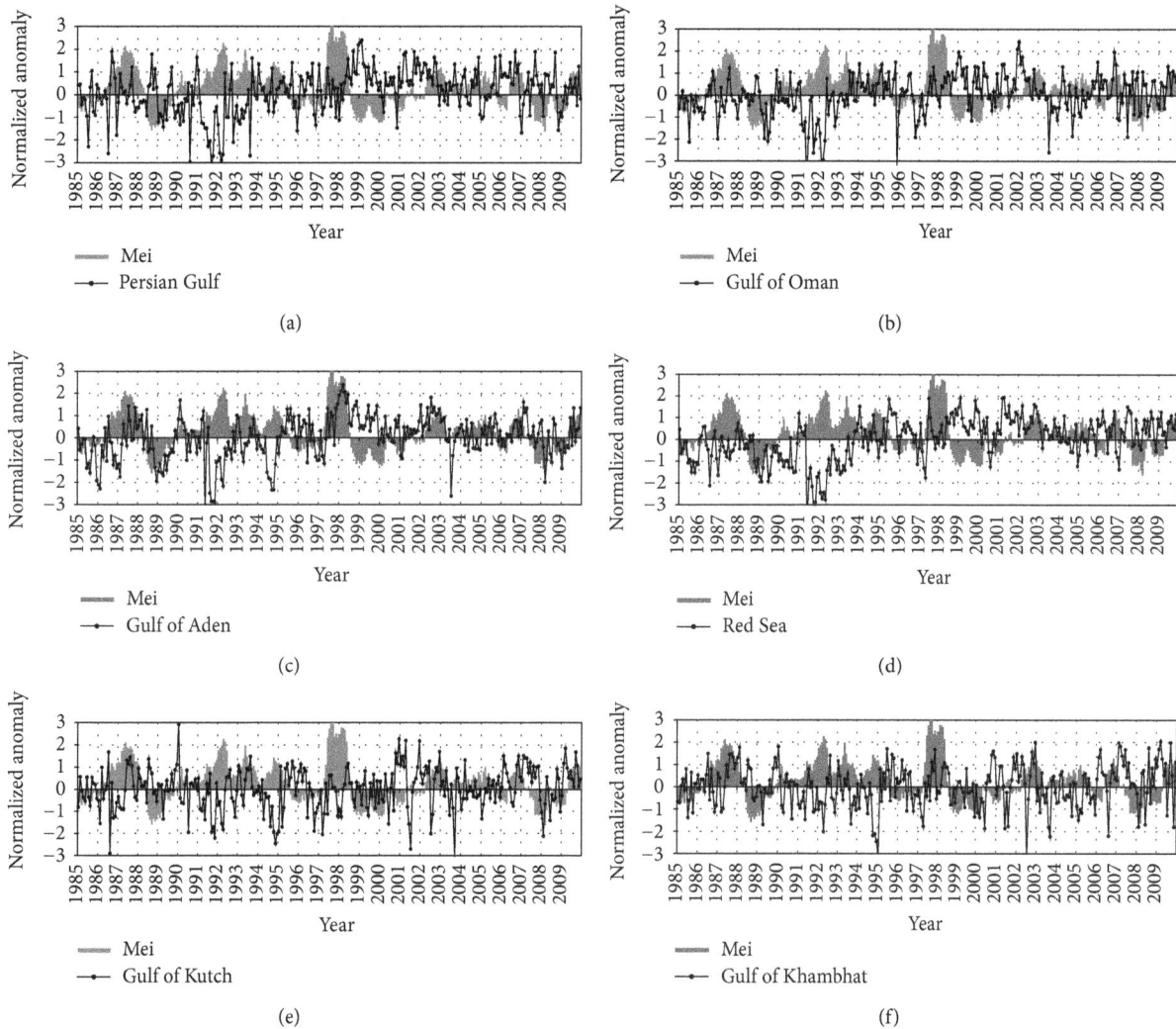

FIGURE 5: Comparison of normalised SST anomaly of (a) Persian Gulf, (b) Gulf of Oman, (c) Gulf of Aden, (d) Red Sea, (e) Gulf of Kutch, and (f) Gulf of Khambhat, from 1985 to 2009 along with Multivariate ENSO Index (MEI).

Gulf of Khambhat. Similarly, the stronger El Niño of 1997-1998 had a major impact on all gulfs except for the Gulf of Kutch. However, a delayed effect was observed in the Persian Gulf, Gulf of Oman, and Red Sea with peaks of normalised anomalies in 1998-1999. The La Niña of 1988-1989 resulted in cooling of Gulf of Oman, Red Sea, and Gulf of Aden but had no influence in the Gulfs of Kutch and Khambhat, whereas the La Niña of 2008 affected all the gulfs except for the Red Sea.

4. Discussion

Time series studies of the oceanographic parameters are useful for understanding seasonal, interannual, and decadal impact of climate change. Numerous investigations have been conducted in the Arabian Sea basin to study the effect of global warming. Rupa Kumar et al. [17] reported that from 1904 to 1994 the Arabian Sea has warmed by 0.5°C. Recent studies have also confirmed about the warming of the

Arabian Sea by 0.16°C/decade from 1971 to 2002 [18] and 0.10°C/decade from 1982 to 2006 [56]. Kailasam and Rao [20] using GISS Surface Temperature observed that there has been an increasing trend of SST, with strong positive SST anomalies (0.8°C) during the period of 1981–2009 over the Northern Indian Ocean.

However, as the gulfs of the Arabian Sea have unique oceanographic features and are quite distinct from the open ocean, it becomes essential to study them individually. These gulfs are single water masses, and therefore studies conducted in few of the coastal areas cannot be generalized for the whole of the gulfs. In the present study, a comparative analysis of SST in different months and seasons was carried out for the eastern and western gulfs of the Arabian Sea and the Red Sea from 1985 to 2009 to study the regional effect of the warming of the Arabian Sea. Annually the SST has increased significantly in all the western gulfs of the Arabian Sea and the Red sea, with maximum increase of normalized SST anomaly in the Red sea (0.64/decade) and the Persian Gulf (0.40/decade).

However, the increase was not significant for the eastern gulfs of the Arabian Sea (Gulf of Kutch and Gulf of Khambhat).

The analysis by Kumar et al. [19] about the anthropogenic induced global warming disrupting the decadal cycle of SST of Arabian Sea after 1995 was also observed in the present study in the eastern gulfs of the Arabian Sea (Gulf of Kutch and Gulf of Khambhat), where the annual SST increased rapidly from 1995 onwards. However, in the western gulfs (Persian Gulf, Gulf of Oman, and Gulf of Aden) and the Red Sea, the disruption of the decadal cycle of SST resulting in a sharp rise of SST was observed from the year 1991-1992. Therefore, a remarkable time lag of (3 years) between the eastern and the western gulfs of the Arabian Sea and the Red sea was observed in the annual increase of SST. There was similarity in the interannual variability pattern of SST amongst the western gulfs and the Red sea on one hand and the eastern gulfs on the other.

Annually for western gulfs of the Arabian Sea, that is, the Persian Gulf, Gulf of Oman, and Gulf of Aden, cooling was observed from 1985 to 1991, and thereafter from 1992 to 2009, warming was found. Whereas in the eastern gulfs of the Arabian Sea, that is, the Gulf of Kutch and Gulf of Khambhat, a significant cooling was observed from 1985 to 1994, and thereafter from 1995 to 2009, a sharp increase in normalized anomaly was found. In the three western gulfs of the Arabian Sea, (Persian Gulf, Gulf of Oman, and Gulf of Aden), the maximum negative normalized SST anomaly was observed in 1991. In the Red Sea the maximum negative normalized SST anomaly was found in 1992 (negative IOD). On the other hand in both the eastern gulfs of the Arabian Sea, maximum negative normalized SST anomaly was observed in 1994. In the western gulfs of Arabian Sea and the Red Sea, 1997-1998, 1999, 2001, 2002, 2006, 2007, and 2009 were the years with high positive normalized SST anomalies, whereas in the eastern gulfs of the Arabian Sea, the years 2002, 2007, and 2009 showed strong positive SST anomalies. Most of these years are reported to be the El Niño years like 1997-98, 2002, 2006, and 2009 which affected the SST variability pattern.

The seasonal SST analysis for the western gulfs of the Arabian Sea and the Red Sea showed an increasing trend of SST in all the four seasons. However, in the eastern gulfs of the Arabian Sea during the SWM season, cooling was observed. During the NEM and the SIM seasons, similar interannual SST variability pattern was observed in the Persian Gulf, Gulf of Oman, and the Red Sea on one hand and Gulf of Aden, Gulf of Kutch and Gulf of Khambhat on the other. During the SWM and FIM seasons similar interannual SST variability pattern was in the eastern gulfs of Arabian Sea on one hand and the western gulfs of Arabian Sea and the Red Sea on the other.

During the NEM season, in the Persian Gulf, Gulf of Oman, and the Red Sea, an alternate cycle of cooling and warming phases was observed from 1986 to 1992, 1992 to 1999, 1999 to 2005, and 2005 to 2009, whereas in the eastern gulfs of the Arabian Sea and the Gulf of Aden, the alternate cycle of cooling and warming phases was observed from 1986 to 1990, 1990 to 1995 (1997 in Gulf of Aden), 1995 to 2001-2002 (2003 in Gulf of Aden), 2001-2002 to 2005, and

2005 to 2009. The Persian Gulf, Gulf of Oman, and the Red Sea exhibited maximum cooling during 1992, owing to the negative IOD. During the NEM season, the years 1987, 1989, and 2008 showed the influence of La Niña, with high negative SST anomalies in the western gulfs of the Arabian Sea and the Red Sea. But in the eastern gulfs of the Arabian Sea, only the influence of La Niña of 2008 was visible. Similarly of all the El Niño years, 1991–1995, 2002, and 2006 resulted in a higher positive SST anomaly in the eastern and the western gulfs of the Arabian Sea and the Red sea.

Similarly, during the SIM season, alternate warming and cooling phases were noticed in the eastern gulfs of the Arabian Sea and the Gulf of Aden from 1985 to 1991-1992, 1991-1992 to 1998, 1998 to 2004-2005 (2003 for Gulf of Aden) and 2004-2005 to 2009. In the eastern gulfs of the Arabian Sea, the maximum positive SST anomaly during SIM was in 2002 that was an El Niño year. Amongst the rest of the western gulfs of the Arabian Sea and the Red Sea, the alternate phases of warming and cooling were observed from 1985 to 1991-1992, 1991-1992 to 2001, 2001 to 2005, and 2005 to 2009. The maximum negative deviation of SST from the climatological mean in the Red Sea, Gulf of Aden, Gulf of Kutch, and Gulf of Khambhat was in 1992, owing to the negative IOD. During the SIM, the influence of El Niño of 1991–1995, 1997-1998, 2002 and 2009 and La Niña of 1987 and 1989 could be seen on the SST anomalies of the eastern and the western gulfs and the Red Sea.

In the SWM season, the periods of warming and cooling included 1985 to 1987-1988, 1987-1988 to 1990-1991, 1991 to 1995 (1998 for the western gulfs and the Red Sea), 1995 to 2003, and 2003 to 2009. The Persian Gulf and the Red Sea showed positive SST anomalies in all the years from 1993 to 2009. The years with strong positive SST anomaly were 1995, 1997, 1998, and 2002 in the Gulf of Aden; 1995, 1997, 1998, 2001, 2007, and 2009 in the Red Sea; 1998, 2002 and 2006 in the Persian Gulf; and 1987 and 1993 in the Gulf of Kutch; and 1987 in the Gulf of Khambhat, of which most of the years like 1995, 1997, 1998, 2002, 2006, and 2009 were the El Niño years. However the influence of La Niña was not seen during SWM season. The years with negative SST anomalies included 1985, 1991, and 1994 in the Gulf of Aden; 1985, 1989, 1990, and 1991 in the Red Sea; 1991 in the Persian Gulf; 1989, 1991, and 2003 in the Gulf of Oman; 2001 and 2003 in the Gulf of Kutch and the Gulf of Khambhat.

During the FIM season, an increasing trend of the normalized anomalies for the eastern and the western gulfs of the Arabian Sea and the Red sea was observed. In the eastern gulfs, the alternate phases of increasing and decreasing SST anomaly included 1985-1986 to 1989-1990, 1989-1990 to 1994, 1994 to 2000, 2000 to 2004, and 2004 to 2009. In the western gulfs of the Arabian Sea and the Red Sea the phases of warming and cooling included 1985 to 1991, 1991 to 1998-1999, 1998-99 to 2003-2004, and 2003-2004 to 2009. The years with high negative SST anomalies during the FIM were 1986, 1988, 1991, and 1994 in the Gulf of Aden, whereas in the Red Sea, these were 1985, 1988, and 1991. Of these 1985 and 1988 were the La Niña years. The years with strong positive SST anomalies during the FIM were 1986, 1998, 2001, 2004, and 2006 in the Persian Gulf; 1999 and 2006 in Gulf of Oman; 1998 and 2002 in Gulf of Aden; and 1998, 2001, and 2002 in the Red Sea. Of these 1986, 1988, 2002, and 2006 were the El Niño years.

The contrasting features between the eastern and western gulfs of the Arabian Sea could be attributed to their difference in oceanographic features, current systems, and also anthropogenic impact. Being enclosed in semiarid surroundings in the northern edges of the Indian subtropical zone, the western gulfs of the Arabian Sea come under the direct influence of multiple atmospheric pressure systems like the impact of the Siberian high pressure system, the El Niño, Southern Oscillation, the North Atlantic Oscillation, and the IOD [57]. The influence of the ENSO and the IOD affecting the interannual variability of SST was found. However, one to one connection was not found. Besides, not all the El Niño or La Niña events had similar impact in the eastern and the western gulfs of the Arabian Sea. The year 1997-1998 was an ENSO dipole year. Hence the extent of warming especially in the western gulfs of the Arabian Sea and the Red Sea was high. Besides, in 1992, a high negative anomaly in the western gulfs and the Red Sea was observed, owing to the strong influence of the negative IOD. During the extended El Niño of 1991–1995, the normalized positive SST anomaly increased drastically in all the gulfs except the Gulf of Kutch and Gulf of Khambhat. Similarly, the stronger El Niño of 1997-1998 had a major impact on all gulfs except for the Gulf of Kutch. The atmospheric temperature over the Persian Gulf including the Gulf of Oman has been reported to be increasing rapidly and is projected to further increase up to 4°C by the end of the century [58]. There are reports of the weakening of the Siberian high pressure system resulting in a decline in the zonal component of wind and hence an increase in surface temperature [59]. Besides, the regional orography of the surrounding land masses which includes deserts and mountains also affects the physical oceanographic processes of the western gulfs of the Arabian Sea. However, the most significant contributor for the sharp rise of SST in the western gulfs of the Arabian Sea that has resulted in disruption of the decadal SST cycle much ahead of that of the Arabian Sea basin and the eastern Gulf is the increasing anthropogenic induced warming. There has been tremendous increase in human activities in the countries surrounding the western gulfs of the Arabian Sea which have contributed to the higher rate of warming along these western gulfs of the Arabian Sea as compared to the Arabian Sea basin and its eastern gulfs.

5. Conclusion

Sea surface Temperature (SST) is one of the key oceanographic parameters, exerting an influential role in many of the meteorological and oceanographic processes. From 1985 to 2009, the SST of the eastern and western gulfs of the Arabian Sea and the Red sea showed a clear cut signal of warming. The rate of increase of SST was highest in the Persian Gulf followed by the Red Sea. From 1991-1992 onwards, the western gulfs of the Arabian Sea and the Red sea exhibited a sharp increase in SST, whereas in the eastern gulfs of the Arabian Sea, the warming was more pronounced from 1995 onwards.

Seasonal differences were also found with respect to increasing temperature. Interannual variability pattern was found to be similar in the annual and seasonal SST amongst the western gulfs and the Red Sea. Similarity in SST variability was also found in the two of the eastern gulfs of the Arabian Sea. Red Sea was found to be warming significantly across all the seasons and months. The largest variation in annual SST was found in the Persian Gulf (CV of 18.02%). The interannual variability in seasonal mean SST was highest in the Persian Gulf in all the seasons except for the SWM season. During SWM season, the maximum variability was observed in the Gulf of Kutch (CV of 10.07%) followed by Gulf of Khambhat (CV of 4.9%). The influence of ENSO and IOD on the SST variability pattern was also found. The impact of rising temperature on other physical and biological parameters of the gulfs of the Arabian Sea and the Red Sea needs to be scrutinized further.

The difference in the oceanographic features of the eastern and western gulfs of the Arabian Sea and the Red Sea, from the open oceans, makes them unique ecosystem. Hence, caution is needed to analyze them individually and not to generalize the findings of the entire basin upon them.

Acknowledgments

The authors are thankful to NASA'S Jet Propulsion Laboratory's Physical Oceanographic Centre and Climate Diagnostic Centre for providing the data.

References

[1] S. Solomon, D. Qin, M. Manning et al., Eds., *Climate Change 2007: The Physical Science Basis. Contribution of Working Group I to the Fourth Assessment Report of the Intergovernmental Panel on Climate Change*, Cambridge University Press, Cambridge, UK, 2007.

[2] A. J. Richardson and D. S. Schoeman, "Climate impact on plankton ecosystems in the Northeast Atlantic," *Science*, vol. 305, no. 5690, pp. 1609–1612, 2004.

[3] M. J. Behrenfeld, R. T. O'Malley, D. A. Siegel et al., "Climate-driven trends in contemporary ocean productivity," *Nature*, vol. 444, no. 7120, pp. 752–755, 2006.

[4] R. Ji, C. S. Davis, C. Chen, D. W. Townsend, D. G. Mountain, and R. C. Beardsley, "Influence of ocean freshening on shelf phytoplankton dynamics," *Geophysical Research Letters*, vol. 34, no. 24, Article ID L24607, 2007.

[5] B. S. Halpern, S. Walbridge, K. A. Selkoe et al., "A global map of human impact on marine ecosystems," *Science*, vol. 319, no. 5865, pp. 948–952, 2008.

[6] J. E. Hansen and S. Lebedeff, "Global trends of measured surface air temperature," *Journal of Geophysical Research: Solid Earth*, vol. 92, no. B1, pp. 345–355, 1987.

[7] J. Hansen, "Public understanding of global climate change," in *Carl Sagan's Universe*, Y. Terzian and E. Bilson, Eds., pp. 247–253, Cambridge University Press, New York, NY, USA, 1997.

[8] C. K. Folland, N. A. Rayner, S. J. Brown et al., "Global temperature change and its uncertainties since 1861," *Geophysical Research Letters*, vol. 28, no. 3, pp. 2621–2624, 2001.

[9] N. A. Rayner, D. E. Parker, E. B. Horton et al., "Global analyses of sea surface temperature, sea ice, and night marine air

temperature since the late nineteenth century," *Journal of Geophysical Research: Atmospheres*, vol. 108, no. D14, p. 4407, 2003.

[10] S. C. Doney, "Oceanography: plankton in a warmer world," *Nature*, vol. 444, no. 7120, pp. 695–696, 2006.

[11] F. Montaigne, "The global fish crisis," *National Geographic*, vol. 211, no. 4, pp. 32–99, 2007.

[12] P. D. Jones and T. M. L. Wigley, "Estimation of global temperature trends: what's important and what isn't," *Climate Change*, vol. 100, pp. 59–69, 2007.

[13] D. W. J. Thompson, J. M. Wallace, P. D. Jones, and J. J. Kennedy, "Identifying signatures of natural climate variability in time series of global mean surface temperature: methodology and insights," *Journal of Climate*, vol. 22, pp. 6120–6141, 2009.

[14] S. A. Henson, J. L. Sarmiento, J. P. Dunne et al., "Is global warming already changing ocean productivity?" *Biogeosciences Discussions*, vol. 6, no. 6, pp. 10311–10354, 2009.

[15] J. Hansen, R. Ruedy, M. Sato, and K. Lo, "Global surface temperature change," *Reviews of Geophysics*, vol. 48, no. 4, Article ID RG4004, 2010.

[16] O. P. Singh, "Recent trends in summer temperatures over the North Indian Ocean," *Indian Journal of Marine Sciences*, vol. 29, no. 1, pp. 7–11, 2000.

[17] R. Rupa Kumar, K. Krishna Kumar, R. G. Ashrit, S. K. Patwardhan, and G. B. Pant, "Climate change in India: observations and model projections," in *Climate Change and India: Issues, Concerns and Opportunities*, P. R. Shukla, S. K. Sharma, and P. Venkata Ramana, Eds., Tata McGraw-Hill Publishing Company Limited, New Delhi, India, 2002.

[18] D. R. Kothawale, A. A. Munot, and H. P. Borgaonkar, "Temperature variability over the Indian Ocean and its relationship with Indian summer monsoon rainfall," *Theoretical and Applied Climatology*, vol. 92, no. 1-2, pp. 31–45, 2008.

[19] S. P. Kumar, R. P. Roshin, J. Narvekar, P. K. D. Kumar, and E. Vivekanandan, "Response of the Arabian Sea to global warming and associated regional climate shift," *Marine Environmental Research*, vol. 68, no. 5, pp. 217–222, 2009.

[20] M. K. Kailasam and R. S. Rao, "Impact of global warming on tropical cyclones and monsoons," in *Global Warming*, S. A. Harris, Ed., 2010, http://www.intechopen.com/books/global-warming/impact-of-global-warming-on-indian-monsoons.

[21] C. R. C. Sheppard and A. L. S. Sheppard, "Corals and coral communities of Arabia," *Fauna of Saudi Arabia*, vol. 12, pp. 3–170, 1991.

[22] C. Sheppard, A. Price, and C. Roberts, *Marine Ecology of the Arabian Region: Patterns and Processes in Extreme Tropical Environments*, Academic Press, London, UK, 1992.

[23] R. F. G. Ormond and S. A. Banaimoon, "Ecology of intertidal macroalgal assemblages on the Hadramout coast of southern Yemen, an area of seasonal upwelling," *Marine Ecology Progress Series*, vol. 105, no. 1-2, pp. 105–120, 1994.

[24] M. Goren and M. Dor, *An Updated Checklist of the Fishes of the Red Sea CLOFRES II*, Academy of Sciences and Humanities, Jerusalem, Israel, 1994.

[25] S. Al-Muzaini and P. G. Jacob, "Marine plants of the Arabian Gulf," *Environment International*, vol. 22, no. 3, pp. 369–376, 1996.

[26] J. M. Kemp, "Zoogeography of coral reef fishes of the Gulf of Aden," *Fauna of Arabia*, vol. 18, pp. 293–321, 2000.

[27] B. Riegl, "Climate change and coral reefs: different effects in two high-latitude areas (Arabian Gulf, South Africa)," *Coral Reefs*, vol. 22, no. 4, pp. 433–446, 2003.

[28] S. L. Coles, *Coral Species Diversity and Environmental Factors in the Arabian Gulf and the Gulf of Oman: A Comparison to the Indo-Pacific Region*, vol. 507 of *Atoll Research Bulletin*, National Museum of Natural History Smithsonian Institution, Washington, DC, USA, 2003.

[29] A. M. Dixit, P. Kumar, L. Kumar, K. D. Pathak, and M. I. Patel, *Economic Valuation of Coral Reef Systems in Gulf of Kachchh. Final Report. World Bank Aided Integrated Coastal Zone Management (ICZM) Project*, Gujarat Ecology Commission, 2010.

[30] PERSGA/GEF, *The Red Sea and Gulf of Aden Regional Network of Marine Protected Areas. Regional Master Plan*, PERSGA Technical Series no. 1, PERSGA, Jeddah, Saudi Arabia, 2002.

[31] C. Sheppard, M. Al-Husiani, F. Al-Jamali et al., "The Gulf: a young sea in decline," *Marine Pollution Bulletin*, vol. 60, no. 1, pp. 13–38, 2010.

[32] W. Gladstone, B. Curley, and M. R. Shokri, "Environmental impacts of tourism in the Gulf and the Red Sea," *Marine Pollution Bulletin*, 2012.

[33] W. Hamza and M. Munawar, "Protecting and managing the Arabian Gulf: past, present and future," *Aquatic Ecosystem Health and Management*, vol. 12, no. 4, pp. 429–439, 2009.

[34] C. Sheppard, "The main issues affecting coasts of the Indian and Western Pacific oceans: a meta-analysis from Seas at the Millennium," *Marine Pollution Bulletin*, vol. 42, no. 12, pp. 1199–1207, 2001.

[35] A. N. Al-Ghadban and A. R. G. Price, "Dredging and infilling," in *The Gulf Ecosystem, Health and Sustainability*, N. Y. Khan, M. Munawar, and A. Price, Eds., pp. 207–218, Bakhuys, Leiden, The Netherlands, 2002.

[36] S.-Y. Chao, T. W. Kao, and K. R. Al-Hajri, "A numerical investigation of circulation in the Arabian Gulf," *Journal of Geophysical Research*, vol. 97, no. 7, pp. 11219–11236, 1992.

[37] R. M. Reynolds, "Physical oceanography of the Gulf, Strait of Hormuz, and the Gulf of Oman—results from the Mt Mitchell expedition," *Marine Pollution Bulletin*, vol. 27, pp. 35–59, 1993.

[38] M. A. Al Saafani and S. S. C. Shenoi, "Water masses in the Gulf of Aden," *Journal of Oceanography*, vol. 63, no. 1, pp. 1–14, 2007.

[39] V. S. Kumar and K. A. Kumar, "Waves and currents in tide-dominated location off Dahej, Gulf of Khambhat, India," *Marine Geodesy*, vol. 33, no. 2-3, pp. 218–231, 2010.

[40] E. Kalnay, M. Kanamitsu, R. Kistler et al., "The NCEP/NCAR 40-year reanalysis project," *Bulletin of the American Meteorological Society*, vol. 77, no. 3, pp. 437–471, 1996.

[41] M. Rouault, B. Pohl, and P. Penven, "Coastal oceanic climate change and variability from 1982 to 2009 around South Africa," *African Journal of Marine Science*, vol. 32, no. 2, pp. 237–246, 2010.

[42] P. C. McClain, "Overview of the satellite data applications in the climate and earth sciences laboratory of NOAA'S National Environmental Satellite, data and information services," in *Proceeding of the US-India Symposium-Cum Workshop on Remote Sensing Fundamentals and Applications*, pp. 1–18, 1985.

[43] S. Krishnaswami and R. Nair, "JGOFS (India)—introduction," *Current Science*, vol. 71, pp. 831–833, 1996.

[44] J. Fang, S. Piao, Z. Tang, C. Peng, and W. Ji, "Interannual variability in net primary production and precipitation," *Science*, vol. 293, no. 5536, p. 1723, 2001.

[45] J. G. Colborn, *Thermal Structure of the Indian Ocean*, IIOE Monograph no. 2, An East West Press, 1975.

[46] R. R. Rao, R. L. Molinari, and J. F. Festa, "Evolution of the climatological near-surface thermal structure of the tropical

Comparative Analysis of Sea Surface Temperature Pattern in the Eastern and Western Gulfs of Arabian Sea and the Red Sea in Recent Past Using Satellite Data

113

Indian Ocean. 1. Description of mean monthly mixed layer depth, and sea surface temperature, surface current, and surface meteorological fields," *Journal of Geophysical Research*, vol. 94, no. 8, pp. 10801–10815, 1989.

[47] T. G. Prasad and M. Ikeda, "A numerical study of the seasonal variability of Arabian Sea high-salinity water," *Journal of Geophysical Research C*, vol. 107, no. 11, pp. 18-1–18-12, 2002.

[48] G. T. Walker, "Correlation in seasonal variations of weather," *Memoirs of the India Meteorological Department*, vol. 24, pp. 333–345, 1924.

[49] N. H. Saji, B. N. Goswami, P. N. Vinayachandran, and T. Yamagata, "A dipole mode in the tropical Indian ocean," *Nature*, vol. 401, no. 6751, pp. 360–363, 1999.

[50] D. W. J. Thompson, J. M. Wallace, P. D. Jones, and J. J. Kennedy, "Identifying signatures of natural climate variability in time series of global-mean surface temperature: methodology and insights," *Journal of Climate*, vol. 22, pp. 6120–6141, 2009.

[51] G. Foster, J. D. Annan, P. D. Jones et al., "Comment on "influence of the Southern Oscillation on tropospheric temperature" by J. D. McLean, C. R. de Freitas, and R. M. Carter," *Journal of Geophysical Research: Atmospheres*, vol. 115, no. D9, Article ID D09110, 2010.

[52] K. Wolter and M. S. Timlin, "Monitoring ENSO in COADS with a seasonally adjusted principal component index," in *Proceedings of the 17th Climate Diagnostics Workshop*, pp. 52–57, NOAA/NMC/CAC, NSSL, Oklahoma Climatological Survey, CIMMS and the School of Meteorology, University of Oklahoma, Norman, Okla, USA, 1993.

[53] P. J. Webster, A. M. Moore, J. P. Loschnigg, and R. R. Leben, "Coupled ocean-atmosphere dynamics in the Indian Ocean during 1997-98," *Nature*, vol. 401, no. 6751, pp. 356–360, 1999.

[54] T. Yamagata, S. K. Behera, S. A. Rao, and N. H. Saji, "Comments on 'dipoles, temperature gradients, and tropical climate anomalies,'" *Bulletin of the American Meteorological Society*, vol. 84, no. 10, pp. 1418–1422, 2003.

[55] M. Nagura and M. Konda, "The seasonal development of an SST anomaly in the Indian Ocean and its relationship to ENSO," *Journal of Climate*, vol. 20, no. 1, pp. 38–52, 2007.

[56] I. M. Belkin, "Rapid warming of large marine ecosystems," *Progress in Oceanography*, vol. 81, no. 1–4, pp. 207–213, 2009.

[57] T. Shinoda, H. H. Hendon, and M. A. Alexander, "Surface and subsurface dipole variability in the Indian Ocean and its relation with ENSO," *Deep-Sea Research Part I*, vol. 51, no. 5, pp. 619–635, 2004.

[58] G. A. Meehl, T. F. Stocker, and W. D. Collins, "Global climate projections," in *Climate Change 2007: The Physical Science Basis. Contribution of Working Group I to the Fourth Assessment Report of the Intergovernmental Panel on Climate Change*, S. Solomon, D. Qin, M. Manning et al., Eds., pp. 80–87, Cambridge University Press, New York, NY, USA, 2007.

[59] S. A. Piontkovski, M. Al-Gheilani, P. B. Jupp, R. A. Al-Azri, and A. K. Al-Hashmi, "Interannual changes in the sea of oman ecosystem," *The Open Marine Biology Journal*, vol. 6, pp. 38–52, 2012.

12

Observation of Oceanic Eddy in the Northeastern Arabian Sea Using Multisensor Remote Sensing Data

R. K. Sarangi

Marine Geo and Planetary Sciences Group, Space Applications Centre (ISRO), Ahmedabad 380 015, India

Correspondence should be addressed to R. K. Sarangi, sarangi74@yahoo.com

Academic Editor: Renzo Perissinotto

An oceanic eddy of size about 150 kilometer diameter observed in the northeastern Arabian Sea using remote sensing satellite sensors; IRS-P4 OCM, NOAA-AVHRR and NASA Quickscat Scatterometer data. The eddy was detected in the 2nd week of February in Indian Remote Sensing satellite (IRS-P4) Ocean Color Monitor (OCM) sensor retrieved chlorophyll image on 10th February 2002, between latitude 16°90′–18°50′N and longitude 66°05′–67°60′E. The chlorophyll concentration was higher in the central part of eddy ($\sim$1.5 mg/m^3) than the peripheral water ($\sim$0.8 mg/m^3). The eddy lasted till 10th March 2002. NOAA-AVHRR sea surface temperature (SST) images generated during 15th February-15th March 2002. The SST in the eddy's center ($\sim$23°C) was lesser than the surrounding water ($\sim$24.5°C). The eddy was of cold core type with the warmer water in periphery. Quickscat Scatterometer retrieved wind speed was 8–10 m/sec. The eddy movement observed southeast to southwest direction and might helped in churning. The eddy seemed evident due to convective processes in water column. The processes like detrainment and entrainment play role in bringing up the cooler water and the bottom nutrient to surface and hence the algal blooming. This type of cold core/anti-cyclonic eddy is likely to occur during late winter/spring as a result of the prevailing climatic conditions.

1. Introduction

Eddies are small features with a big impact. They are where a lot of ocean physics happen and are an integral part of our climate system. Coastal eddies have a major role in regulating the weather near the shore and they are important for fisheries. In the open ocean, eddies bring nutrient-rich cold water up to the surface and are an important part of the global carbon cycle [1]. These eddies are rotating masses of water that have broken off from a strong front. Eddies are crucial to the transport of heat, momentum, trace chemicals, biological communities, the oxygen, and nutrients relating to life in the sea [2]. They are also active in air-sea interaction, both through response to weather and in shaping the patterns of warmth that drives the entire atmospheric circulation. Measurements of ocean color and the fate of light in the ocean are extremely useful for describing biological dynamics in surface waters [3–5]; thus the oceanographic communities has made a substantial commitment to remote sensing of ocean color from space [6, 7]. Regional scale ocean color imageries are particularly useful for space-borne studies of ocean margins because of the importance and dominance of mesoscale features (upwelling plumes, eddies, filaments, and river plumes) in continental shelf and slope waters [8]. The oceanic and coastal processes rapidly alter the optical properties of waters and these effects get manifested in the color of the water [9].

Recent study over South China Sea describes the three-dimensional dynamic structure of eddy on the basis of the analyses of *in situ* current, hydrographic measurements, and concurrent satellite observations. That is important in the development, sustenance, and dissipation of cold eddies in the ocean [10]. Oceanic circulation, stratification, and eddies have been studied using scatterometer data [11]. A cyclonic eddy was observed around the mooring location of the Marine Optical Buoy (MOBY) offshore of Lanai Island, Hawaii. The satellite observations such as the sea surface temperature (SST) and the chlorophyll concentration were analyzed and the cold core cyclonic eddy displayed a significant increase in surface chlorophyll [12]. Eddies are thus usually associated with high biological activity, as reported in several papers during the past decade [13–20]. However, the eddies role in the biological and biogeochemical processes remain enigmatic, partially because of the challenges of *in situ* eddy measurements due to the spontaneity in eddy generation and present technological limitations [21]. Chlorophyll-a (Chl-a) concentration, an index of phytoplankton biomass, is the most important property of the marine ecosystem. Remote sensing images of ocean color, converted into Chl-a concentration, provide a window into the ocean ecosystem with synoptic scales. It is a promising approach for understanding the oceanic biological and physical processes and for monitoring ocean waters [22–24].

The Arabian Sea is unique among low-latitude sea by terminating at latitude of $25°N$ and being under marked continental influence. It is the region of monsoon and its biological production is affected by the physical processes altering the vertical flux of nutrients in the mixed layer [25]. Satellite-based observations on ocean color provides a tool to monitor the biological productivity in terms of phytoplankton concentration. Remote sensing images of ocean color, converted into chl-a concentration, provide a window into the ocean ecosystem with synoptic scales [2]. The biological processes are being manifested by the physical events in the sea, like their circulation patterns and other surface features caused due to wind and ocean currents. These physical processes are relevant to ocean color studies. Remote sensing technique offers a practical tool in overcoming the problems like regular *in situ* observations of physical and biological processes linked to surface ocean and relevant data analysis [26, 27].

2. Objectives of the Study

(i) Analysis of the IRS-P4 OCM data for the northeastern Arabian Sea during the period February-March 2002 and detection of eddy using chlorophyll images and monitoring the phases of the eddy;

(ii) analysis of the NOAA-AVHRR data for the northeastern Arabian Sea during the period February-March 2002 to retrieve Sea Surface Temperature (SST) images and correlate the OCM derived chlorophyll images with SST images;

(iii) analyze the NASA Quickscat scatterometer retrieved wind speed and wind vector data for the above period over northeast Arabian Sea and interpret the direction and magnitude of the eddy.

3. IRS-P4 OCM Sensor

The ocean color monitor (OCM) of the Indian Remote Sensing satellite IRS-P4 is optimally designed for the estimation of chlorophyll in coastal and oceanic waters, detection and monitoring of phytoplankton blooms, studying the suspended sediment dynamics, and the characterization of the atmospheric aerosols. IRS-P4 satellite carrying onboard remote sensing payload Ocean Color Monitor (OCM) was launched successfully by Polar Satellite Launch Vehicle (PSLV) on May 26, 1999 from Sriharikota, India. The

TABLE 1: Technical characteristics of IRS-P4 OCM payload.

Spectral range	404–882 nm
no. of channels	8
	Channel 1: 404–423 (340.5)
	Channel 2: 431–451 (440.7)
	Channel 3: 475–495 (427.6)
Wavelengths range (nm) and Signal to noise ratio (SNR)	Channel 4: 501–520 (408.8)
	Channel 5: 547–565 (412.2)
	Channel 6: 660–677 (345.6)
	Channel 7: 749–787 (393.7)
	Channel 8: 847–882 (253.6)
Satellite altitude (km)	720
Spatial resolution (m)	360 × 236
Swath (km)	1420
Repeativity	2 days
Quantisation	12 bits
Equatorial crossing time	12 noon
Along track steering (to avoid sunglint)	20°

technical specifications of the OCM sensor are mentioned in Table 1 [28].

4. OCM Data Processing

The retrieval of ocean color parameters, such as phytoplankton pigment (chlorophyll-a) in oceanic waters, involves two major steps, the first being atmospheric correction of visible channels to obtain normalized water leaving radiances in shorter wavelengths and second application of the bio-optical algorithm for retrieval of phytoplankton pigment concentration.

4.1. Atmospheric Correction of the IRS-P4 OCM Imagery. In the ocean remote sensing, the signal received at the satellite altitude is dominated by radiances contribution through atmospheric scattering processes and only 8–10% signal corresponds to oceanic reflectance. Therefore, it has been mandatory to correct the atmospheric effect to retrieve any quantitative parameter from space. The OCM scenes were corrected for atmospheric effects of Rayleigh and aerosol scattering using an approach called long wavelength atmospheric correction method. The approach used the two near infra red channels at 765 and 865 nm to correct for the contribution of molecular and aerosol scattering in visible wavelengths at 412, 443, 490, 510, and 555 nm [29–31]. The water leaving radiances derived from atmospheric correction procedure was used to compute chlorophyll-a pigment concentration.

4.2. Chlorophyll Algorithm. A number of bio-optical algorithms for retrieval of chlorophyll have been developed to relate measurements of ocean radiance to the *in situ* concentrations of phytoplankton pigments. O'Reilly and Maritorena et al. [32] proposed an empirical algorithm (also known as Ocean Chlorophyll 2 or OC2) and operated successfully on SeaWiFS ocean color data. This algorithm captures the inherent sigmoid relationship. The algorithm operates with five coefficients and has following mathematical form:

$$C = -0.071 + 10^{(0.319 - 2.336*X + 0.879*X^2 - 0.135*X^3)}, \quad (1)$$

where C is chlorophyll concentration in mg/m^3 and $X = \log_{10}[R_{rs}490/R_{rs}555]$. R_{rs} is remote sensing reflectance. This algorithm has been presently used for generating the chlorophyll maps, using IRS-P4 OCM-derived water leaving radiances. The chlorophyll retrieval accuracy obtained is within ±30% [33].

5. Dataset Used

Fourteen date chlorophyll images were generated using the above mentioned procedure from the cloud-free passes of IRS-P4 OCM sensor during the period February-March 2002 covering the offshore water in the northeast Arabian Sea. The consecutive cloud-free passes were selected during February 10, 2002 to March 10, 2002. The chlorophyll images were geometrically corrected and gridded with 3.0 degree latitude and longitude interval. The path 9 and row 13, 14 scenes of IRS-P4 OCM data were processed and chlorophyll images were generated for each day and two pass images were mosaiced. Similarly, NOAA-AVHRR data were processed during the above period to retrieve and visualize the SST images. The Quickscat data product received from PO.DAAC, JPL/NASA website were processed and plotted to retrieve the wind speed and wind vector products during February 15, 2002 to March 15, 2002.

6. Methodology for SST Retrieval

The SST retrieval using AVHRR Channels 4 and 5 was performed using following three steps.

6.1. Radiometric Calibration. It was done to convert the channel 4 and 5 digital data into brightness temperatures. It was assumed that the output of each channel (in counts) is a linear function to sensed radiances. The function is defined as

$$N = G * X + I, \quad (2)$$

where N is the radiance of target at count value X, G and I are gain and intercept, respectively. The brightness temperature (T_b) was obtained using Planck's radiation equation:

$$T_b = \frac{C_2 n}{\ln[1 + (C_1 n^3/N)]}, \quad (3)$$

where, $C_1 = 1.1910659 * 10^{-5}$ Cm·K and $C_2 = 1.1438833$ Cm·K.

6.2. SST Computation. Using McClain's algorithm [34–36]

$$SST = 3.6569 * T_{11} - 2.6705 * T_{12} - 268.92, \quad (4)$$

where, T_{11} and T_{12} are the brightness temperatures for the central wavelength 11 and 12 μm of the channels 4 and 5. The NOAA-AVHRR retrieved SST has been validated for Indian Ocean region and found to be within well accuracy range, ±0.5°C [37].

6.3. Geometric Correction. It is performed using the Ground Control Points (GCPs) resampled from the master image to the AVHRR image and Mercator projection is applied.

7. Results and Discussion

The eddy was detected in the 2nd week of February in IRS-P4 OCM derived chlorophyll image on 10th February 2002 covering latitudes 16°90′–18°50′N and longitudes 66°05′–67°60′E (Figure 1). During this period, several phytoplankton bloom features were seen in the northeast Arabian Sea. The chlorophyll concentration was found to be higher in the central part of the eddy ($\sim$1.5 mg/m^3) than the peripheral water around the eddy ($\sim$0.8 mg/m^3) as observed in the image. Analysis has been done with three locations along the eddy region (Eddy core, 19.58°N and 68.27°E; eddy periphery, 19.60°N and 67.94°E and outside eddy, 19.00°N and 69.21°E). There has been observation of increase in the chlorophyll concentration (1.10 mg/m^3) in the eddy core and in the periphery region as compared to the outer region of eddy (Figure 2). With the movement of the eddy, the outer region observed to be spread along with eddy region and chlorophyll concentration in late stage indicates the dissipation of eddy. The chlorophyll concentration decreases to 0.14 mg/m^3 (Figure 2). The eddy appeared as a circular concentric structure which lasted for a month period till 10th March 2002. The eddy was spread for about 150 km diameter. The eddy appeared to be cold core type as the central part was more productive may be due to lesser temperature than its peripheral water. The eddy was at its peak during 22nd February–6th March 2002 and then subsequently its size and diameter reduced. The surrounding water showed the disappearance of bloom patches.

To study the relationship of the eddy with sea surface temperature (SST), NOAA Advanced very high-resolution radiometer (AVHRR) sensor-retrieved SST images were studied during the period 15th February 2002–15th March 2002 (Figure 3). The SST images detected the eddy's appearance with lower temperature in the eddy's center ($\sim$23°C) than the water surrounding the eddy ($\sim$24.5°C). There was a significant SST gradient of 1.5°C observed in the SST images, around the eddy. The feature was consistent during a month period as compared to the IRS-P4 OCM-retrieved chlorophyll images. So, the eddy was confirmed to be cold core type with the warmer surrounding water in periphery. The movement of eddy appeared to be in clockwise direction and it was an anticyclonic eddy. These results were being interpreted with NASA Quickscat scatterometer retrieved wind vector data (Figure 4). The average wind speed was about 10 cm/sec

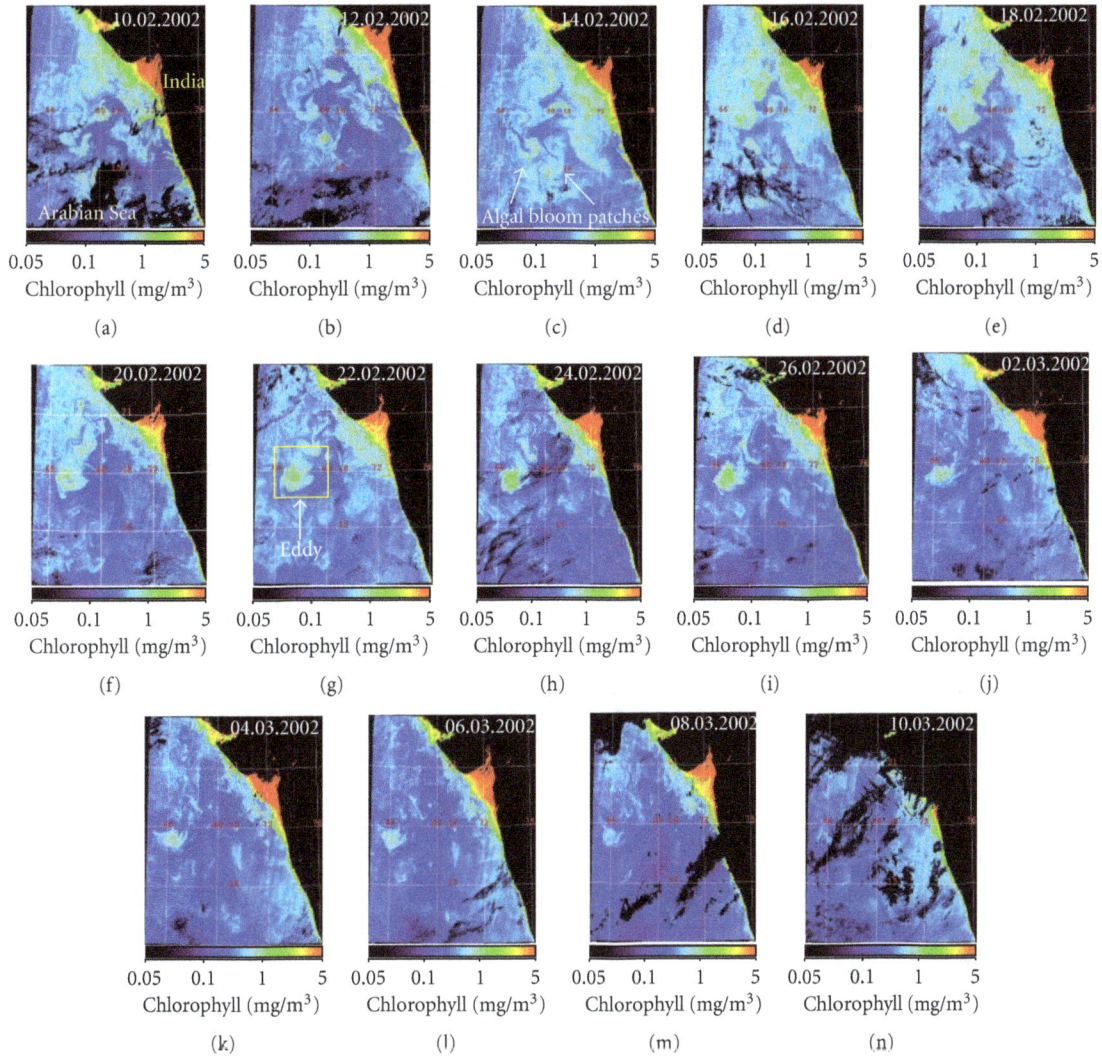

FIGURE 1: IRS-P4 OCM-derived chlorophyll images for February-March 2002 showing the eddy in different stages in the northeast Arabian Sea.

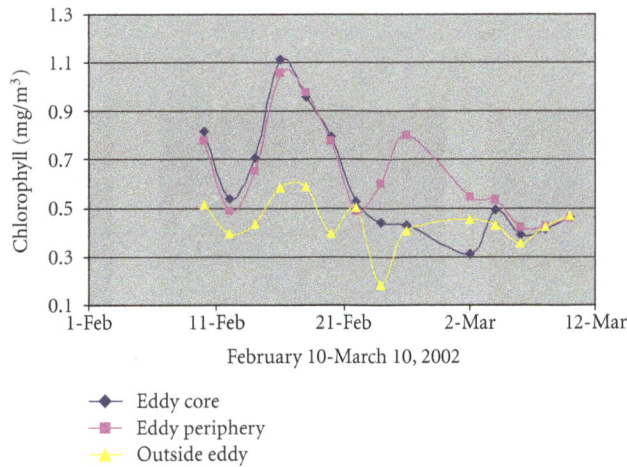

FIGURE 2: Observation of OCM derived chlorophyll concentration along the eddy region.

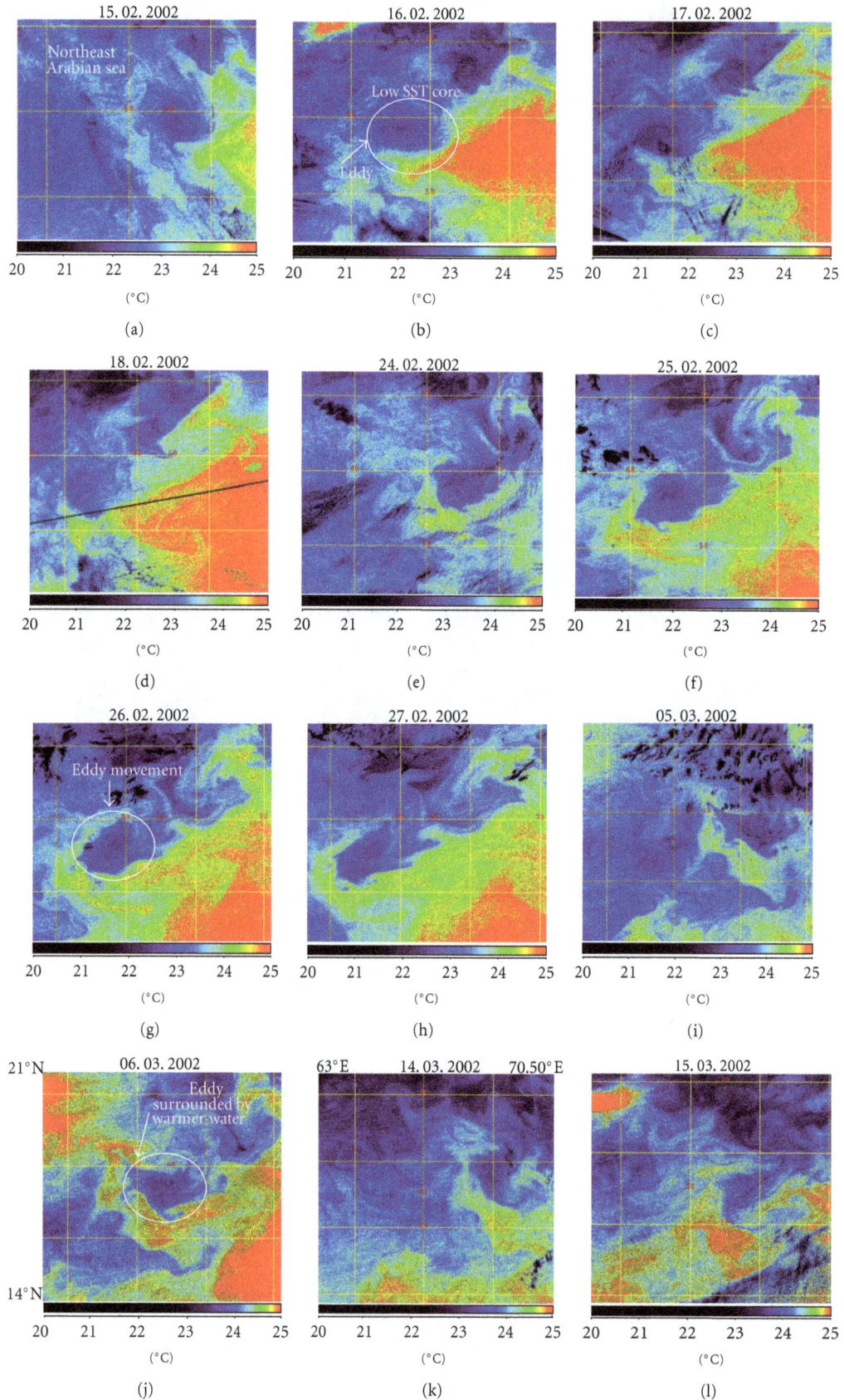

FIGURE 3: NOAA-AVHRR retrieved Sea Surface Temperature (SST) images showing oceanic eddy and its subsequent stages in the Arabian Sea during February-March 2002.

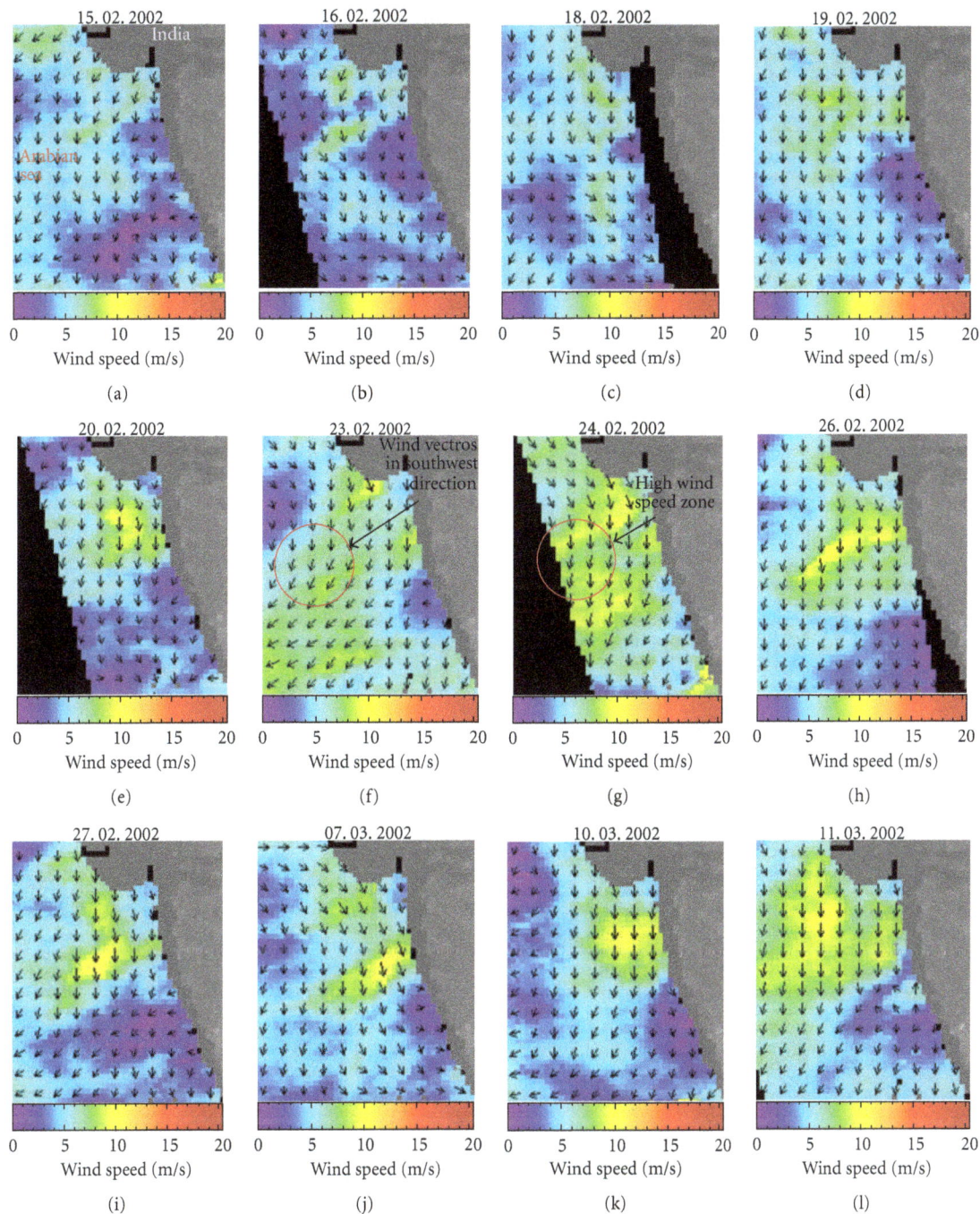

FIGURE 4: NASA-Quickscat retrieved wind vector data over northern Arabian Sea during February-March 2002.

along the eddy. Wind speed was seen high around 8–10 m/sec in the eddy and surrounding zone, but the wind vector was not seen prominently. As the eddy was of small scale and of lower magnitude, the wind vector of comparatively lower resolution has not made much impact and was not identified in the 360 meter high-resolution OCM–derived eddy region 65–67°E latitude and 17-18°N longitude. Still the high wind speed (8–10 m/sec) has been observed around the above zone in the study area, northeast Arabian Sea. The

wind vector showed its movement in southeast to southwest direction, which might have helped in churning the surface water and eddy formation. The increase in the chlorophyll concentration (0.80 to 1.50 mg/m^3) at the core of eddy was evident, where the SST was decreasing (24.5°C to 23°C) and the wind speed was high (8–10 m/sec). These were the prevailing conditions and observations along the eddy. The link between the chlorophyll, SST, and wind speed/vector has been understood and hence the validity of biological productivity in the northeast Arabian Sea, during this winter cooling period of northeast monsoon.

The northern Arabian Sea is one of the most biologically productive ocean regions [38]. Dynamics and thermodynamics of the surface layer of the Arabian Sea are dominated by the monsoon-related annual cycle of air-sea momentum and heat fluxes [39]. The northern Arabian Sea is renowned for complicated flow pattern consisting of several eddies [40] and the water mass structure in the Arabian Sea is complex [41]. During the northeast monsoon (November–February), the winds blow from the northeast and have maximum wind stress magnitudes about 2 dyne·cm^{-2} [39]. The circulation pattern consists of several eddies and meanders, with a pronounced anticyclonic eddy around 24°N and 64°E. In general, eddies appear to have appreciable deep vertical extension [42]. The eddy circulation is attributed to the influence of bottom topography, which is marked by depressions and rises [43]. Upwelling in the eastern Arabian Sea occurs during southwest monsoon during July–September [39]. During the winter, the northeast monsoon during February, the convection process tends to generate eddies and brings up bottom water with nutrient rich water to make the region cooler and productive [44] and even the wind plays a major role to enhance the cooling process and algal blooming [45]. So, the eddy has relation to convection and wind pattern as well, so those eddies can be termed as convective eddies.

The winter convection in the northeastern Arabian sea leads to elevated pigment values, as does even modest mixing during southwest (summer) monsoon. This is due to winter cooling due to high evaporation, mixing, and convection processes leading to the nutrient injection to the surface layer. During winter the northern Arabian Sea (especially north of 10°N) experiences a net heat loss, about 140 W/m^2 [44]. The cool dry continental air brought by the prevailing north-east monsoon winds enhances evaporation, leading to surface cooling [38]. Apart from the cooling due to evaporation, the decrease in insolation (net short wave radiation) from 220 W/m^2 during October to about 160 W/m^2 during January also contributes to the further cooling of the surface waters [46]. Thus, the observed reduction in SST (by about 4°C) [47] and creation of a deep mixed layer depth (MLD), in the northern Arabian Sea during winter is forced by combination of enhanced evaporation and decrease in the insolation. In fact, this starts during December and persists till the end of February [44]. Accordingly, the northern Arabian Sea, north of 15°N, experiences cooling and densification, which leads to sinking and convective mixing and injects nutrients into the surface layers from thermocline region. Off India, the bloom occurs towards the beginning of the February because the mixed layer detrains earlier than the other regions. This occurs when the mixed layer detrains after a period of entrainment, during which the layer is thick enough to inhabit phytoplankton growth. Detrainment occurs when there is a decrease in turbulent mixing (when the wind weakens or there is surface heating). The deeper mixed layer which often being present after winter time cooling. As fluid detrains from the mixed layer, a fossil layer is created that thickens in time. Detrainment blooms tend to be highly productive because detrainment results in

a thin mixed layer, which intensifies the depth-averaged light intensity and favors phytoplankton growth. The bloom-forming features of the phytoplankton due to eddy observed in the present study are in good agreement with the previous works carried out by several researchers in the same regions, along the northern Arabian Sea during the winter monsoon period [44, 48]. This type of cold core/anticyclonic eddy is likely to occur regularly during late winter/spring as a result of the prevailing climatic conditions at that time of the year.

Entrainment of water reduces the surface water column temperature [45]. During summer monsoon (July–September), upwelling/Ekman pumping/lateral advection processes are mostly relevant to the eastern Arabian Sea. But, during the winter/northeast monsoon (January–March), it is the convection and entrainment/detrainment processes forced by intraseasonal wind induces the eddy, as evident from the current study. It is observed that in the northeast Arabian Sea the eddy is formed due to convective processes from bottom and water column, during the winter monsoon period in February. The processes like detrainment and entrainment play role in bringing up the cooler water and the bottom nutrient to surface and hence the algal blooming and increase in the chlorophyll concentration. The information on oceanic eddies are very much required in studies like acoustic propagation, optimum ship route planning, fishery forecasting and zoning, delineating good/bad monsoon years, heat transport, and so forth.

8. Conclusion

The present study on the detection and monitoring the phases of eddy in the Arabian Sea water gives an insight into the utilization of multisensor satellites (IRS-P4 OCM, NOAA-AVHRR, NASA-Quickscat) data-retrieved multiparameters like the chlorophyll, SST, and wind vector field respectively. The eddy is well understood as a cold core type from the IRS-P4 OCM multidate chlorophyll images interpretation and is well supported from the NOAA-AVHRR-retrieved SST images information. The rotational movement of the eddy was appearing in clockwise direction in chlorophyll images, which was confirmed with the information obtained from the Quickscat-retrieved wind vectors, as clockwise moving anticyclonic eddy. The increase in the chlorophyll concentration at the core of eddy was evident, where the SST was decreasing and the wind speed was high. The link between the three parameters (chlorophyll, SST and wind) has been understood and the enhancement in the biological productivity has been resulted with effect of the cold core eddy and its churning phases. The eddy observed to be regulated by the processes like convection and hence the entrainment and detrainment to enhance the surface water chlorophyll with the bottom water nutrients. So, the detailed knowledge about the eddies, their types, circulation patterns, and finally their role in regulating the marine living resources up to the secondary production in terms of the increase in fish catch details in the Arabian Sea water, in different regions and seasons are essential.

Acknowledgments

The authors are thankful to the Director of, Space Applications Centre for providing necessary guidance and facilities for carrying out the work. The Quickscat wind vector data provided by PO.DAAC, JPL/NASA are sincerely acknowledged.

References

[1] Yi and Chao, http://www.jpl.nasa.gov/releases/2002/release_2002_175.html.

[2] D. Tang, H. Kawamura, and A. J. Luis, "Short-term variability of phytoplankton blooms associated with a cold eddy in the northwestern Arabian Sea," *Remote Sensing of Environment*, vol. 81, no. 1, pp. 82–89, 2002.

[3] C. S. Yentsch, "The influence of phytoplankton pigments on the colour of sea water," *Deep Sea Research*, vol. 7, no. 1, pp. 1–9, 1960.

[4] C. J. Lorenzen, "Extinction of light in the ocean by phytoplankton," *Journal du Conseil International pour l'Exploration de la Mer*, vol. 34, pp. 262–267, 1972.

[5] R. C. Smith, J. Marra, M. J. Perry et al., "Estimation of a photon budget for the upper ocean in the Sargasso Sea," *Limnology & Oceanography*, vol. 34, no. 8, pp. 1673–1693, 1989.

[6] J. Aiken, G. F. Moore, and P. M. Holligan, "Remote sensing of oceanic biology in relation to global climate change," *Journal of Phycology*, vol. 28, no. 5, pp. 579–590, 1992.

[7] B. G. Mitchell, "Coastal Zone Color Scanner retrospective," *Journal of Geophysical Research*, vol. 99, no. 4, pp. 7291–7292, 1994.

[8] J. A. Yoder, "Chapter 12 An overview of temporal and spatial patterns in satellite-derived chlorophyll-a imagery and their relation to ocean processes," *Elsevier Oceanography Series*, vol. 63, pp. 225–238, 2000.

[9] R. K. Sarangi, P. Chauhan, and S. R. Nayak, "Phytoplankton bloom monitoring in the offshore water of northern Arabian Sea using IRS-P4 OCM satellite data," *Indian Journal of Marine Sciences*, vol. 30, no. 4, pp. 214–221, 2001.

[10] J. Hu, J. Gan, Z. Sun, J. Zhu, and M. Dai, "Observed three-dimensional structure of a cold eddy in the southwestern South China Sea," *Journal of Geophysical Research*, vol. 116, Article ID C05016, pp. 1–11, 2011.

[11] C. Chavanne, P. Flament, R. Lumpkin, B. Dousset, and A. Bentamy, "Scatterometer observations of wind variations by oceanic islands: implications for wind driven ocean circulation," *Journal of Remote Sensing*, vol. 28, pp. 466–474, 2002.

[12] C. Dong, T. Mavor, F. Nencioli et al., "An oceanic cyclonic eddy on the lee side of Lanai Island, Hawaii," *Journal of Geophysical Research C*, vol. 114, no. 10, Article ID C10008, 13 pages, 2009.

[13] T. D. Dickey, F. Nencioli, V. S. Kuwahara et al., "Physical and bio-optical observations of oceanic cyclones west of the island of Hawai'i," *Deep-Sea Research Part II*, vol. 55, no. 10–13, pp. 1195–1217, 2008.

[14] V. S. Kuwahara, F. Nencioli, T. D. Dickey, Y. M. Rii, and R. R. Bidigare, "Physical dynamics and biological implications of Cyclone Noah in the lee of Hawai'i during E-Flux I," *Deep-Sea Research Part II*, vol. 55, no. 10–13, pp. 1231–1251, 2008.

[15] S. L. Brown, M. R. Landry, K. E. Selph, E. Jin Yang, Y. M. Rii, and R. R. Bidigare, "Diatoms in the desert: Plankton community response to a mesoscale eddy in the subtropical North Pacific," *Deep-Sea Research Part II*, vol. 55, no. 10–13, pp. 1321–1333, 2008.

[16] M. R. Landry, M. Decima, M. P. Simmons, C. C. S. Hannides, and E. Daniels, "Mesozooplankton biomass and grazing responses to Cyclone *Opal*, a subtropical mesoscale eddy," *Deep-Sea Research Part II*, vol. 55, no. 10–13, pp. 1378–1388, 2008.

[17] C. R. Benitez-Nelson, R. R. Bidigare, T. D. Dickey et al., "Mesoscale eddies drive increased silica export in the subtropical Pacific ocean," *Science*, vol. 316, no. 5827, pp. 1017–1021, 2007.

[18] R. R. Bidigare, C. Benitez-Nelson, C. L. Leonard et al., "Influence of a cyclonic eddy on microheterotroph biomass and carbon export in the lee of Hawaii," *Geophysical Research Letters*, vol. 30, no. 6, article 1318, 4 pages, 2003.

[19] R. D. Vaillancourt, J. Marra, M. P. Seki, M. L. Parsons, and R. R. Bidigare, "Impact of a cyclonic eddy on phytoplankton community structure and photosynthetic competency in the subtropical North Pacific Ocean," *Deep-Sea Research Part I: Oceanographic Research Papers*, vol. 50, no. 7, pp. 829–847, 2003.

[20] M. P. Seki, J. J. Polovina, R. E. Brainard, R. R. Bidigare, C. L. Leonard, and D. G. Foley, "Biological enhancement at cyclonic eddies tracked with GOES thermal imagery in Hawaiian waters," *Geophysical Research Letters*, vol. 28, no. 8, pp. 1583–1586, 2001.

[21] T. D. Dickey and R. R. Bidigare, "Interdisciplinary oceanographic observations: the wave of the future," *Scientia Marina*, vol. 69, supplement 1, pp. 23–42, 2005.

[22] J. A. Yoder, C. R. McClain, G. C. Feldman, and W. E. Esaias, "Annual cycles of phytoplankton chlorophyll concentrations in the global ocean: a satellite view," *Global Biogeochemical Cycles*, vol. 7, no. 1, pp. 181–193, 1993.

[23] H. Kawamura and OCTS Team, "OCTS mission overview," *Journal of Oceanography*, vol. 54, no. 5, pp. 383–399, 1998.

[24] D. L. Tang, I. H. Ni, D. R. Kester, and F. E. Müller-Karger, "Remote sensing observations of winter phytoplankton blooms southwest of the Luzon Strait in the South China Sea," *Marine Ecology Progress Series*, vol. 191, pp. 43–51, 1999.

[25] K. Banse and D. C. English, "Geographical differences in seasonality of CZCS-derived phytoplankton pigment in the Arabian Sea for 1978–1986," *Deep-Sea Research Part II*, vol. 47, no. 7-8, pp. 1623–1677, 2000.

[26] G. Zibordi, F. Parmiggiani, and L. Alberotanza, "Application of aircraft multispectral scanner data to algae mapping over the venice lagoon," *Remote Sensing of Environment*, vol. 34, no. 1, pp. 49–54, 1990.

[27] P. Lavery, C. Pattiaratchi, A. Wyllie, and P. Hick, "Water quality monitoring in estuarine waters using the Landsat thematic mapper," *Remote Sensing of Environment*, vol. 46, no. 3, pp. 268–280, 1993.

[28] R. R. Navalgund and A. S. Kiran Kumar, "IRS-P4 Ocean Color Monitor (OCM)," Technical Note, 1999.

[29] H. W. Gordon and M. Wang, "Retrieval of water-leaving radiance and aerosol optical thickness over the oceans with SeaWiFS: a preliminary algorithm," *Applied Optics*, vol. 33, no. 3, pp. 443–452, 1994.

[30] M. Mohan, P. Chauhan, M. Raman et al., "Initial results of MOS validation experiment over the Arabian sea, atmospheric correction aspects," in *Proceedings of the 1st International Workshop on MOS-IRS and Ocean Color*, pp. 1–11, DLR, Berlin, Germany, 1997.

[31] P. Chauhan, C. R. C. Nagur, M. Mohan, S. R. Nayak, and R. R. Navalgund, "Surface chlorophyll-a distribution in Arabian Sea and Bay of Bengal using IRS-P4 Ocean Color Monitor (OCM) satellite data," *Current Science*, vol. 80, pp. 40–41, 2001.

[32] J. E. O'Reilly, S. Maritorena, B. G. Mitchell et al., "Ocean color chlorophyll algorithms for SeaWiFS," *Journal of Geophysical Research C*, vol. 103, no. 11, pp. 24937–24953, 1998.

[33] P. Chauhan, M. Mohan, R. K. Sarngi, B. Kumari, S. Nayak, and S. G. P. Matondkar, "Surface chlorophyll a estimation in the Arabian Sea using IRS-P4 Ocean Colour Monitor (OCM) satellite data," *International Journal of Remote Sensing*, vol. 23, no. 8, pp. 1663–1676, 2002.

[34] E. P. McClain, "Overview of satellite data applications in the climate and earth sciences laboratory of NOAA's National Environment al Satellite, data and information services," in *Proceedings of the US-Indo Symposium-cum-Workshop on Remote Sensing Fundamentals and Applications*, p. 247, Ahmedabad, India, 1985.

[35] E. P. McClain, W. G. Pichel, and C. C. Walton, "Comparative performance of AVHRR-based multichannel sea surface temperatures," *Journal of Geophysical Research*, vol. 90, no. 6, pp. 11587–11601, 1985.

[36] C. C. Walton, "Nonlinear multichannel algorithms for estimating sea surface temperature with AVHRR satellite data," *Journal of Applied Meteorology*, vol. 27, pp. 115–124, 1988.

[37] M. V. Rao, A. Narendra Nath, T. D. Suhasini, D. Jogendra Nath, and K. H. Rao, "A comparison of satellite MCSST with ship measured SST in the North Indian Ocean," *Geocarto International*, vol. 12, no. 2, pp. 13–22, 1997.

[38] M. Madhupratap, S. P. Kumar, P. M. A. Bhattathiri et al., "Mechanism of the biological response to winter cooling in the northeastern Arabian Sea," *Nature*, vol. 384, no. 6609, pp. 549–552, 1996.

[39] S. R. Shetye, A. D. Gouveia, and S. S. C. Shenoi, "Circulation and water masses of the Arabian Sea," in *Biogeochemistry of the Arabian Sea*, D. Lal, Ed., pp. 9–25, Indian Academy of Sciences, Phototypeset at Thomson Press, New Delhi, India, 1994.

[40] E. Böhm, J. M. Morrison, V. Manghnani, H. S. Kim, and C. N. Flagg, "The Ras al Hadd Jet: remotely sensed and acoustic Doppler current profiler observations in 1994-1995," *Deep-Sea Research Part II*, vol. 46, no. 8-9, pp. 1531–1549, 1999.

[41] J. M. Morrison, L. A. Codispoti, S. Gaurin, B. Jones, V. Manghnani, and Z. Zheng, "Seasonal variation of hydrographic and nutrient fields during the US JGOFS Arabian Sea Process Study," *Deep-Sea Research Part II*, vol. 45, no. 10-11, pp. 2053–2101, 1998.

[42] S. Z. Qasim, "Oceanography of the northern Arabian Sea," *Deep Sea Research Part A*, vol. 29, no. 9, pp. 1041–1068, 1982.

[43] V. K. Das, A. D. Gouveia, and K. K. Varma, "Circulation and water characteristics on isanosteric surfaces in the northern Arabian Sea," *Indian Journal of Marine Sciences*, vol. 9, pp. 156–165, 1980.

[44] S. Prasanna Kumar and T. G. Prasad, "Winter cooling in the northern Arabian Sea," *Current Science*, vol. 71, no. 11, pp. 834–841, 1996.

[45] K. Banse and D. C. English, "Geographical differences in seasonality of CZCS-derived phytoplankton pigment in the Arabian Sea for 1978–1986," *Deep-Sea Research Part II*, vol. 47, no. 7-8, pp. 1623–1677, 2000.

[46] S. Hastenrath and P. J. Lamb, *Climatic Atlas of the Indian Ocean. Part I: Surface Climate and Atmospheric Circulation*, University of Wisconsin Press, Madison, Wis, USA, 1979.

[47] S. Hastenrath and L. L. Greischar, *Climatic Atlas of the Indian Ocean, Part III: Upper–Ocean Structure*, The University of Wisconsin Press, Madison, Wis, USA, 1989.

[48] K. Banse and C. R. McClain, "Winter blooms of phytoplankton as observed by the Coastal Zone Color Scanner," *Marine Ecology Progress Series*, vol. 34, pp. 201–211, 1986.

13

Eocene (Lutetian) Shark-Rich Coastal Paleoenvironments of the Southern North Sea Basin in Europe: Biodiversity of the Marine Fürstenau Formation Including Early White and Megatooth Sharks

C. G. Diedrich

PaleoLogic, Nansenstraße 8, D-33790 Halle, Germany

Correspondence should be addressed to C. G. Diedrich, cdiedri@gmx.net

Academic Editor: Roberto Danovaro

The Fürstenau Formation (Lutetian, Paleogene, Eocene) is based on type sections near Fürstenau in Germany (central Europe) and is built of 22 meter thick marine glauconitic and strongly bioturbated sands, clays, and a vertebrate-rich conglomerate bed. The conglomerate layer from the Early Lutetian transgression reworked Lower Cretaceous, and Paleogene marine sediments. It is dominated by pebbles from the locally mountains which must have been transported by an ancient river in a delta fan. Marine reworked Lower Cretaceous and Paleogen pebbles/fossils, were derived from the underlying deposits of northern Germany (= southern pre North Sea basin). The benthic macrofauna is cold upwelling water influenced and non-tropical, and medium divers. The vertebrate fish fauna is extremely rich in shark teeth, with about 5,000 teeth per cubic meter of gravel. The most dominant forms are teeth from sand shark ancestors *Striatolamia macrota*, followed by white shark ancestors *Carcharodon auriculatus*. Even teeth from the magatooth shark ancestor *Carcharocles sokolovi* are present in a moderately diverse and condensed Paleogene fish fauna that also includes rays, chimaeras, and more then 80 different bony fish. Fragmentary turtle remains are present, and few terrestrial vertebrates and even marine mammals with phocids, sirenians and possibly whales.

1. Introduction

Early research in the Dalum fossil locality in the Fürstenau region of north-west Germany (Figure 1(a)) concentrated largely on the stratigraphy and the fossil shark tooth content [1]. A preliminary analysis of the conglomerate components has also been published by Bartholomäus [2]. The site has only been excavated by private collectors, who have been active there since 1971 [2]. Whereas aragonite and calcite fossils at Dalum have all been dissolved away, they are still preserved in a sand pit at Osteroden (Figure 1(b)) from which 11,000 fish otoliths have previously been analysed by Schwarzhans [3]. The extremely rare teeth of terrestrial mammals from the Dalum and Osteroden sites were the focus of a paper by Franzen and Mörs [4].

The Paleogene (Middle Eocene) marine localities at Dalum and Osteroden (Figures 1(a) and 1(b)), in the south-ern pre-North Sea basin, represent an important environmental bridge between marine, coastal, deltaic swamp, and terrestrial faunas (Figure 1(c)). The marine-dominated fossils at Dalum also offer valuable information on the paleobiodiversity of the central European tropical terrestrial faunas during the Eocene.

Other important marine Eocene sites in Europe are at Helmsleben in Germany, and at Oosterzele and Balegem in Belgium (Figure 1(c)), which are also well known for their enormous quantities of shark teeth [5] and which also contain rare remains of terrestrial mammals [4]. The marine Paris Basin to the south of the Artois submarine swell (Figure 1(c)) is also one of the classic fossil regions of Europe in which many of the shark species presented herein were first found, as described in the monograph by Lériche [6], as well as by other authors. Finally, the Lower Eocene London Clay Formation in England, for example, on the Isle of Sheppey

(a)

(b)

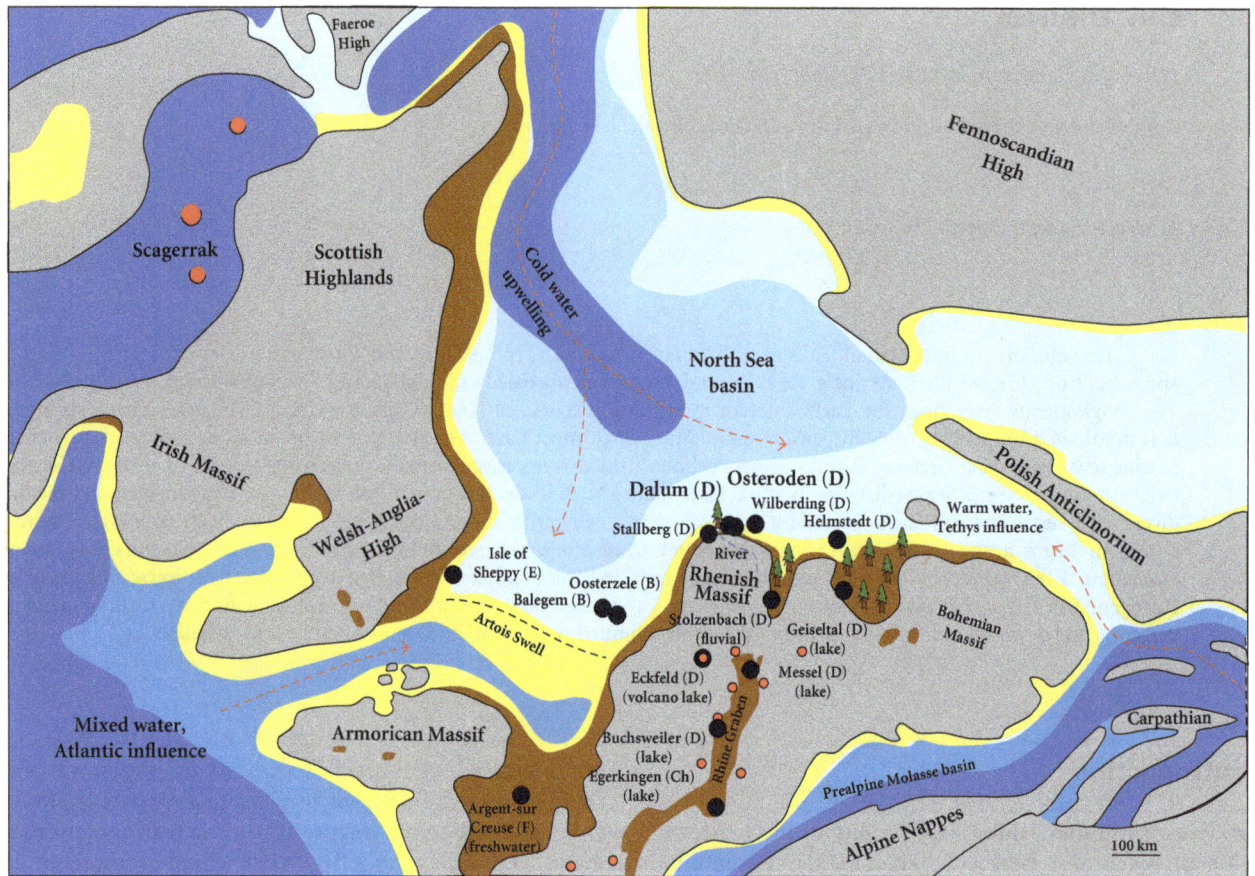
(c)

▨ Mainland/Islands	▨ Shallow marine (mainly limestones)
▨ Brackish-limnic	▨ Deep marine (mainly clays)
▨ Coastal beach zones	● Volcano
▨ Very shallow marine (mainly sands)	● Vertebrate fossil site
▨ Shallow marine (clays/sands)	♣ Swamp coal areas

FIGURE 1: (a) The Dalum site near Fürstenau and (b) the Osteroden sand pit near Bippen, in north-western Germany. (c) Paleogeography of important Eocene marine and terrestrial vertebrate fossil sites of Europe (compiled from [4, 5, 9–14]).

Eocene (Lutetian) Shark-Rich Coastal Paleoenvironments of the Southern North Sea Basin in Europe: Biodiversity
of the Marine Fürstenau Formation Including Early White and Megatooth Sharks

125

(Figure 1(c)), represents near-shore, fresh-water-influenced, coastal paleoenvironments that contain overlapping terrestrial and marine faunal remains [7].

The Eocene terrestrial fauna (Figure 1(c)) was made famous by the abundance of early mammal, bird, reptile, anuran, and fish skeletons, as well as botanical specimens and insects, from the German at UNESCO World Heritage Messel sites at Geiseltal (an Early to Middle Eocene lake site) [8] and at Messel (Middle Eocene) [9]. The Eckfeld Maar volcano lake site has also yielded important fossil material of terrestrial fauna and flora from the Middle Eocene [10], as have the Eocene coal mine at Stolzenbach [11] and the Swiss mammal site at Egerkingen, which is of Upper Eocene of age [12].

This study presents preliminary stratigraphic and sedimentological analyses focussed on the marine, terrestrially influenced, vertebrate-rich gravels at Dalum, and includes a preliminary statistically based overview of the site with regard to the quality and quantity of its fossils. This information is used for a paleoenvironmental reconstruction resulting in the compilation of a new paleogeographical map, and to provide an overview of the (mainly marine) biodiversity, and marine palaeocurrents on the southern pre North Sea basin.

2. Material and Methods

In May and June of 2011 the UNESCO-supported "Geo and Naturpark TERRA.Vita" near Osnabrück (NW Germany) started a systematic excavation and research project (under the leadership of the PaleoLogic private research institute) at the Dalum forest site near Fürstenau (Figure 1(a)) and at a second locality in the Osteroden sand pit, Bippen (Figure 1(b)). The eventual objective, following the completion of the research program, was to develop and promote the Shark Center at Bippen, which is a visitor and research center within the UNESCO-supported "Geo and Naturpark TERRA.Vita" of north-west Germany.

In May 2011 a 4-day field program using heavy machinery was completed at the Dalum and Osteroden sites (Figures 1(a) and 1(b)). Stratigraphic cross-sections were prepared at both of these sites (making use of 5-meter drillholes at Osteroden) indicating a combined 22-meter thick marine section (Figure 3).

A total of 180 cubic meters of conglomeratic material was excavated from the Dalum site for sieving, of which only 250 buckets (10 litres each) were sieved during the following two weeks (i.e., only 0.1% of the recovered material). It was only possible to excavate 10 cubic meters of material at the Osteroden site, which has not yet been sieved (other than for a preliminary sample) because of the intense weathering and damage shown by all of the fossils, and the large degree of root dissolution in the teeth.

Attention has therefore been focussed on the Dalum gravel, which was sieved into two different fractions of +4 mm and −4 to +1 mm, in order to obtain a good variety of gravel types (Figures 4–5), invertebrate remains (Figures 6–8), and vertebrate remains (Figures 9–15), and to facilitate the gravel pebble analysis. Only material greater then 4 mm is presented

herein, which is adequate for qualitative and quantitative gravel analyses and for analysis of the main vertebrate fauna. In total, 14,437 fossils were recovered and studied, of which 95% are shark teeth and only 5% (747 specimens) are teeth from other types of animals (Figure 5).

More than half of the 13,683 shark teeth recovered were identifiable, but lower proportions of the ray teeth and fin spines and the bone fish teeth and bones. The finer material (less than 4 mm) appears (after sieving a preliminary sample) to contain many additional species, which will be recovered and examined during the coming years from the retained fine fraction of the gravel. A separate, more detailed fish analysis (covering all material >1 mm) will also be the subject of future research, after much more material has been recovered to allow a more complete picture of the fish and invertebrate paleobiodiversity in the pre-North Sea basin during the Eocene.

All the material described and illustrated herein from the 2011 excavations (14,440 fossils in total) is now housed in the Shark Center, Bippen (SCB).

3. Geology and Sedimentology

3.1. Glacial Tectonics. The Tertiary sediments in north-west Germany were deformed and compressed as a result of Middle Pleistocene glacial tectonics [13]. The Ankumer range in particular is the result of proglacial folding due to a Saalian glacier, as indicated by a U-shaped moraine that extends as far as the north-western part of the Wiehengebirge mountain range near Bramsche (Figure 2(a)) [15]. The Eocene sediments at the Dalum (Figure 2(c)) and Osteroden sites are therefore strongly folded. The fold axis at Dalum is oriented east to west (parallel to the main glacial fold axis, cf. Figure 2(c)), with a "saddle" that dips at about 10–15° towards the east. In a road cutting about 50 meters to the east of the new sampling trench and parallel to it (Figure 2(c)) the top of the saddle structure is missing in the road section [1] having been eroded during the Saalian and further the Weichselian Ice Age. At the new section in the forest instead the top of the saddle is preserved and the convex deformed sediment layers that would normally be expected are more or less horizontal (Figure 2(c)), indicating that the saddle has been compressed from above. At Osteroden the deformation is more complex, and in the recent pit the gravel layers dip at about 30–60° to the north-east. This is also parallel to the main axis of glacial deformation (Figure 2(a)).

3.2. Paleogene Stratigraphy, Sedimentology, and Dating in North-West Germany. No firmly dated early Tertiary (Paleocene) deposits are known in north-west Germany [13].

Early Eocene (Ypresian) marine sediments have, however, been dated by microfossils such as diatoms, radiolarians, and foraminifers [16, 17], and are known from a clay pit at Wilberding, near Damme in north-west Germany [18], and also from another clay pit at Stallberg near Emsbüren (Figure 1(c)) which contains shark remains and reworked iron-carbonate concretions from the Lower Cretaceous [19]. These gray-green, glauconitic marine clays of Wilberding and

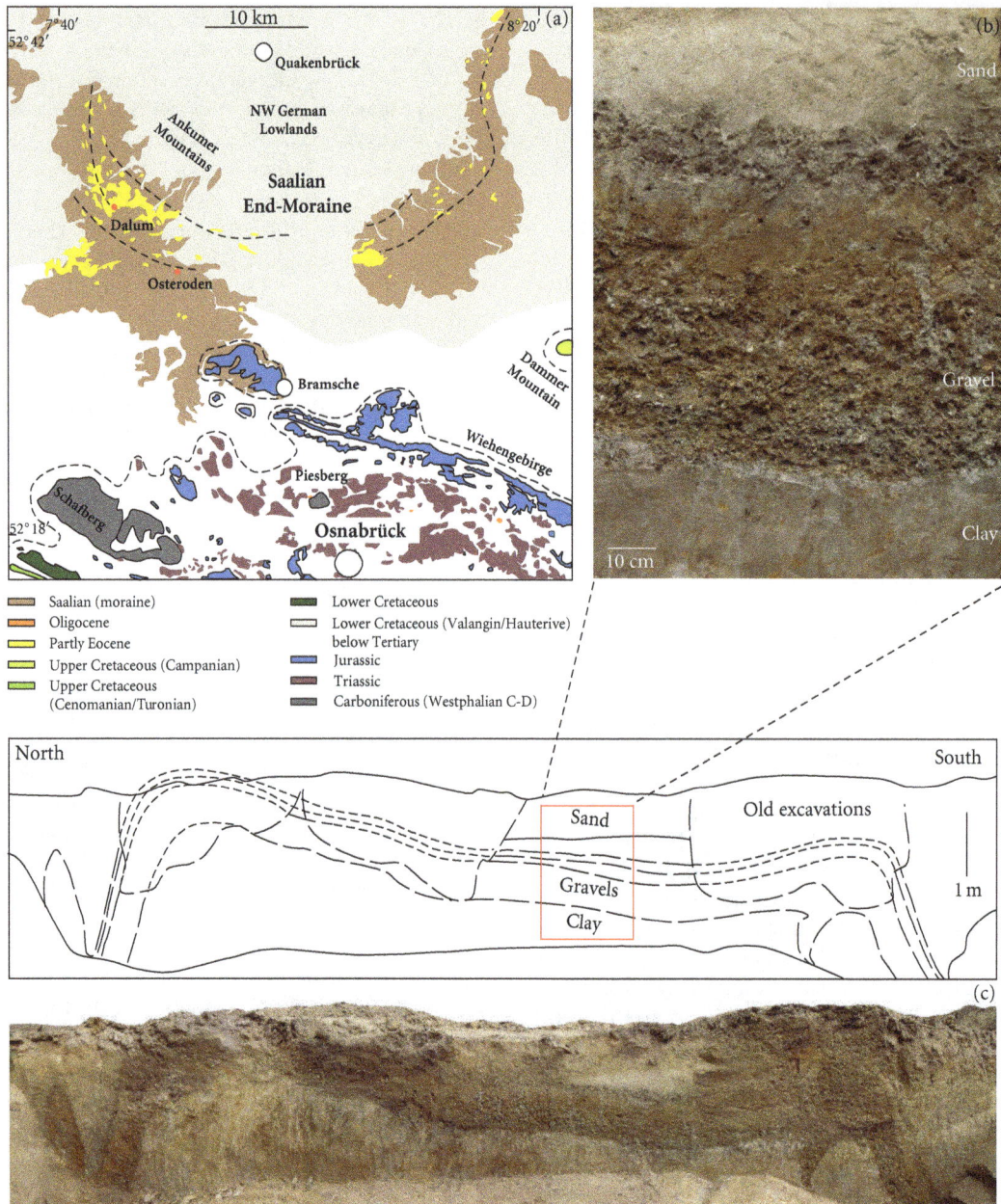

FIGURE 2: Glacial tectonics and distribution of Tertiary deposits as result of (a) the Saalian Ice Age end-moraine forming the Ankumer Mountains (geology redrawn and simplified from [15]), including the Eocene sites at Dalum and Osteroden; (b) north-south section through the gravels at Dalum; (c) photo of the same section at Dalum, showing the Eocene marine deposits, which where deformed into a saddle-like structure (well visible in the Eocene gravel layer) by glacial compression.

Stallberg containing phosphorite nodules (phosphorites), that appear to be of Lower Eocene age, represent the earliest known marine Tertiary sediments in north-west Germany. The Stalberg section also contained many mixed layers 1 mm to 40 mm thick of volcanic ash (tuff) and marine sediments, which are explained as being the result of volcanic ash clouds from the marine strait Skagerrak (possibly also from the Eifel volcanos, e.g., the Eckfelder Maar, cf. Figure 1(c)) [20]. This was a tectonically active period that was also responsible for the Rhine Graben structure of central Europe (Figure 1(c)) [20], and it was also the time when the Messel and Geiseltal fossil sites were formed in Germany, together with the Stolzenbach site and the coastal swamps of the Rhenish Massif (Figure 1(c)).

Middle Eocene (Lutetian) marine deposits for the Dalum and Osteroden sites are presented herein, and a new formation name introduced. A compilation of the two sections from Dalum and Osteroden (Figure 3) has allowed the dating of these sediments, and also enabled sedimentological and paleoenvironmental interpretations to be made.

Eocene (Lutetian) Shark-Rich Coastal Paleoenvironments of the Southern North Sea Basin in Europe: Biodiversity of the Marine Fürstenau Formation Including Early White and Megatooth Sharks

127

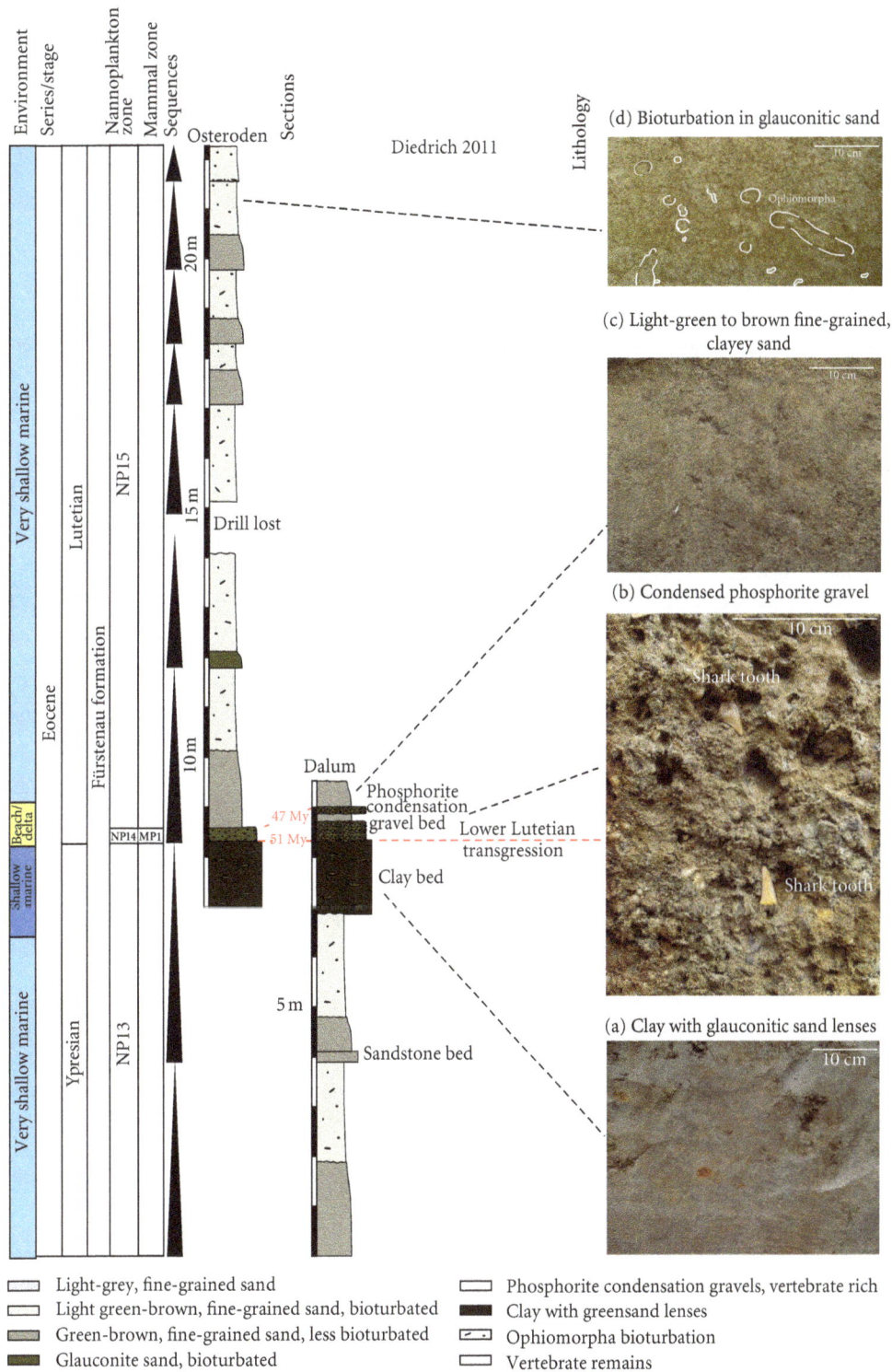

FIGURE 3: Stratigraphy and sedimentology of the Eocene Fürstenau Formation vertebrate sites at Osteroden and Dalum (NW Germany). (a) Clay with glauconitic sand lenses. (b) Condensed phosphorite gravel bed with abundant vertebrate remains, especially shark teeth. (c) Light-green to brown fine-grained, clayey sand, at the base of the sequence. (d) Bioturbated glauconitic sands, with *Ophiomorpha* and other burrows (top of sequence).

3.3. The New "Fürstenau Formation" (Ypresian to Lutetian, Early to Middle Eocene)

3.3.1. Type Stratum. Fürstenau Formation, Early to Middle Eocene, after the nearby city of Fürstenau.

3.3.2. Type Localities. The Dalum site near Fürstenau and the Osteroden sand pit near Bippen, north-west Germany.

3.3.3. Description. The Fürstenau Formation is exposed over a thickness of about 22 meters (Figure 3), the lower part of which (Early Eocene, Ypresian) consists of slightly glauconitic marine sands, with thinly bedded sandstone layers above (Dalum site). These are followed by 1 to 2 meters of thick gray-green clays with lenses of glauconitic sand (possibly still Early Eocene), the top of which have been eroded by a conglomerate bed 40 cm thick at Osteroden and 80 cm thick at Dalum (Figure 2(b)). These conglomerates can be dated herein into the basal Lutetian (basal Middle Eocene). They are fully decalcified at Dalum but are carbonatic at Osteroden where they contain calcareous shells of marine invertebrates, as well as aragonitic fish otoliths in places. The middle Middle Eocene marine glauconitic sands (NP 15 Zone) are composed of at least seven parasequence sets, starting with green, clay-rich, fine-grained sands and ending with green-gray, glauconitic, highly bioturbated (mainly *Ophiomorpha*), medium-grained sands (Figures 3, 7(1–3)). The gravels and the macrofauna from the Fürstenau Formation are described below.

4. Analysis of the Condensed Lutetian Phosphorite Conglomerate Bed

4.1. Pebble Types and Origin. The degree of rounding in the pebbles varies (Figure 4), but this property is not very useful for the Carboniferous quartzite pebbles as they are rederived Paleozoic fluvial clasts that were already rounded. Further analysis of the degree of rounding and estimates of transport distances are therefore unlikely to succeed, at least for that particular fraction of the pebbles.

Other pebble groups such as most of the flint pebbles, however, do indicate long-distance transport over a long period of time within a marine environment, while those comprised of local sediments, especially carbonates, have clearly only travelled relatively short distances.

The qualitative rock analysis reveals more then 30 different types of rock within the conglomerates from Dalum and Osteroden (Figure 4), for which provenances cannot always be determined. It has, however, been possible to identify the main rock types and their original ages (Table 1) which are important for an understanding of terrestrial or limnic influences, marine paleocurrents, and the erosional history of the region, and have allowed a new paleogeographical map to be drawn up (Figure 5).

4.1.1. Paleozoic Sedimentary Clasts. Clear to white and pale red or milky, well-rounded quartz pebbles (Figures 4(1) and 4(4)) are from the Westphalian C-D sandstone series

TABLE 1: Dalum gravel analysis based on 3,927 pebbles (>4 mm).

Age	Rock type	Number of clasts
Paleogene (Early/Middle Eocene)	Phosphorite concretions	600
	Phosphorite burrows	1
	Phosphorite fossils	1
	Fine-grained white sandstone	10
	Vertebrate remains (mainly shark teeth)	150
Upper Cretaceous (Campanian-Maastrichtian)	Flint	10
Lower Cretaceous (Valanginian-Hauterivian)	Iron-carbonate concretion: rock fragments	980
	Iron-carbonate concretion: burrows	50
	Iron-carbonate concretion: ammonite fragments	10
Jurassic (Bathonian-Kimmeridgian)	Oolitic iron ores	1
	Oolitic carbonates	1
	Silica-iron-carbonate concretions	1
Carboniferous (Westphalian C-D)	Quartzite	2,100
	Quartzite conglomerates	10
Devonian	Black lydite	1
Volcanic rocks (age unclear)	Red volcanic rocks	1

of the Upper Carboniferous, which outcrops mainly within the Schafberg horst structure, but also at the lesser extent Piesberg and Hüggel horsts near Osnabrück [21]. These quartzite pebbles were already well rounded by previous transport during the Paleozoic and similar well-rounded quartzite pebbles can be seen in the conglomerates of the Piesberg and Schafberg horsts. These pebbles have therefore been redeposited after travelling only short distances, and do not provide any evidence of long-distance transport during the Paleogene, to their secondary deposition site.

4.1.2. Paleozoic Volcanic Clasts. Clasts of red volcanic rocks (Figures 4(11–12)) are only present as small pebbles (<3 cm) and their origin remains unclear. They were probably carried southwards from the Fennoscandian High by upwelling currents, because no such volcanic rock types are known to the south in the northern Rhenish Massif.

4.1.3. Lower Cretaceous Reworked Fossils and Clasts. The most abundant reworked fossils are ammonite fragments (Figures 4(27) and 4(31)), especially single chambered fragments (see the "cat paw" specimen in Figure 4(28)), which are all found in carbonate-iron concretions that originated from the Lower Cretaceous claystones of northern Germany. Most abundant are fragments of orthicon ammonites (Figures 4(27) and 4(28)).

Eocene (Lutetian) Shark-Rich Coastal Paleoenvironments of the Southern North Sea Basin in Europe: Biodiversity of the Marine Fürstenau Formation Including Early White and Megatooth Sharks

129

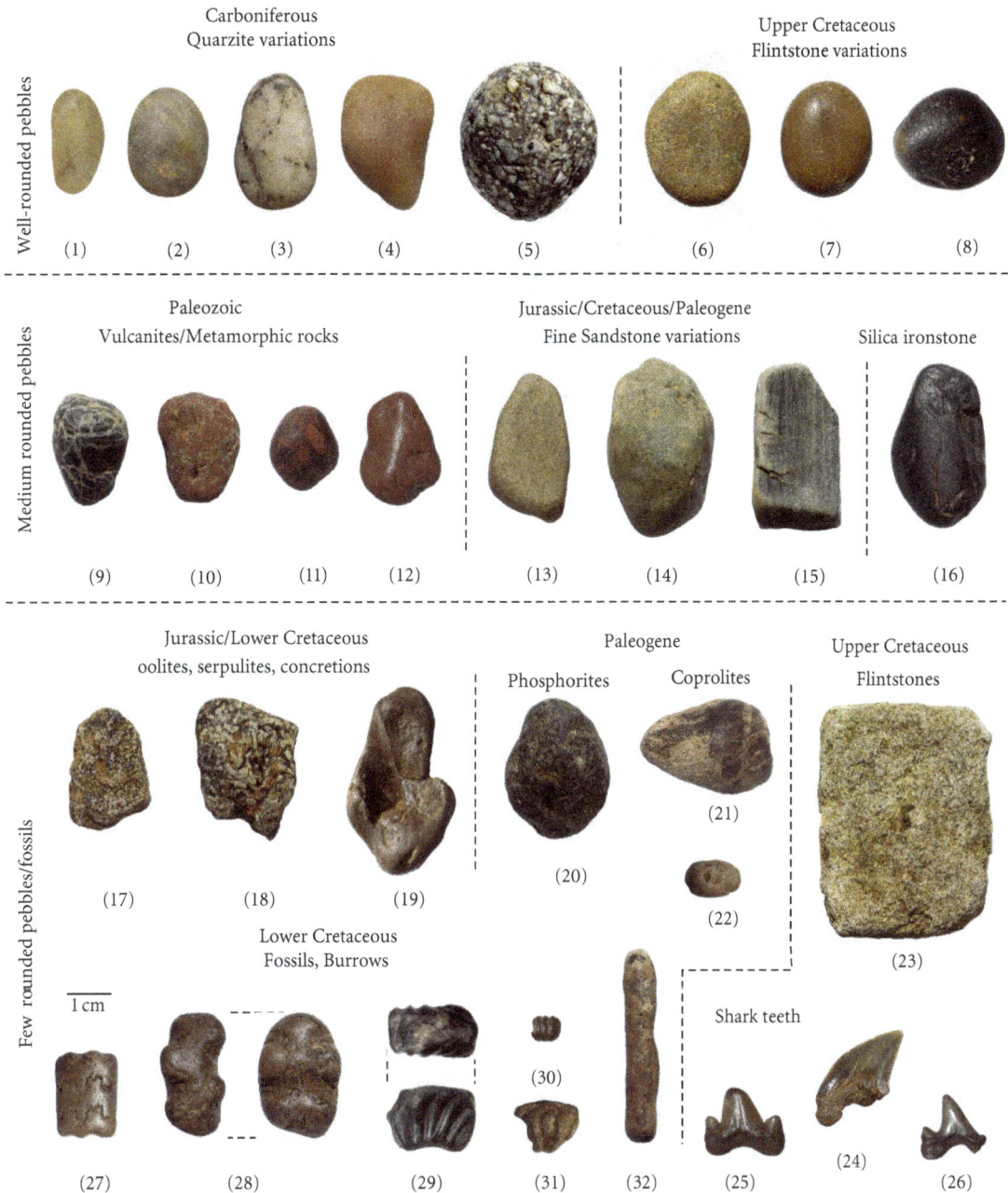

FIGURE 4: Middle Eocene (basal Lutetian) gravel types. (1–3(?4)) Carboniferous (Westphalian C-D, originating from the Schafberg and Piesberg mountains) quartzite variants which were already well rounded in the Paleozoic, and have only been transported a short distance since then (SCB no. Sedim-2). (5) Carboniferous conglomerate (Westphalian C-D), originating from the Schafberg and Piesberg mountains (SCB no. Sedim-3). (6–8) Upper Cretaceous (Campanian-Maastrichtian) flints (SCB no. Sedim-6), consisting of Campanian "Rijckholt" flint (6-7) from the Aachen region and northern Netherlands (SCB no. Sedim-6a) and "Maas-Egg," black flint (8) from the Maastrichtian of the northern Netherlands (SCB no. Sedim-6b). (9) Paleozoic metamorphite of uncertain origin (SCB no. Sedim-10), (10–12) red Paleozoic volcanic rocks of uncertain origin, possibly from Scandinavia (SCB no. Sedim-13), (13) Paleogene white fine-grained sandstone of uncertain origin (SCB no. Sedim-9), (14-15) Paleogen glauconitic fine-grained sandstones (SCB no. Sedim-11), and (16) Jurassic silica/iron-carbonate concretion from the Osnabrücker Bergland (SCB no. Sedim-17). (18) Jurassic or Lower Cretaceous serpulid (SCB no. Sedim-4). (19) Iron-carbonate concretion fragment with orthicon (*Bochianites*) ammonite negatives from the Valanginian/Hauterivian, Lower Cretaceous (SCB no. Sedim-18). (20) Middle Eocene phosphorite nodule (SCB no. Sedim-7). (21-22) Eocene phosphatic shark coprolites (SCB no. Sel-890, 898). (23) Upper Cretaceous (Campanian) platy flint (SCB no. Sedim-8). (24) Shark tooth from *Squalicorax kaupi*, reworked from the Campanian, Upper Cretaceous (SCB Dal-Sel-901), labial. (25) Shark tooth from *Cretolamna appendiculata*, reworked from the Santonian-Campanian, Upper Cretaceous (SCB Dal-Sel-882), labial. (26) Shark tooth from *Cretolamna appendiculata,* reworked from the Campanian, Upper Cretaceous (SCB Dal-Sel-881), labial. (27) Lower Cretaceous (Upper Valanginian) orthicon ammonite fragment of (*Bochianites* sp. (SCB no. Sedim-18). (28) Lower Cretaceous (Upper Valanginian) orthicon ammonite fragment, a so-called "cat paw" (SCB no. Sedim-17). (29) Lower Cretaceous (Lower Hauteriveian) ammonite fragment of *Endemoceras* sp. (SCB no. Sedim-22). (30) Lower Cretaceous (Upper Hauteriveian) ammonite fragment of *Aegocrioceras* sp. (SCB no. Sedim-22). (31) Lower Cretaceous (Upper Valanginian) ammonite fragment of *Oostrella* sp. (SCB no. Sedim-18). (32) Lower Cretaceous elongated burrow preserved in iron-carbonate concretion (SCB no. Sedim-43).

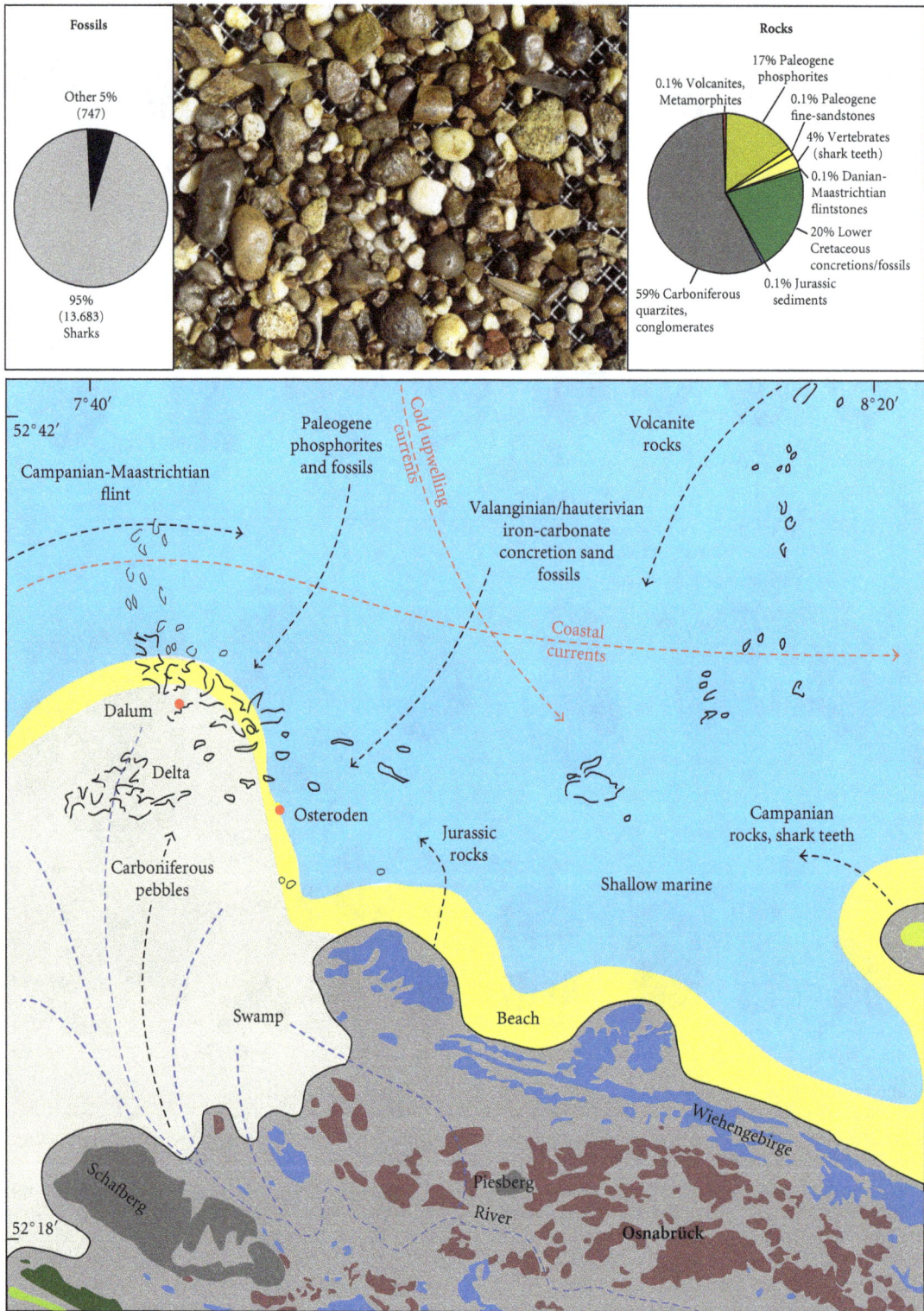

Figure 5: Basal Middle Eocene (Early Lutetian) gravel: quantitative analysis and transportation direction reconstruction. The main Paleozoic components (Carboniferous quartz) must have been transported by an ancient river running between today's Wiehengebirge and Teutoburger Wald mountain ranges, ending in a delta. Other components (Jurassic material) are from the Wiehengebirge range, whereas Lower Cretaceous pebbles and ammonite fossils are from north Germany. Flints were accumulated from different directions (Maastricht/Aachen area, and Dammer Berge; Material no. SCB Dal-Sedim-1–16).

The ammonite fragments, which are also found in the Lower Cretaceous claystones of north-west Germany [22], range in age from Valanginian to Hauterivian (Lower Cretaceous) and include *Bochianites* sp. and *Oosterella* sp. (Upper Valanginian), *Endemoceras* sp. and *Leopoldia* sp. (Lower Hauterivian), and *Aegocrioceras* sp. (Upper Hauterivian). All ammonites must have already been previously reworked during the Lower Eocene [19], and again during the basal Middle Eocene transgression. Rock types with similar (medium-brown) preservation to these reworked fossils contain abundant fossilized burrow casts (Figure 4(32)) and must have originated from the Lower Cretaceous marine sediments of north-west Germany. These can be distinguished from Eocene burrow casts since the latter are different and occur in phosphatic concretions (cf. Figure 7(2–5)).

4.1.4. Upper Cretaceous Clasts.

The Upper Cretaceous clasts are carbonates from the Teutoburger Wald (Cenomanian to Santonian) and Dammer Berg (Campanian) regions of Germany (Figure 2(a)) that weather easily (except for the Campanian flint nodules, but these are absent from the Cenomanian to Santonian). The platy flints, which show little rounding, (Figure 4(23)) possibly originated from the Dammer Berge region. The Maastrichtian marine carbonate deposits in the studied region are not represented but typical black flint from further west in Belgium is common, including the "Maas-Eggs" (Figure 4(8)) that can sometimes be found in the Dalum gravels. Other red-to-brown flint variants resemble the "Rijckholt" flint (Figures 4(6) and 4(7)), typical of the Campanian limestones in northern Holland and on the Dutch-German border. All of these flint pebbles in the Dalum gravels (except for the platy flints) are well rounded and seem to have been transported to north-west Germany by marine paleocurrents from the west (Figure 5).

4.1.5. Jurassic Clasts.

The black silica-ironstone concretions found in these gravels (Figure 4(16)), which resemble black Paleozoic lydite, are from the Middle Jurassic Soninnien beds and are well known in the Osnabrücker Bergland, as well as also being found in reworked Quaternary deposits of the region [23]. Rare, typically red, oolites (iron ores, Figure 4(17)) originate from the Middle Jurassic Macrocephaly oolith beds (also know as "Portha-Oolith or Wittekindsflöz") of the central part of the Wiehengebirge mountain range [24] (Figure 5) and, as with the carbonates, do not survive over long transport distances. In total, the sedimentary clasts from the Jurassic of the Osnabrücker Bergland, and in particular the Wiehengebirge mountain range, are poorly represented and comprise less then 1% of the total number of clasts (Figure 5).

4.1.6. Lower to Middle Eocene Clasts.

The main hard rock types represented from the Lower and Middle Eocene are the phosphorite nodules (Figure 4(20)) that have been reworked from the soft Eocene marine sands (including most of the marine vertebrate remains that accumulated in the Dalum and Osteroden condensed phosphorite nodule gravels), and possibly some white fine-grained sandstones (Ypresian,

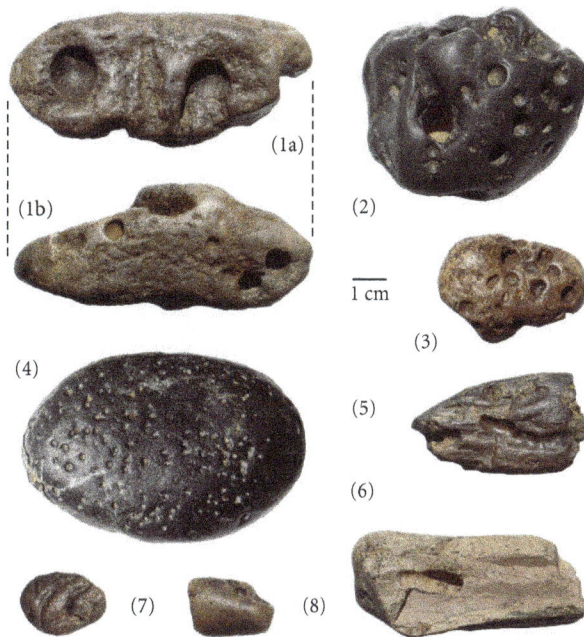

FIGURE 6: Bored and drilled beach pebbles from the Middle Eocene (Lutetian) gravels at Dalum. Only softer rock types, such as iron-carbonate concretions (Valanginian-Hauterivian, Lower Cretaceous) and phosphorite nodules, or wood (Paleocene-Eocene, Paleogene) were drilled by bivalves or were chemically dissolved by sponges. (1) Large *Gastrochaenolites* boreholes in reworked Lower Cretaceous iron-carbonate concretion pebble (SCB no. Dal-Invert-2). (2) Small *Gastrochaenolithes* boreholes in reworked Eocene phosphorite (SCB no. Dal-Invert-1). (3) Small *Gastrochaenolithes* boreholes in reworked Lower Cretaceous iron-carbonate concretion pebble (SCB no. Dal-Invert-3). (4) Small boreholes made by *Clione*-type sponges in reworked Eocene phosphorite (SCB no. Dal-Invert-4). (5) Boreholes of the bivalve *Teredo* sp. in Paleogene (Eocene) silicified driftwood (SCB no. Dal-Invert-6). (6) Boreholes of the bivalve *Teredo* sp. in silicificated Eocene driftwood pieces (SCB no. Dal-Invert-7). (7) Boreholes in reworked Lower Cretaceous ammonite fragment, iron-carbonate concretion (SCB no. Dal-Invert-5a). (8) Small *Gastrochaenolithes* boreholes in reworked Lower Cretaceous orthicon ammonite fragment (*Bochianites*), iron-carbonate concretion (SCB no. Dal-Invert-5b).

Figure 4(13)). The more polished pebbles appear to have come from the older Lower Eocene (Paleocene) horizons that have been reported from other Paleogene sites in north-west Germany [13] and have later been reworked and bored by organisms during the Middle Eocene (Figures 6(2) and 6(4)), while the unpolished phosphorites (Figure 4(20)) may be younger (Middle Eocene) or else have not travelled as far. Also less-transported are sporadic phosphatic shark coprolites, some of which have been reworked from earlier Eocene horizons (Figures 4(21) and 4(22)).

The quantitative rock-type analysis (Figure 5) is based on 3,927 pebbles greater than 4 mm (Table 2, Figure 5). The main rock types identified were abundant Carboniferous (Westphalian C-D) quartzites and (less common) quartzite conglomerate clasts (59%), Lower Cretaceous (Valagin/Hauterive) iron-carbonate concretions and fossils

TABLE 2: Fossil types at the Dalum site, based on 14,440 specimens in material >4 mm.

Fossil type	Amount
Plants	173
Burrows/boreholes in pebbles	184
Burrows (refilled, phosphatic)	43
Bivalves (fragments)	61
Gastropods	25
Crustaceans	29
Anthozoans	4
Chimaera	15
Rays	206
Sharks	13,690
Turtles	5
Crocodiles	3
Mammals	2

(20%), and Paleogene (Paleocene) Eocene phosphorite nodules (15%). These three main rock types derived from different directions and have been reworked from sediments of different ages (Figure 5). While the Carboniferous material must have been washed in from the south east by a river passing the Schafberg and reaching into a fluvial delta system, the Lower Cretaceous pebbles and fossils were washed opposite south and south east towards the coast, together with the Paleogene phosphorite and fossil material (Figure 5). All other rock types and components are less well represented and not important to the reconstruction of paleocurrents and erosional history.

4.2. *Fossil Content.* In total, 14,437 fossils were recovered from the Dalum site excavation; these are listed by fossil type in Table 2. Nearly all of the Eocene invertebrate remains are only preserved in the reworked phosphorites; the only fossils preserved by silica replacement are the shells of larger oysters (with silica rings) and the solitary corals. The phosphatic coprolites, the silicified wood, all the bones, and in particular the teeth, were the most resistant specimens.

4.2.1. *Invertebrates.* The invertebrate fauna is very incomplete and largely represented by phosphorite steinkerns of shells and burrow casts that survived the enrichment processes in the Middle Eocene gravels of Dalum.

Drills and Boreholes. Drills and boreholes in pebbles [25] are quite common in pebble-rich beach Eocene paleoenvironments also found in the Oligocene pebble-rich beaches of north-west Germany [26]. The mechanical bivalve drills and chemical sponge boreholes in the Eocene pebbles from the Dalum site have only been made in soft to moderately hard rock types, and in particular in 85% of the reworked Lower Cretaceous iron-carbonate nodules (as in the Oligocene beach pebbles from the Osnabrück Bergland region) [26] and ammonite fragments; they are less common in Eocene phosphorites. The two typical ichnites of *Gastrochaenolites* isp. produced by bivalves (Figures 6(1–3), 6(7) and 6(8)), and

Clione sponge burrows (Figure 6(4)), are common in Lower Cretaceous iron-carbonate concretion pebbles and lesser in Eocene phosphorite pebbles. *Teredo* bivalve boreholes are also present, but only abundant in driftwood (Figures 6(5) and 6(6)).

Anthozoans. Corals without reef structures are known throughout the southern North Sea basin, represented by the solitary coral *Paracyathus caryophyllus*. The few specimens of this coral that have been recovered have been preserved by metasomatic silica replacement (Figure 8(12)). This solitary, nonhermatypic coral is also found in Oligocene shallow marine, beach pebble paleoenvironments [26, 27].

Crustaceans. These are common in the Eocene North Sea basin [28, 29], but in the Dalum material they are only preserved in the reworked phosphorite concretions. The remains of *Hoploparia* sp. (Figures 7(6–8)) are significant as these lobsters could be largely responsible for the high degree of bioturbation in the glauconitic sands (Osteroden section, cf. Figures 3(d), 7(1)) and may have been responsible for the *Ophiomorpha* isp. burrows (Figure 7(1)), fragments of which are also found (as phosphorite concretion casts) within the gravels (Figures 7(2) and 7(3)). Smaller burrows, especially the y-shaped forms (Figures 7(4) and 7(5)), may also be related to smaller crustaceans. Crab remains are more common and at least three different species can be identified from their pincers and carapaces: *Xanthilithes* sp. (Figures 7(11) and 7(12)), *Zanthopsis* (Figure 7(9)), and *Litoricola/Portunites* (Figure 7(10)).

Molluscs. Bivalves have been identified following the published works on the Belgian [30, 31] and English [7] sites. In the Dalum material almost all of these have been destroyed because of their aragonite shells, with only a few infaunistic forms such as *Arca* sp. and *Striarca* sp. (Figure 8(9)) preserved in phosphorite concretions, some epifaunistic forms such as *Pteria* sp. (Figure 8(10)), and the more abundant large oyster *Pycnodonte* sp. (only shell fragments preserved by metasomatic silica replacement), which seem to have settled on the gravels of the pebble beaches. The gastropods are more diverse with at least eight species represented (all reworked Eocene phosphorite steinkerns, Figures 8(1–8)), although most of the specimens are from the infaunistic *Natica* sp.; epifaunistic forms include *Sconsia* sp., *Cyprea* sp., *Tectonatica* sp., and *Xenophora* sp.

4.2.2. *Vertebrates.* The chimera, shark, and ray material was compared to the main important works on Paleocene-Eocene sharks of Europe [5, 7, 32–39]. The vertebrates remains are dominated by shark teeth (95%), followed by remains of rays (2%); the rest are chimera, bony fish, reptile, and (extremely rare) mammal remains (Table 2, Figure 5).

Chimera. Only one chimera species, *Edaphodon bucklandi* Agassiz, 1843, has been recorded to date; it is represented by jaw fragments (Figure 14(19)).

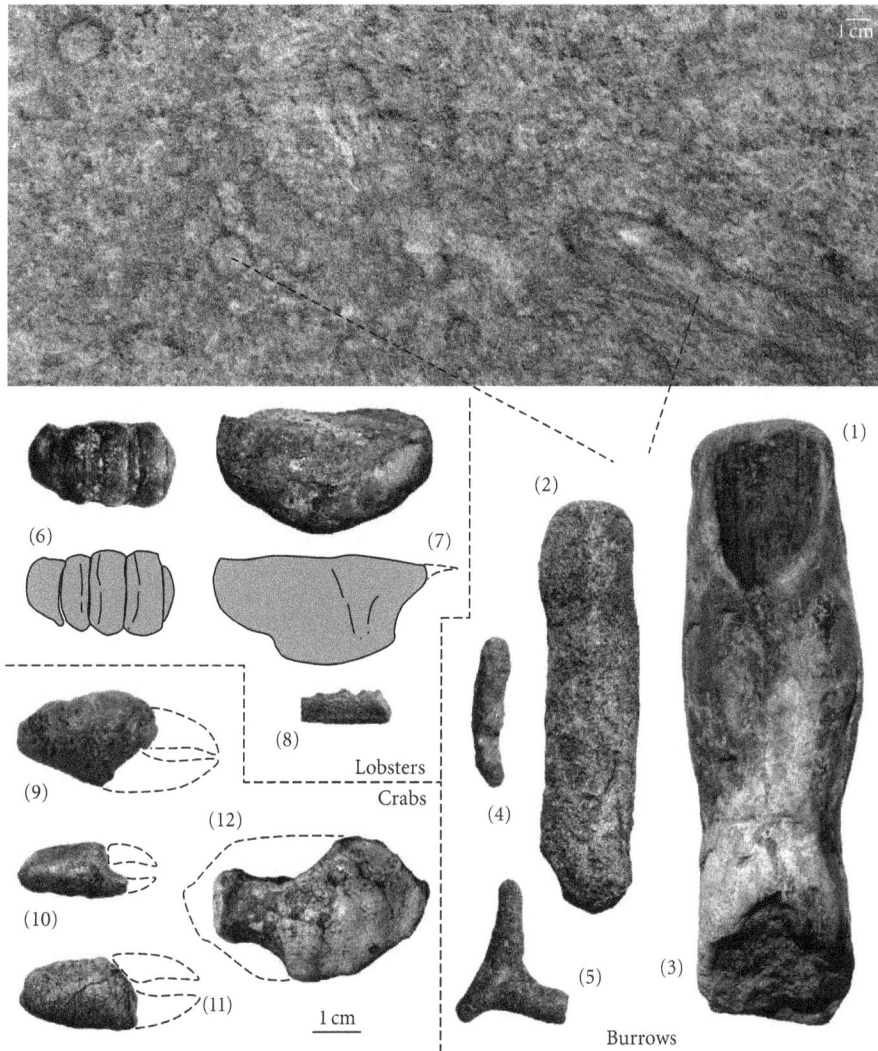

FIGURE 7: Crustacean remains and burrows from the Middle Eocene (Lutetian) gravels at Dalum. (1) Burrows by crustaceans, with large burrows of *Ophiomorpha* isp. and other smaller burrows in bioturbated glauconitic sands of the Fürstenau Fm. at the Osteroden sand pit (see also burrows in Figure. 3(d)). (2) Refilled *Ophiomorpha* isp. burrow fragment, phosphorite concretion (SCB no. Invert-41). (3) Encrusted *Ophiomorpha* isp. burrow fragment, phosphorite concretion (SCB no. Invert-38). (4) Refilled burrow fragment, phosphorite concretion (SCB no. Invert-40). (5) Triaxonic burrow fragment, phosphorite concretion (SCB no. Invert-39). (6) Abdomen fragment from the lobster *Hoploparia* sp., phosphorite concretion (SCB no. Invert-14), dorsal. (7) Carapace of the lobster *Hoploparia* sp., phosphorite concretion (SCB no. Invert-16), lateral. (8) Lower pincer of the lobster *Hoploparia* sp., phosphorite concretion (SCB no. Invert-13), lateral. (9) Crab *Zanthopsis* large right pincer, phosphorite concretion (SCB no. Invert-18), anterior. (10) Crab *Litoricola/Portunites* right pincer, phosphorite concretion (SCB no. Invert-17), anterior. (11) Crab *Xanthilites* sp. large right pincer, phosphorite concretion (SCB no. Invert-20), anterior. (12) Crab *Xanthilites* sp. carapace in phosphorite concretion (SCB no. Invert-19), dorsal.

Rays. These are the second most common type of fish remains, represented mainly by broken teeth, but also by fin spines. At least four species can be recognised from their teeth including *Aetobatus irregularis* (Agassiz, 1843), *Leidybatis jugosus* (Leidy, 1876), *Myliobatis dixoni* (Agassiz, 1843), and one large unidentified species (Figures 14(14–17)), while *Pristis lethami* Galeotti, 1837, is represented by several rostral tooth fragments (Figure 14(18)).

Sharks. Out of the total of 13,690 Paleogene shark teeth recovered from the Dalum site, 6,946 (more then 50%) have been identified at a species level (Table 3). This large

number of identifications is a result of the large quantity of material and many complete and comparable teeth recovered, from different jaw positions in different species. The preliminary qualitative analysis herein includes only 19 of the larger Paleogene shark species and five ray species (>4 mm material), which can be found listed in Table 3 together with their time ranges and abundances. Preliminary sampling of the material below 4 mm indicates twice as much fish biodiversity, including of course the smaller species whose teeth are, however, quite rare.

Large teeth from *Otodus obliquus* (Agassiz, 1843; Figure 9(1–3)) are all incomplete and mostly very rounded,

FIGURE 8: Mollusc and anthozoa remains from the Middle Eocene (Lutetian) gravels at Dalum. (1) Gastropod *Natica* sp., phosphorite steinkern (SCB no. Invert-25), dorsal and lateral. (2) Gastropod *Sconsia* sp., phosphorite steinkern (SCB no. Invert-27), dorsal and lateral. (3) Gastropod *Cyprea* sp. phosphorite steinkern (SCB no. Invert-29), dorsal and lateral. (4) Gastropod *Tectonatica* sp., phosphorite steinkern (SCB no. Invert-26), lateral. (5) Gastropod *Volutospina* sp. phosphorite steinkern (SCB no. Invert-30), lateral. (6) Gastropod *Littoriniscala* sp., phosphorite steinkern (SCB no. Invert-32), lateral. (7) Gastropod *Xenophora* sp. phosphorite steinkern (SCB no. Invert-28), dorsal and lateral. (8) Gastropod *Falsifusus* sp. phosphorite steinkern-negative (SCB no. Invert-31), lateral. (9) Bivalve *Striarca* sp. phosphorite steinkern (SCB no. Invert-34), lateral. (10) Bivalve *Pteria* sp., phosphorite steinkern (SCB no. Invert-35), lateral. (11) Oyster *Pycnodonte* sp., shell fragment, metasomatically replaced, silicified (SCB no. Invert-36), lateral. (12) Anthozoa *Paracyathus caryophyllus*, metasomatically replaced, silicified (SCB no. Invert-37), lateral and dorsal.

but the sharp-edged teeth illustrated herein would appear to be more or less autochthonous (also rare sharp edged in abundant *Striatolamia* teeth), which is significant because this species now appears to extend from the Paleocene into the lower Middle Eocene in Europe, but not into the middle or upper parts of the Middle Eocene. All of the large teeth from *Carcharocles subserratus* (Agassiz, 1843; Figure 9(4-10)) are rounded, indicating most probably in all cases the reworking of horizons that appear to extend into the upper Paleocene or Lower Eocene. However, this material may partly belong to the most abundant type of white shark, *Carcharodon auriculatus* (Blainville, 1818), which is represented by both enrolled teeth and sharp serrated teeth, from various jaw positions (Figures 10(1–8)). The largest and most robust teeth recovered from the Dalum site are from the megatooth shark lineage—*Carcharocles sokolovi* (Jaeckel, 1895)—which is not well defined, and is rare in any material. All of the megatooth shark ancestor material will be discussed further in the future, when more teeth are available (only 13 have been recovered to date, and only two of these are in good condition: Figures 10(9) and 10(10)) and can be better distinguished from the other serrated teeth. Another, more abundant, species is *Isurus praecursor* (Lériche, 1905), with material from all jaw positions (Figure 11(1–10)). *Jaeckelotodus* is represented by two species, *J. trigonalis* (Jaeckel, 1895) from which all teeth are reworked (Figures 11(11–15)), and *J. robustus* (Lériche, 1921) which has both reworked and sharp-edged specimens (Figures 11(16–18)). The most abundant shark is the sand shark ancestor *Striatolamia macrota* (Agassiz, 1843), with teeth present from all the main jaw positions (Figure 12) and including teeth from juveniles right through to teeth from mature individuals. Only 0.1% of the teeth are sharp edged with little evidence of any transport. Smaller

species are also abundant, such as *Sylvestrilamia teretidens* (White, 1931; Figures 13(1–12)), and *Hypotodus verticalis* (Agassiz, 1843; Figures 13(13–19)). Less abundant are teeth from *Anomotodon novus* (Winkler, 1874; Figures 13(20–22)) which are all reworked, as is also largely the case with teeth from *Brachycarcharias lerichei* (Casier, 1946; Figures 14(1–6)). This latter species sometimes also has sharp-edged teeth, such as *Carcharias* sp. (Figures 14(7) and 14(8)). Rare teeth from some other species are also present (Table 3, Figures 14(9) and 14(13)).

Bone Fish. Some vertebrae and teeth, mainly from *Phylodus* sp. (Figure 14(20)) and *Pycnodus* sp. (Figure 14(21)) and some rostrum fragments from *Cylindracanthus* sp. (Figure 14(22)) provide only a poor representation of the diverse fish fauna, which is normally estimated from the otoliths, but these are absent from Dalum.

Terrestrial Vertebrates. Only single vertebrate teeth have been described from Early to Middle Eocene marine deposits around the world [4, 7, 40], but large pachyostotic rib fragments from Dalum appear to be attributable to early marine Sirenian mammals when compared to Oligocene material [41], and these rare Middle Eocene marine mammal remains will be described in a future publication. The terrestrial vertebrate remains recovered from the Dalum material are of rare crocodile teeth, osteoderm fragments from *Diplocynodon* (Figures 15(1) and 15(2)), and freshwater turtle remains of *Trionyx* (Figure 15(3)), all of which are suggestive of freshwater lakes or swamps and therefore a possible deltaic environment.

Some bone fragments are impossible to identify but appear to relate to mammals, judging by their spongious nature. A distal humerus fragment (Figure 15(4)) appears to

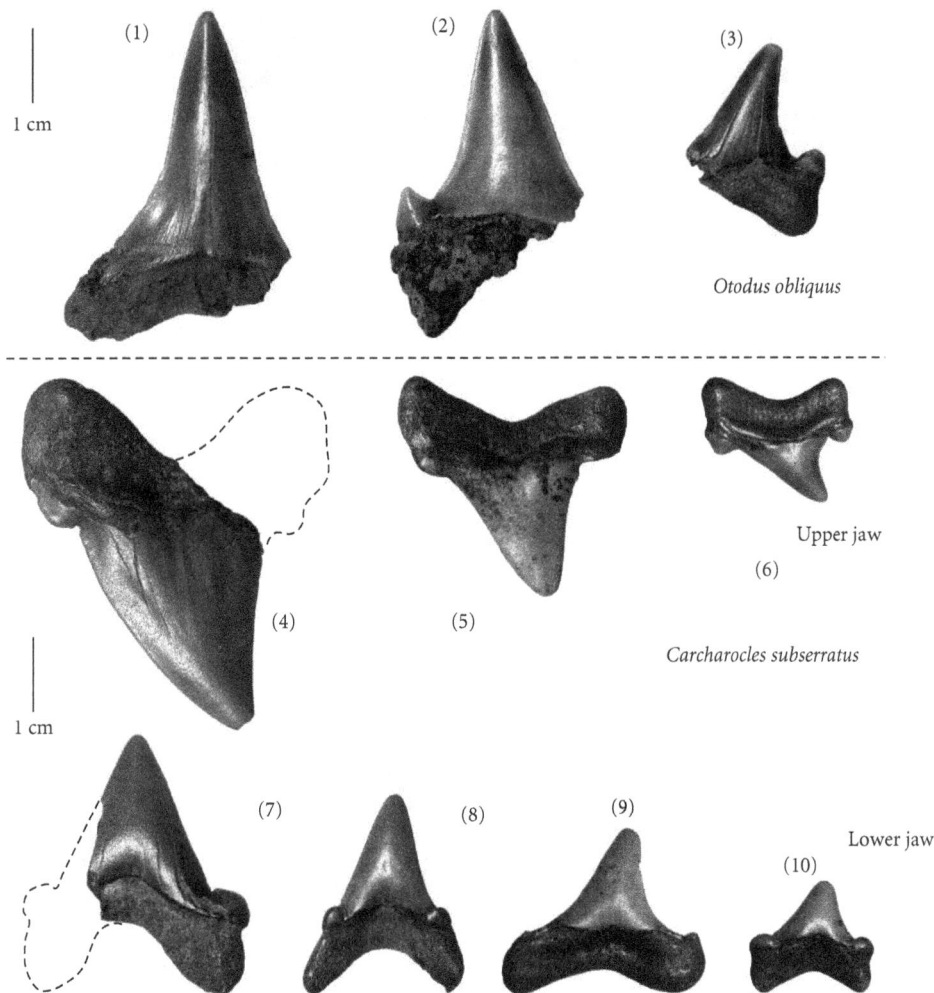

FIGURE 9: Shark teeth from lamniform (Odontidae) sharks (from different jaw positions) from the Middle Eocene (Lutetian) gravels at Dalum. (1 3). *Otodus obliquus* (Agassiz, 1843): (1) lower jaw anterior tooth (SCB no. Scl-856), (2) lower jaw anterolateral tooth (SCB no. Sel-857), and (3) lateral tooth (SCB no. Sel-858). (4–10). *Carcharocles subserratus* (Agassiz, 1843): (4) upper jaw anterior tooth (SCB no. Sel-869), (5) upper jaw lateral tooth (SCB no. Sel-870), (6) upper jaw posterolateral tooth (SCB no. Sel-872), (7) lower jaw anterior tooth (SCB no. Sel-873), and (8) Lower jaw anterolateral tooth (SCB no. Sel-874), (9) lower jaw lateral tooth (SCB no. Sel-876), (10) lower jaw posteriolateral tooth (SCB no. Sel-875; all in labial view).

belong to a medium-sized marine mammal, and a femur fragment (Figure 15(5)) may belong to a larger mammal.

5. Discussion

5.1. Stratigraphy and Dating. The sediments from the 22-meter-thick sections at Dalum and Osteroden have a different lithology from the Lower to Middle Eocene Brussel and Lede Sand Formations of the Belgian Oosterzele and Balegem marine deposits [42]. They also contain less carbonate and no storm shell beds, but only massive clay and conglomerate layers, and hence a new formation name is proposed for these deposits: the Fürstenau Formation.

At the Dalum site only the older (Ypresian) Eocene layers are present, with initial marine sequences possibly overlain by a massive clay bed, although precise dating of this part of the section is not yet available. The conglomerate layer can be dated by the terrestrial mammal teeth (MP 11 zone), as at Balegem, in Belgium [4, 43], and by shark teeth indicating the presence or absence of certain species (e.g., *O. obliquus*). The more carbonate-influenced Middle Eocene conglomerates (two horizons) of the Lede Sand Formation at Balegem and Osterzele [44] have been described as being of Middle Lutetian age, and may therefore possibly be slightly younger than those from northern Germany [1, 5]. The condensed phosphorite gravel beds at Dalum and Osteroden may be isochronous with the two bivalve-enriched condensed beds in the middle Middle Eocene Lede Sand Formation of Belgium [42, 45]. The conglomerates at Dalum and Osteroden also appear to be isochronous with the Lower Geiseltalium, and would therefore be of a similar age to the Messel and older Geiseltal mammal faunas [9].

The marine clay-sand parasequences of the Fürstenau Formation have been dated by nannoplankton into the

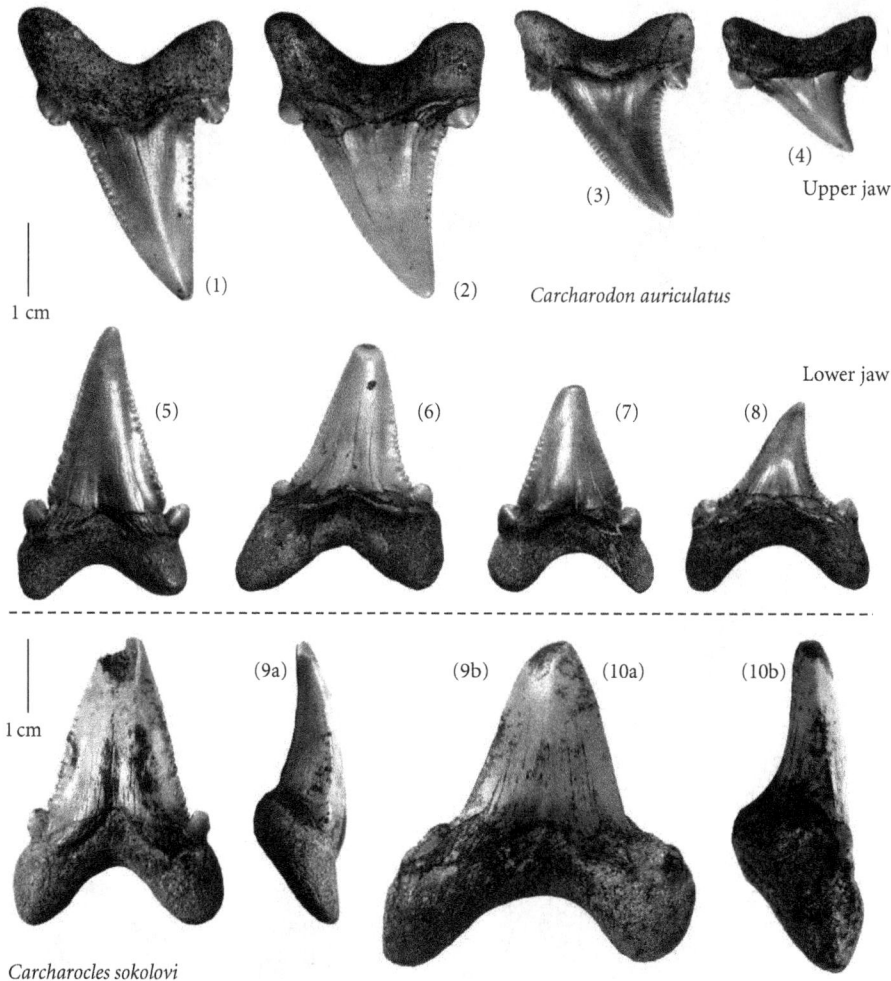

FIGURE 10: Shark teeth from lamniform (Carcharidae) sharks (from different jaw positions) from the Middle Eocene (Lutetian) gravels at Dalum. (1–8) *Carcharodon auriculatus* (Blainville, 1818): (1) upper jaw anterior tooth (SCB no. Sel-26), mirrored, (2) upper jaw anterolateral tooth (SCB no. Sel-25), mirrored, (3) upper jaw anterior lateral tooth (SCB no. Sel-27), mirrored, (4) upper jaw posterior lateral tooth (SCB no. Sel-65), mirrored, (5) lower jaw anterior tooth (SCB no. Sel-4), mirrored, (6) lower jaw anterolateral tooth (SCB no. Sel-5), mirrored, (7) lower jaw anterior lateral tooth (SCB no. Sel-7), (8) lower jaw posterior lateral tooth (SCB no. Sel-66; (1-2(a)) labial and (b) lateral, (3–10), labial views). (9-10) *Carcharocles sokolovi* (Jaeckel, 1895): (1) lower anterior tooth (SCB no. Sel-85) and (2) lower anteriolateral tooth (SCB no. Sel-88).

middle Middle Eocene (Lutetian, NP 14 zone [46, 47]) and appear to be nonisochronous with the shark-tooth-rich layers at Balegem (NP 15 zone) [43, 48]. The lithology of the newly introduced Fürstenau Formation differs from that of the marine localities in the southern North Sea and Paris Basins in having a much lower carbonate content, an almost complete absence of sandstone beds, and an absence of the bivalve (dominated by the bivalve *Megacardita*) and gastropod shell beds.

At least two shark species appear to be from reworked Upper Cretaceous deposits, represented by some teeth from *Cretolamna appendiculata* and a single tooth from *Squalicorax kaupi* (Figures 4(24) and 4(26)). Upper Paleocene and Lower Eocene tooth material appears to be particularly well represented by *Carcharocles obliquus* and *Otodus subserratus*. While *O. obliquus* is typically present in the Late Paleocene and Lower Eocene of Europe (cf. London Clay in: [7]) it is represented here by sharp-edged teeth that show little signs

of transport, and so its time range may extend into the Early Lutetian. In contrast, *O. obliquus* and *C. appendiculata*, are both absent from the diverse selachian fauna of Middle Lutetian age in Belgium [5]. The *Carcharodon auriculatus* teeth from Dalum and Osteroden in Germany, and those from Balegem and Oosterzele in Belgium, differ in their serrations, especially on the lateral cusps which are more strongly developed in the younger material from Belgium [5]. The shark fauna from the Dalum and Osteroden gravels appears to be a little younger than the Middle Eocene Belgian faunas, which supports the dating of the Dalum and Osteroden gravels (but not all the teeth therein) into the Early Lutetian transgression (Early Middle Eocene [48], during which older Paleogene tooth material was reworked and concentrated.

5.2. Biodiversity and Facies. The mollusc population seems to be represented by forms from the sandy floor of a shallow

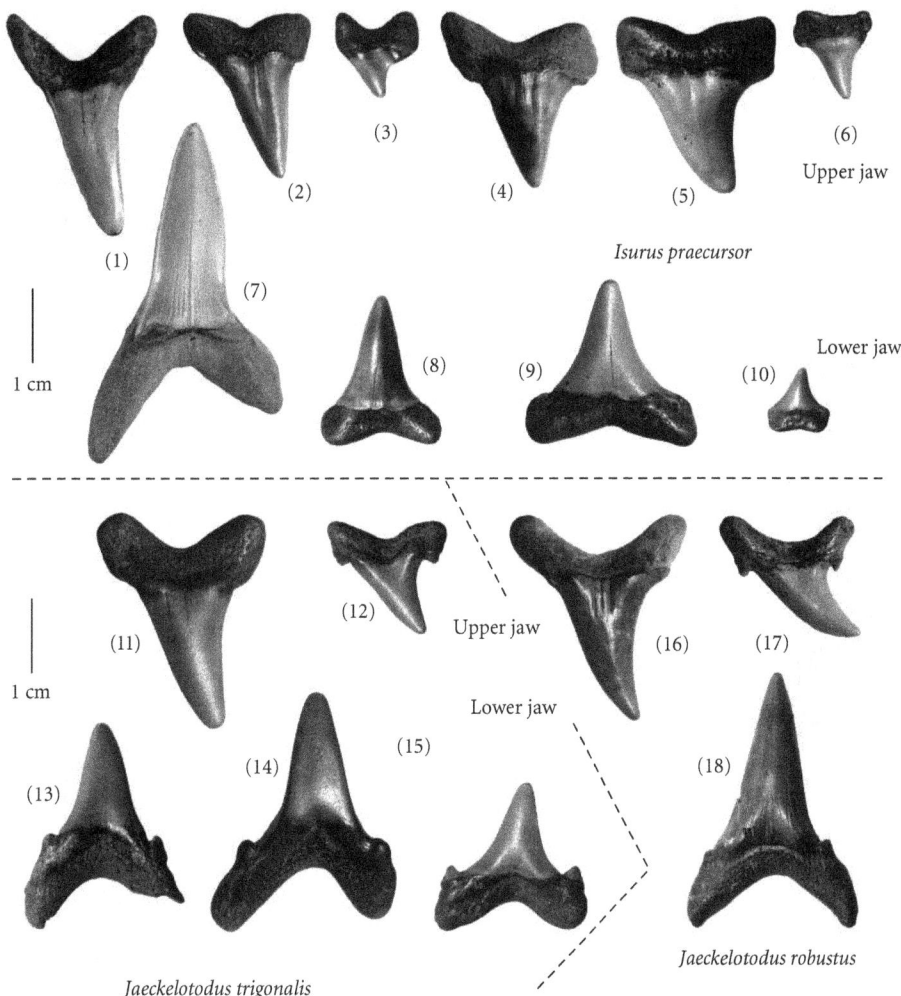

FIGURE 11: Shark teeth from lamniform (Lamnidae, Jaeckelotodontidae) sharks (from different jaw positions) from the Middle Eocene (Lutetian) gravels at Dalum. (1–10) *Isurus praecursor* (Lériche, 1905): (1) first upper anterior tooth (SCB no. Sel 101), 2. upper jaw anterior tooth (SCB no. Sel-117), (3) second upper jaw tooth (SCB no. Sel-112),(4) upper jaw anterolateral tooth (SCB no. Sel-124), (5) upper jaw anterior lateral tooth (SCB no. Sel-123), (6) upper jaw posterior lateral tooth (SCB no. Sel-125), (7) first lower jaw anterior tooth (SCB no. Sel-97), (8) second lower jaw anterolateral tooth (SCB no. Sel-109), (9) lower jaw middle lateral tooth (SCB no. Sel-104), (10) lower jaw distal lateral tooth (SCB no. Sel-108). (11–15) *Jaeckelotodus trigonalis* (Jaeckel, 1895): (11) upper jaw anterior tooth (SCB no. Sel-157), (12) upper jaw lateral tooth (SCB no. Sel-156), (13) lower jaw first anterior tooth (SCB no. Sel-160), (14) lower jaw anterolateral tooth (SCB no. Sel-158), (15) lower jaw lateral tooth (SCB no. Sel-159). (16–18) *Jaeckelotodus robustus* (Lériche, 1921): (16) upper jaw anterolateral tooth (SCB no. Sel-131), (17) upper jaw lateral tooth (SCB no. Sel-137), and (18) lower jaw anterior tooth (SCB no. Sel-129; all in labial view).

sea, although only rudimentary taphonomic comparisons have been made to the Belgian and English sites [7, 30, 31]. There is anyway too little material available at this stage for a model to be made of the autochthonous benthic fauna on the sandy sea floor. The invertebrate fauna of the Middle Eocene beach gravel paleoenvironment is expected to be restricted, with a dominance of oysters [26], as has been demonstrated for the Oligocene pebble beaches of northwest Germany. Oysters such as *Pycnodonte* appear to have been attached to the gravels, but proof is yet to be found in the fragmented material from Dalum. The solitary corals are also in same state of preservation as the oysters and would not have survived any significant amount of transport, which means that they are also likely to have been attached to the gravels. A restriction of the marine invertebrate benthic

fauna must also be expected due to the freshwater influence, but further analysis of the Osteroden site, with its well-preserved mollusc assemblages (Felker *pers. com.*) is still required.

The vertebrates are dominated by the ancestral sand shark *Striatolamia macrota* (Figure 12, Table 3), which appears to have also had nursery schools in those shallow waters, as indicated by the presence of teeth from all sizes of animals including many very small teeth from young individuals. Most interesting in the shark research are two different "white and megatooth shark" ancestors, one with massive and the other with slim tooth morphologies: the megatooth shark ancestor *Carcharocles sokolovi* (the massive-toothed form) [49], and the great white shark ancestor *Carcharodon auriculatus* (the smaller form) [49].

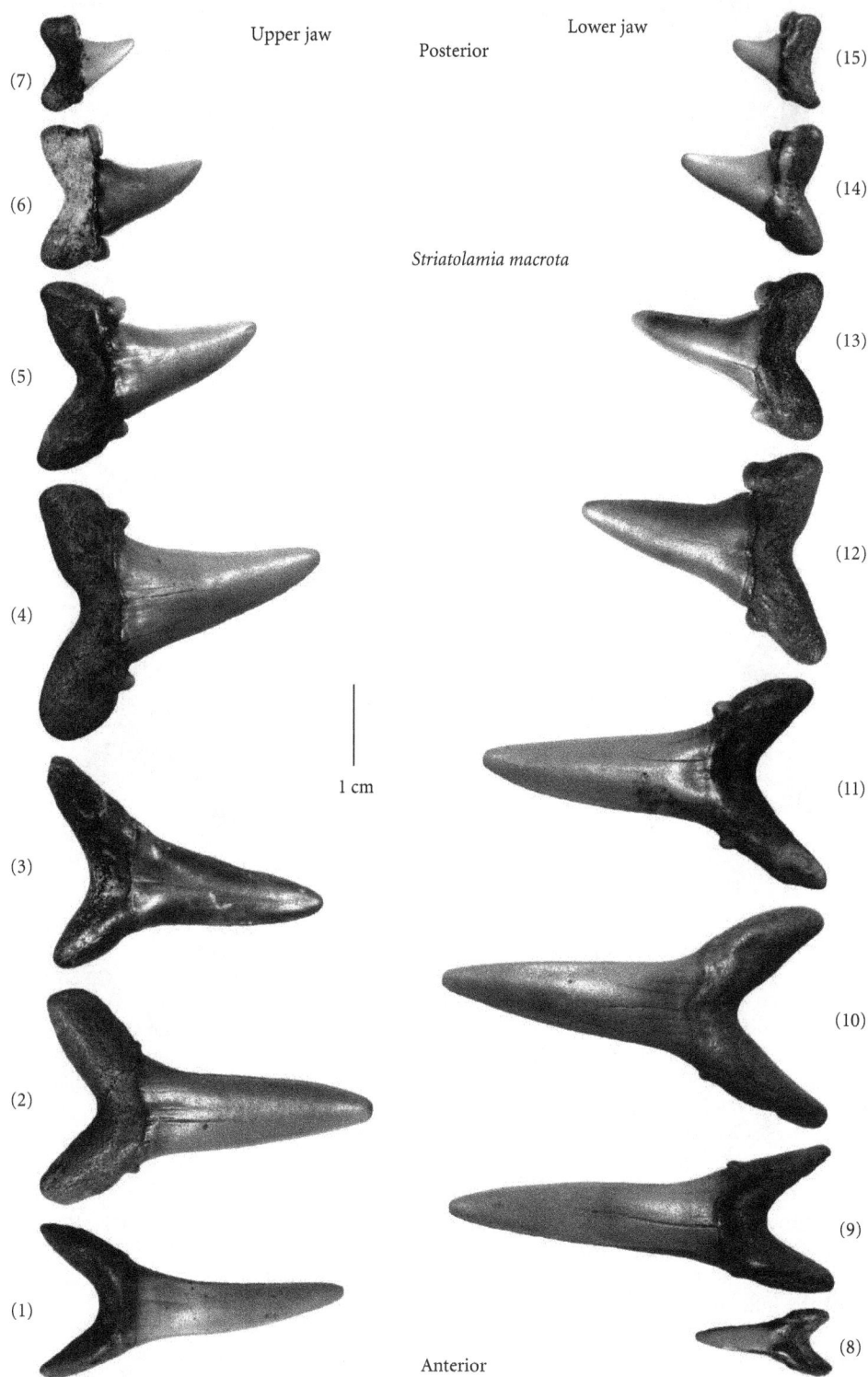

FIGURE 12: Shark teeth from lamnid (Odontaspididae) *Striatolamia macrota* (Agassiz, 1843) sharks (from different jaw positions) from the Middle Eocene (Lutetian) gravels at Dalum. (1) upper jaw first tooth (SCB no. Sel-195), (2) upper jaw second tooth (SCB no. Sel-179), (3) upper jaw third tooth (SCB no. Sel-212), (4) upper jaw fifth tooth (SCB no. Sel-195), (5) upper jaw middle lateral tooth (SCB no. Sel-172), (6) upper jaw middle lateral tooth (SCB no. Sel-171), (7) upper jaw posteriolateral tooth (SCB no. Sel-220), (8) lower jaw first (symphyseal) tooth (SCB no. Sel-167), (9) lower jaw second tooth (SCB no. Sel-168), (10) lower jaw third tooth (SCB no. Sel-169), (11) lower jaw fourth tooth (SCB no. Sel-170), (12) lower jaw anterior lateral tooth (SCB no. Sel-217), (13) lower jaw middle lateral tooth (SCB no. Sel-218), (14) lower jaw distal lateral tooth (SCB no. Sel-219), and (15) lower jaw posterolateral tooth (SCB no. Sel-214; and all in labial view).

Eocene (Lutetian) Shark-Rich Coastal Paleoenvironments of the Southern North Sea Basin in Europe: Biodiversity
of the Marine Fürstenau Formation Including Early White and Megatooth Sharks

139

TABLE 3: Shark species from the Dalum site based on the number of identifiable teeth >4 mm.

Sharks	Quantity	Time range
Striatolamia macrota	5,901	Paleocene-Eocene
Carcharodon auriculatus	455	Middle/Upper Eocene
Isurus praecursor	179	Lower/Middle Eocene
Sylvestrilamia teretidens	95	Eocene
Hypotodus verticalis	64	Eocene
Carcharocles subserratus	48	Lower Eocene
Otodus obliquus	47	Paleocene-Lower Eocene
Jaeckelotodus robustus	46	Lower/Middle Eocene
Brachycarcharias lerichei	36	Eocene
Jaeckelotodus trigonalis	33	Lower/Middle Eocene
Carcharocles sokolovi	13	Middle Eocene-Lower Oligocene
Carcharias sp.	12	Eocene
Anomotodon novus	5	Eocene
Isurolamia affinis	4	Eocene
Galeocerdo latidens	3	Eocene
Pachygaleus lefevrei	2	Eocene
Xiphodolamia ensis	1	Eocene
Usakias wardi	1	Eocene
Odontaspis winkleri	1	Eocene
Total	6,946	
Selachii Indet.	6,737	
Total number of teeth	13,683	

Two different "large serrated shark" forms can therefore be shown (from the well-dated and large quantities of teeth) to have already developed by the basal Middle Eocene, both of which will need to be further analysed in greater detail on the basis of the material from Dalum in order to further define the ancestry of the white/megatooth sharks and their evolutionary record, which remains as yet unclear, controversial, and a matter of some discussion [49, 50]. The Dalum material includes intermediate forms between *Otodus obliquus* and *Carcharocles sokolovi*, such as *Carcharocles subserratus* whose teeth seem to have all been reworked from Lower Eocene horizons, as have most of the *O. obliquus* teeth. The origin of the white/megatooth shark lineage will be investigated using about 1,000 serrated shark teeth from the two sites at Dalum and Osteroden, which is possibly the largest collection in general [51], and for the Eocene period in particular.

5.3. Paleogeographic Reconstruction and Paleoenvironment. Analysis of conglomerate pebbles and the reworked Mesozoic fossils has allowed a detailed paleogeographic map of the Osnabrück region, north of the Rhenish Massif, to be prepared, revising the former less refined models [13, 16, 20]. Previous pebble analysis work by Bartholomäus [2], in which the iron-carbonate clasts, ammonite fragments, and several pebble types were not recognised, has been considerably extended herein.

Large quantities of Paleozoic quartzite material (from Carboniferous sandstones and conglomerates) must have been transported, mainly from Schafberg area but also from the Piesberg area (see Carboniferous horst structures and Westphal C-D sandstones [21]), which appears to have been the result of major erosion under a tropical climate by an ancient (hypothetically reconstructed) river that flowed during the Middle Eocene between the mountain chains that today form the Wiehengebirge and Teutoburger Wald ranges (Figure 5). This river also may also have transported Paleozoic gravels (including lydite) from the Rhenish Massif. Only the hard silica-rich rock types could have survived long-distance transport. With such a river and a tropical climate an estuarine or deltaic situation can be expected at its mouth [52], as has been reconstructed herein (on a preliminary basis) using only the two Dalum and Osteroden sections and their differences in gravel thickness. Dalum would have been more proximal (80 cm thickness) than Osteroden (40 cm thickness) in the deltaic fan (see [52]). More sections and drill cores from the Fürstenau area will need to be studied in the future in order to obtain an improved understanding of this paleoenvironment. The conglomerates can anyway not be simply treated as tempestites or as being of purely transgressive origin [1, 13], as they are demonstrated herein to be a combination of marine transgressive and fluvial deltaic deposits. The Lower Cretaceous (Valanginian/Hauterivian) iron-carbonate concretions, the smaller burrows, and the many ammonite fragments, with all softer rock types showing burrows and boreholes from Eocene times (Figure 5), were transported during the Early Eocene (Ypresian [19]) and subsequently reworked during the Middle Eocene from the Early Eocene marine deposits, or possibly even reworked twice, and finally accumulated north of the Wiehengebirge mountain range (to the north-west of Osnabrück) in the southern pre-North Sea basin coastal pebble facies that were mixed with deltaic deposits (Figure 5).

The region north-west of Fürstenau (i.e., the Dalum/Osteroden region) represents a deltaic, fresh-water-influenced, low-carbonate, siliciclastic glauconite sand to pebble facies in the shallow marine coastal environment of the southern pre-North Sea basin, to the north of the Rhenish Massif. In contrast, the Belgian (Balegem/Osterzele) region was more carbonate influenced and was located in shallow marine areas between the southwestern pre-North Sea and the Paris Basins [44]. The Osterzele and Balegem facies had similarities with the Middle Eocene facies from north-west Germany, but had less terrestrial influences whereas in the Lower Eocene London Clay Formation the terrestrial input was even higher into the coastal paleoen- vironments [14].

6. Conclusion

The Middle Eocene (Lutetian) transgressive conglomerates from Dalum and Osteroden in north-west Germany are a mixture that includes reworked Lower Cretaceous (Valanginian-Hauterivian) marine claystones, with the derived ammonite fragments and pebbles making up 20% of the total number of clasts. About 80% of these reworked Lower Cretaceous pebbles contain burrows or boreholes (e.g., *Gastrochaenolithes* boreholes made by bivalves). Boreholes have also been left in driftwood by the bivalve *Teredo*.

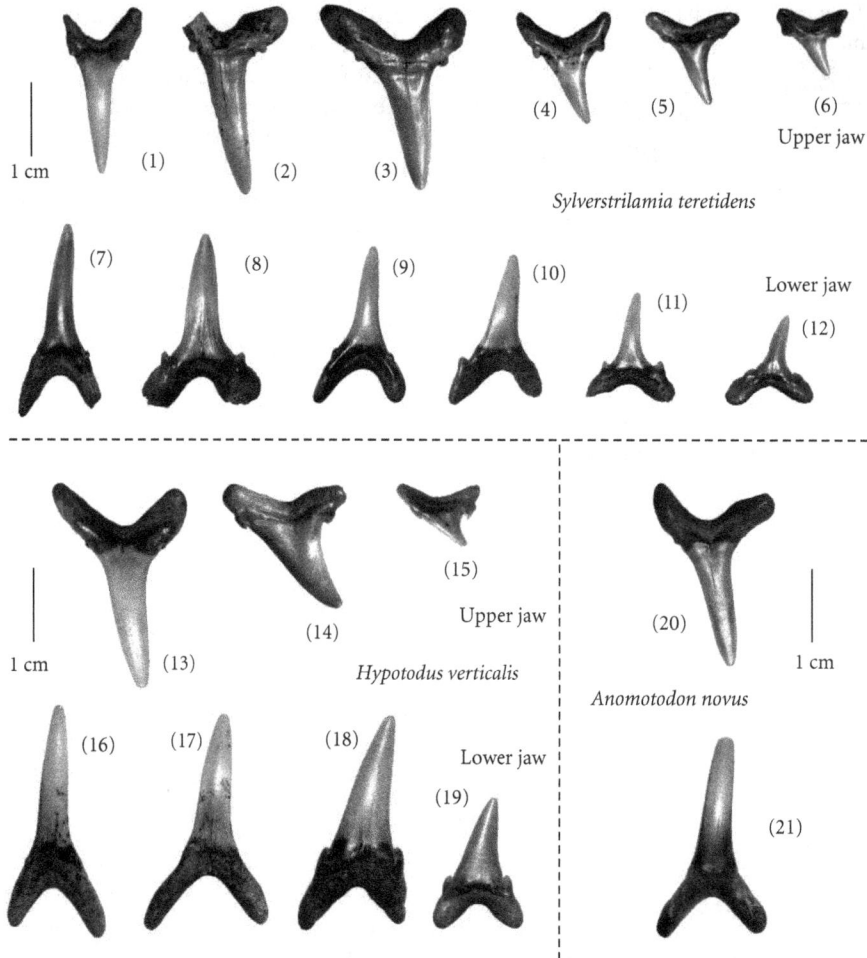

FIGURE 13: Shark teeth from lamniform (Odontaspididae, Mitsukurinidae) sharks (from different jaw positions) from the Middle Eocene (Lutetian) gravels at Dalum. (1–12). *Sylvestrilamia teretidens* (White, 1931): (1) upper jaw first anterior tooth (SCB no. Sel-832), (2) upper jaw second anterior tooth (SCB no. Sel-833), (3) upper jaw third anterior tooth (SCB no. Sel-834), (4) upper jaw middle lateral tooth (SCB no. Sel-835), mirrored, (5) upper jaw middle lateral tooth (SCB no. Sel-836), mirrored, (6) upper jaw posteriolateral tooth (SCB no. Sel-837), (7) lower jaw second anterior tooth (SCB no. Sel-838), (8) lower jaw third anterior tooth (SCB no. Sel-839), (9) lower jaw anterior tooth (SCB no. Sel-840), (10) lower jaw anterolateral tooth (SCB no. Sel-841), (11) lower jaw middle lateral tooth (SCB no. Sel-843),and (12) lower jaw posterior lateral tooth (SCB no. Sel-845), mirrored. (13–19). *Hypotodus verticalis* (Agassiz, 1843): (13) upper jaw anterior tooth (SCB no. Sel-823), (14) upper jaw middle lateral tooth (SCB no. Sel-824), (15) upper jaw posteriolateral tooth (SCB no. Sel-825), (16) lower jaw second anterior tooth (SCB no. Sel-827), (17) lower jaw third anterior tooth (SCB no. Sel-828), (18) lower jaw fourth anterior tooth (SCB no. Sel-829), (19) lower jaw middle lateral tooth (SCB no. Sel-826). (20–22) *Anomotodon novus* (Winkler, 1874): (20) upper jaw anterior tooth (SCB no. Sel-818), and (21) Lower jaw anterior tooth (SCB no. Sel-819), (22) lower jaw anterior tooth (SCB no. Sel-; 820) all in labial view).

In addition, Paleogene phosphorite nodules sometimes contain benthic marine invertebrate fauna (and also burrow casts). These were transported southward to the pebble beach coast, as were all the Lower to basal Middle Eocene vertebrate remains and, in particular, the shark teeth that are highly concentrated within these gravels, with a density of about 5,000 per cubic meter of gravel.

The other main components of the conglomerate are Paleozoic (Carboniferous) siliciclastic sandstones, and the (mostly well-rounded) quartzites that make up 59% of the pebbles, but these must have been transported by river and come from the opposite direction, deriving from the mainland to the south. An ancient river between the

Wiehengebirge and Teutoburger Wald mountain ranges must have flowed to the west of Osnabrück and the Carboniferous Schafberg where a river passed to the west of the main quartz source. Even the resistant sandstone rocks were affected by weathering in the harsh tropical climate. A "Fürstenau-Delta" is postulated, with a conglomerate thickness of about 80 cm in Dalum (proximal) and 40 cm in Osteroden (distal), suggesting a deltaic fan close to the western end of the mountain chains on the Rhenish Massif mainland. The coastline during the Early Eocene (Ypresian) was a little further south, and later moved some tens of kilometres north of the Wiehengebirge mountain range, where it was during the Middle Eocene (Lutetian).

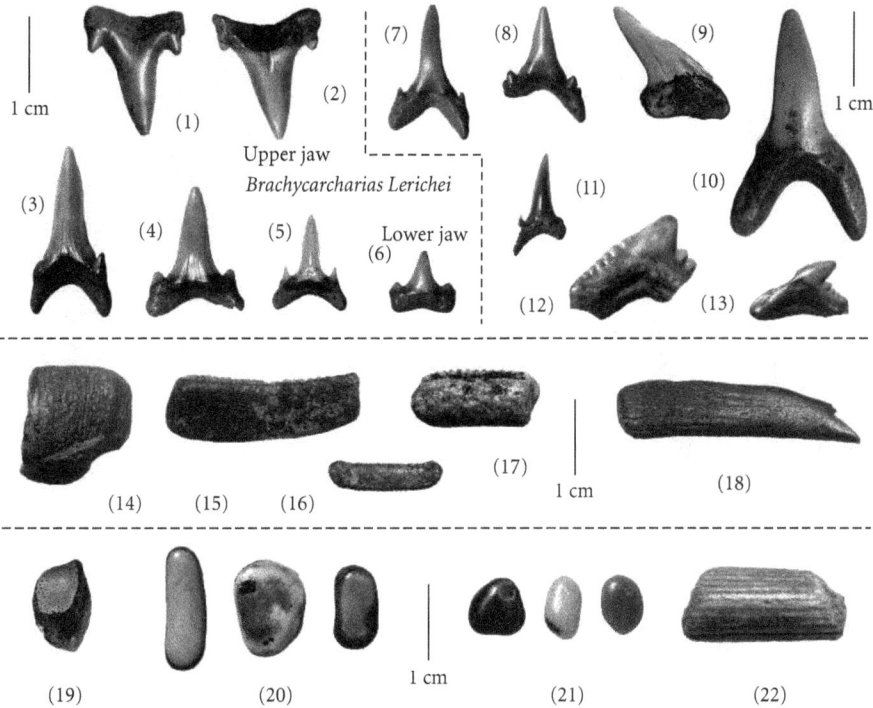

FIGURE 14: Shark teeth from rare shark species, rays, chimaera, and bony fish remaining from the Middle Eocene (Lutetian) gravels at Dalum. 1–6. *Brachycarcharias lerichei* (Casier, 1946): (1) upper jaw second anterior tooth (SCB no. Sel-849), (2) upper jaw third anterior tooth (SCB no. Sel-850), (3) lower jaw first anterior tooth (SCB no. Sel-851), (4) lower jaw third anterior tooth (SCB no. Sel-848), (5) lower jaw lateral tooth (SCB no. Sel-852), (6) lower jaw posterolateral tooth (SCB no. Sel-853). 7-8. *Carcharias* sp.: (7) lower jaw anterolateral tooth (SCB no. Sel-878), (8) lower jaw lateral tooth (SCB no. Sel-879). (9) Xiphodolamia ensis Leidy, 1877 upper jaw second tooth (SCB no. Sel-859). (10) *Isurolamna affinis* (Casier, 1946) lower jaw first anterior tooth (SCB no. Sel-865). (11) *Otodus winkleri* Lériche, 1905 upper jaw anterolateral tooth (SCB no. Sel-883). (12) *Galeocerdo eaglesomei* (White, 1955) tooth (SCB no. Sel-2). (13) *Pachygaleus lefevrei* (Dammeries, 1891) tooth (SCB no. Sel-1). (14) Tooth from a large unidentified ray (SCB no. Bat-7). (15) *Aetobatis irregularis* (Agassiz, 1843) tooth fragment (SCB no. Bat-4). (16) *Leidybatis jugosus* (Leidy, 1876) tooth (SCB no. Sel-902). (17) *Myliobatis dixoni* (Agassiz, 1843) tooth fragment (SCB no. Bat-6). (18) *Pristis lathami* Galeotti, 1837 rostral tooth (SCB no. Bat-9). (19) *Edaphodon bucklandi* Agassiz, 1843 (SCB no. Chim-1). (20) *Phylodus* sp. teeth (SCB no. Tel-1). (21) *Pycnodus* sp. three lateral row teeth (SCB no. Tel-2). (22) *Cylindracanthus* sp. rostrum fragment (SCB no. Tel-8). (1–13 labial view).

The northern Rhenish Massif (the Westphalian Cretaceous Basin/Bay) was not flooded during a medium high stand of the Early to Middle Eocene pre-North Sea Basin, nor was the area between the Wiehengebirge and Teutoburger Wald mountain ranges subjected to marine flooding. The area between the Wiehengebirgs and Teutoburger Wald mountain ranges was, however, subjected to flooding later during the Oligocene (Latdorfian to Chattian), when the old river drainage system of Eocene times was flooded and destroyed.

The shark fauna is mainly dominated by the sand shark *Striatolamia macrota*, for which a full range of ages is well represented. *Otodus obliquus* is also another common large shark. A "great white" shark ancestor is represented by teeth from *Carcharodon auriculatus* and a megatooth shark ancestor by the larger and more massively rooted teeth of *Carcharocles sokolovi*. Other moderately common sharks include *Isurus praecursor*, *Jaeckelotodus robustus*, and *J. trigonalis*. The *Otodus obliquus*, and *C. subserratus* teeth, in particular, appear to indicate reworking of older Early Eocene marine sediments, which are known to exist in northwest Germany. All of the other large shark species (teeth >4 mm), out of a total of 19 that have been identified, are more rare. The larger sharks show a moderate degree of biodiversity and do not indicate tropical marine water temperatures, nor does the absence of coral reefs and the low carbonate production, which is in contrast to the terrestrial tropical forest mammal fauna biodiversity of the Rhenish Massif Mainland.

Acknowledgments

The excavation field project was sponsored by the "Geo- and Naturpark TERRA.Vita" (H. Escher), and the Landkreis, Osnabrück. The private company "PaleoLogic" was responsible for the field work and sponsoring of the scientific research. Special thanks are due to A. Felde from the Reuter company for his excellent work with the backhoe, and to the company for its full support. The Kuhlhoff Education Center leaders Mr. A. Bruns and Mr. W. Hollermann, and the mayor of Bippen, Mr. H. Tolsdorf, are thanked for their managements to install the field sieving laboratory and for any support. Finally, the author would also like to

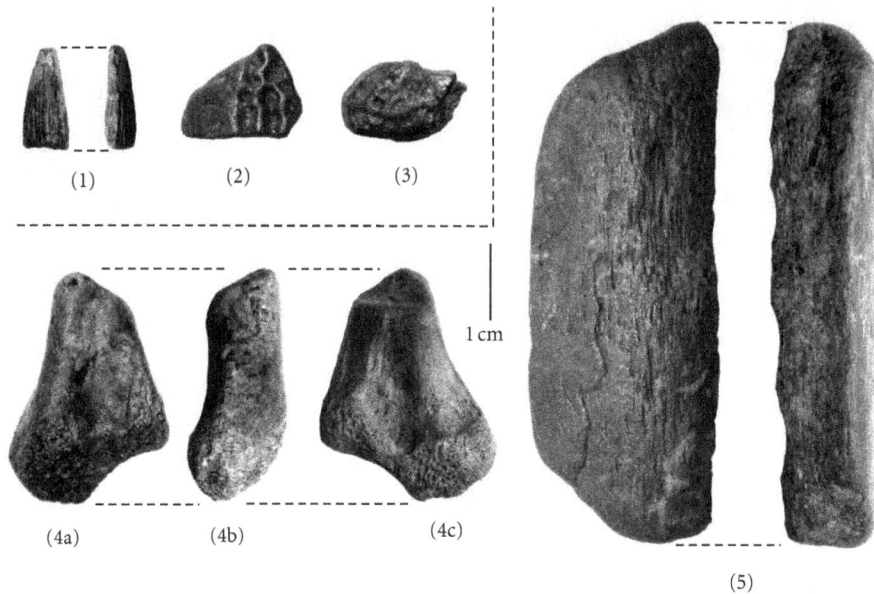

FIGURE 15: Reptile and mammal remains form the Middle Eocene (Lutetian) gravels at Dalum. (1) Crocodile tooth from *Diplocynodon* (SCB no. Rept-1), labial and lateral. (2) Crocodile osteoderm fragment from *Diplocynodon* (SCB no. Rept-2), dorsal. (3) Turtle *Trionyx* osteoderm fragment (SCB no. Rept-3), dorsal. (4) Distal joint of a humerus from an unidentified medium-sized mammal (SCB no. Mam-1), (a) dorsal, (b) lateral, and (c) ventral. (5) Femur fragment from an unidentified large mammal (SCB no. Mam-2), dorsal and lateral.

thank H. Felker, a local hobby paleontologist, who provided support, helped with sieving the material, and allowed access to his large collection (obtained over 25 years from both localities (Dalum and Osteroden) for comparisons.

References

[1] F. Hocht, "Eine Lagerstätte kreidezeitlicher und paläogener Chondrichthys-Reste bei Fürstenau (Niedersachsen)," *Osnabrücker Naturwissenschaftliche Mitteilungen*, vol. 6, pp. 35–44, 1979.

[2] W. A. Bartholomäus, "Ein kondensiertes Knochenlager im marinen Alttertiär von Fürstenau/SW-Niedersachsen," *Fossilien*, vol. 2, no. 93, pp. 79–82, 1993.

[3] W. Schwarzhans, "The otoliths from the Middle Eocene of Osteroden near Bramsche, north-western Germany," *Neues Jahrbuch fur Geologie und Palaontologie*, vol. 244, no. 3, pp. 299–369, 2007.

[4] J. L. Franzen and T. Mörs, "Das nördlichste Vorkommen paläogener Säugetiere in Europa," *Palaeontologische Zeitschrift*, vol. 81, no. 4, pp. 447–456, 2007.

[5] G. Van den Eeckhaut and P. De Schutter, "The Elasmobranch fauna of the Lede Sand Formation at Osterzele (Lutetian, Middle Eocene of Belgium)," *Palaeofocus*, vol. 1, pp. 1–57, 2009.

[6] M. Lériche, "L'Eocene des Bassins Parisiennes et Belge, Livret-guide de la réunion extraordinaire de la Société Géologique de France," *Bulletin de la Société Géologique de France*, vol. 4, no. 12, pp. 692–807, 1912.

[7] D. Rayner, T. Mitchel, M. Rayner, and F. Clouter, *London Clay Fossils of Kent and Essex. Medway Fossil and Mineral Society*, Caxton and Holmesdale Press Ltd, Kent, Ohio, USA, 2009.

[8] J.L. Franzen, "The biostratigraphic and palaeoecologic significance of the Middle Eocene locality Geiseltal near Halle (German, Democratic Republic)," *Münchner Geowissenchaftliche Abhandlungen A*, vol. 10, pp. 93–100, 1987.

[9] S. Schaal and W. Ziegler, Eds., *Messel—ein Schaufenster in die Geschichte der Erde und des Lebens*, Waldemar Kramer, Frankfurt, Germany, 1988.

[10] F. W. Negendank, G. Irion, and J. Linden, "Ein eozänes Maar bei Eckfeld nordöstlich Manderscheid (SW-Eifel)," *Mainzer Geowissenschaftliche Mitteilungen*, vol. 11, pp. 157–172, 1982.

[11] H. H. Schleich, "Neue Reptilfunde aus dem Tertiär Deutschlands. 13. Schildkröten- und Krokodilreste aus der eozänen Braunkohle des Untertargebaues Stolzenbach bei Borken (Hessen)," *Courier des Forschungsinstitutes Senckenberg*, vol. 173, pp. 79–101, 1994.

[12] J.-L. Hartenberger, "Les Mammifêres d'Egerkingen et l'histoire des faunes de l'Eocène d'Europe," *Bulletin de la Société Geologique France, Serie VII*, vol. 12, no. 5, pp. 886–893, 1970.

[13] H. Hiltermann, "Tertiär," in *Geologie des Osnabrücker Berglandes*, H. Klassen, Ed., pp. 463–494, Osnabrück Germany, 1984.

[14] C. King, "The stratigraphy of the London Clay Formation and Virginia Formation in the coastal sections of the Isle of Sheppy (Kent, England)," *Tertiary Research*, vol. 5, no. 3, pp. 121–160, 1984.

[15] P. Mangelsdorf, *Geologische Wanderkarte Landkreis Osnabrück*, Landkreis Osnabrück, Hanover, Germany, 1984.

[16] F. Bettenstaedt, "Paläogeographie des nordwestdeutschen Tertiärs," *Erdöl und Tektonik in NW Deutschland*, vol. 1941, pp. 143–172, 1949.

[17] F. Bettenstaedt, "Tertiär Norddeutschlands," *Leitfossilien der Mikropaläontologie*, vol. 1962, pp. 339–379, 1962.

[18] J. Rohling, "Beiträge zur Stratigraphie und Tektonik des Tertiärs in Südoldenburg," *Decheniana A*, vol. 100, pp. 1–103, 1941.

[19] K. Staesche and H. Hilterman, "Mikrofaunen aus dem Tertiär NW Deutschlands," *Abhandlungen der Reichsanstalt für Bodenforschung Neue Folge*, vol. 201, pp. 1–26, 1940.

[20] P. A. Ziegler, *Geological Atlas of Western and Central Europe*, Shell, Amsterdam, The Netherlands, 1990.

[21] K.-H. Josten, K. Köwing, A. Rabitz et al., "Oberkarbon," in *Geologie des Osnabrücker Berglandes*, H. Klassen, Ed., pp. 7–77, Osnabrück, Germany, 1984.

[22] E. Kempe, *Die Tiefe Unterkreide im Vechte-Dinkel-Gebiet*, Dobler-Druck, Alfeld,Germany, 1992.

[23] C. Diedrich, "Kieselgeoden als wichtiger Rohstoff für die Beilproduktion im Neolithikum von Nordwestdeutschland—erste Nachweise von Produktionsplätzen der Trichterbecherkultur," pp. 1–36, 2004, www.jungsteinsite.de/.

[24] W. Weitschat, "Lias und Dogger," in *Geologie des Osnabrücker Berglandes*, H. Klassen, Ed., pp. 335–385, Osnabrück, Germany, 1984.

[25] M. Bertling and K. Hermanns, "Autochthone Muschelbohrungen im Neogen des Rheinischen Braunkohlenreviers und ihre sedimentologische Bedeutung," *Zentralblatt für Geologie und Paläontologie I*, vol. 1995, no. 1-2, pp. 33–44, 1996.

[26] C. Diedrich, "Palaeoecology, facies and stratigraphy of the shallow marine macrofauna from the Upper Oligocene (Palaeogene) in the southern Pre- North-Sea Basin of Astrup near Osnabrück (NW Germany)," *Cainozoic Research*. In press.

[27] M. Berndt and J. Welle, "Ahermatype Korallen aus dem Oberoligozän der Schächte Rheinberg und Sophia Jacoba 8 (Niederrheinische Bucht)," *Neues Jahrbuch für Geologie und Paläontologie Abhandlungen*, vol. 198, no. 1-2, pp. 47–68, 1995.

[28] W. J. Quayle, "English Eocene Crustacea (Lobsters and Stomatopods)," *Tertiary Research*, vol. 30, no. 5, pp. 581–612, 1987.

[29] J. S. H. Collins and J. Saward, "Three new genera and species of crabs from the Lower Eocene London Clay of Essex, England," *Bulletin of the Mitzunami Fossil Museum*, vol. 33, no. 2006, pp. 67–76, 2002.

[30] M. Gilbert, "Les Bivalvia du Lédien (Eocène ,oyen supérieur) de la Belgique I, Paleotaxodonta, cryptodonta, Pteriomorpha," *Bulletin d l'Institut royal des Sciences Naturelles de Belgique/Mededelingen van het Koninklijk Belgish Instituut voor Natuurwetenschappen*, vol. 51, no. 3, p. 1, 1975.

[31] M. Gilbert, "Les Bivalvia du Lédien (Eocène ,oyen supérieur) de la Belgique II. Heterodonta, et Anomalodesmata," *Bulletin d l'Institut Royal des Sciences Naturelles de Belgique/Mededelingen van het Koninklijk Belgish Instituut voor Natuurwetenschappen*, vol. 52, no. 5, pp. 1–80, 1980.

[32] O. Jaeckel, *Die Eozänen Selachier aus Monte Bolca*, 1894.

[33] O. Jaeckel, "Unter-Tertiäre Selachier aus Südrussland," *Mémoires du Comité Géologique de St.-Pétersbourg*, vol. 9, pp. 19–35, 1895.

[34] M. Lériche, "Contribution à lètude des poisons fossils du Nord de la France et des regions voisines," *Mémoires de la Société Geologique du Nord*, vol. 5, pp. 1–430, 1906.

[35] E. Casier, "Contributions à l'étude des poisons fossiles de la Belgique VII—les Pristidés Eocènes," *Bulletin de l'Institut Royal des Sciences Naturelles de Belgique/Mededelingen van het Koninklijk Belgisch Instituut voor Natuurwetenshappen*, vol. 25, no. 19, pp. 1–52, 1949.

[36] E. Casier, "Fauna ichthyologique du London Clay," *British Museum Natural History London*, vol. 1966, pp. 1–496, 1966.

[37] H. Cappetta, *Chondrichthyes II. Mesozoic and Cenozoic Elasmobranchii. Handbook of Paleoichthyology*, Gustav Fischer, Stuttgart, Germany, 1987.

[38] H. Cappetta and D. Nolf, "Révision de quelques Odontaspididae (Neoselachii: Lamniformes) du Paléocène et de l'Eocène du Bassin de la mer du Nord," *Bulletin de l'Institut Royal des Sciences Naturelles de Belgique/Mededelingen van het Koninklijk Belgisch Instituut voor Naturwetenshappen, Sciences de la Terre/Aardwetenshappen*, vol. 75, pp. 237–266, 2005.

[39] S. Adnet, "Nouvelles faunes de Sélachiens (Elasmobranchii, neoselachii) de l'Eocène moyen des Landes (Sud-Ouest France)," *Implications Dans la Connaissance des Communautés de Sélachians Dèaux Profonds. Palaeo Ichthyologica*, vol. 10, pp. 1–128, 2006.

[40] S. G. Lucas, "Fossil mammals and the Paleocene/Eocene series boundary in Europe, North Amrica, and Asia," in *Late Paleocene-Early Eocene Climatic and Biotic Evolution*, M.-P. Aubry, S. Lucas, and W. A. Berggren, Eds., pp. 451–500, Columbia University Press, New York, NY, USA, 1998.

[41] C. Diedrich, "The food of the miosiren *anomotherium langenwieschei* (Siegfried)—indirect proof of seaweed or seagrass by xenomorphic oyster fixation structures in the Upper Oligocene (Neogene) of the Doberg, Bünde (NW Germany) and comparisons to modern *Dugong dugon* (Müller) feeding strategies," *Senckenbergiana Maritima*, vol. 38, no. 1, pp. 59–73, 2008.

[42] J. Hermann, E. Steurbaut, and N. Vandenberghe, "The boundary between the Middle Eocene Brussel Sand and the Lede Sand Formations on the Zaventem-Nederokkerzeel area (Northeast of Brussels, Belgium)," *Geologica Belgica*, vol. 3, no. 3-4, pp. 213–255, 2000.

[43] T. Smith, B. De Wilde, and E. Steurbaut, "Primitive equoid and Tapiroid mammals: keys for interpreting the Ypresian/Lutatian transition in Belgium," *Bulletin van het Koninklijk Belgisch Institut voor Natuurwetenshappen, Aardwetenschappen, Supplement*, vol. 74, pp. 165–175, 2004.

[44] B. Fobe and V. Spiers, "Sedimentology and facies distribution of the Lede Formation (Eocene) in Belgium and northern France," *Contributions to Tertiary and Quaternary Geology*, vol. 29, no. 1-2, pp. 9–20, 1992.

[45] P. Jacobs and E. Sevens, "Middle Eocene sequence stratigraphy in the Balegem quarry (Western Belgium, Southern Bigth North Sea)," *Bulletin de la Société belge de Géologie/Bulletin van de Belgische Vereniging voor Geologie*, vol. 102, no. 1-2, pp. 203–213, 1994.

[46] M.-P. Aubry, "Palaeocene/Eocene boundary events in space and time," *Israel Bulletin of Earth Sciences*, vol. 44, pp. 1–253, 1998.

[47] H.-P. Lauterbacher, J. R. Ali, H. Brinkhuis et al., "The paleogene period," in *A Geological Time Scale*, G. M. Gradstein, J. G. Ogg, A. Smith et al., Eds., pp. 384–408, Cambridge University Press, Cambridge, UK, 2004.

[48] W. A. Berggren and M.-P Aubry, "Late Paleocene-Early Eocene NW European and North sea magnetostratigraphic correlation network. A sequence stratigraphic approach," in *Correlation of the Early Palaeogene in Northwestern Europe*, O. R. B. Knox, R. C. Corfiled, and R. E. Dunay, Eds., vol. 101, pp. 309–352, 1996.

[49] A. Martin, "Systematics of the Lamnidae and the Origination Time of *Carcharodon carcharias* inferred from the comparative analysis of mitochondrial DNA sequences," in *Great White Sharks: The Biology of Carcharodon Carcharias*, A. P. Kimley and D. G. Ainley, Eds., pp. 67–78, Academic Press, San Diego, Calif, USA, 1996.

[50] S. P. Applegate and L. Espinosa-Arrubarrena, "The fossil history of *Carcharodon* and its possible ancestor, *Cretolamna*: a study in tooth identification," in *Great White Sharks—The Biology of Carcharodon Carcharias. Ainley*, A. P. Klimley and G. David, Eds., pp. 19–36, Academic Press, San Diego, Calif, USA, 1996.

[51] D. J. Long and B. M. Waggoner, "Evolutionary relationships of White Shark: a phylogeny of lamniform sharks based on dental morphology," in *Great White Sharks—The Biology of Carcharodon Carcharias*, A. P. Klimley and G. David, Eds., pp. 37–47, Academic Press, San Diego, Calif, USA, 1996.

[52] H. Füchtbauer and G. Müller, *Sedimente und Sedimentgesteine*, Schweizbart' sche Verlagsbuchhandlung, Stuttgart, Germany, 1970.

14

C-Band Polarimetric Coherences and Ratios for Discriminating Sea Ice Roughness

Mukesh Gupta, Randall K. Scharien, and David G. Barber

Centre for Earth Observation Science, Department of Environment and Geography, Clayton H. Riddell Faculty of Environment, Earth, and Resources, University of Manitoba, Winnipeg, MB, Canada R3T 2N2

Correspondence should be addressed to Mukesh Gupta; mukesh_gupta@umanitoba.ca

Academic Editor: Grant Bigg

The rapid decline of sea ice in the Arctic has resulted in a variable sea ice roughness that necessitates improved methods for efficient observation using high-resolution spaceborne radar. The utility of C-band polarimetric backscatter, coherences, and ratios as a discriminator of ice surface roughness is evaluated. An existing one-dimensional backscatter model has been modified to two-dimensions (2D) by considering deviation in the orientation (i.e., the slopes) in azimuth and range direction of surface roughness simultaneously as an improvement in the model. It is shown theoretically that the circular coherence (ρ_{RRLL}) decreases exponentially with increasing surface roughness. The crosspolarized coherence (ρ_{HHVH}) is found to be less sensitive to surface roughness, whereas the copolarized coherence (ρ_{VVHH}) decreases at far-range incidence angles for all ice types. A complete validation of the adapted 2D model using direct measurements of surface roughness is suggested as an avenue for further research.

1. Introduction

Arctic sea ice is going through a rapid decline [1, 2]. Thinner first-year ice (FYI) is replacing multiyear ice, leaving an ice cover, which is more sensitive to deformation and changes in atmospheric and ocean forcing. Increased open water and marginal ice zones (MIZs), due to the enhanced mobility of a relatively thinned pack ice, are further susceptible to increases in surface roughness and greater surface roughness variability [3]. Greater surface roughness in the MIZ is of importance due to higher rates of heat flux [4] and momentum [5] exchanges occurring across the ocean-sea ice-atmosphere interface, greater biological productivity [6], and potential limitations imposed on ship navigation. Although the literature contains information on how the MIZ responds to wind and wave forces, it is necessary to investigate the electromagnetic (EM) response of the MIZ to facilitate satellite-based observations. Satellite-based observation is necessary due to the scarcity of surface observations in a MIZ, as well as the difficulties in collecting physical measurements due to the instability and roughness of the ice floes.

The use of polarimetric synthetic aperture radar (pol-SAR) represents a promising approach for satellite-based monitoring of surface roughness and, concurrently, discriminating sea ice types within a MIZ. A pol-SAR records the amplitude and phase information of backscattered energy for four transmit-receive polarizations (HH, HV, VH, and VV), thereby facilitating the derivation of the full polarimetric response of the target. It is recognizable that the diversity in polarization achievable by pol-SARs or even by dual-polarization SAR systems provides more complete inference of target features (e.g., sea ice) than conventional, single channel SARs. Furthermore, recently launched pol-SARs are capable of higher spatial resolution (<10 m) imaging, leading to enhanced potential for monitoring complex ice environments.

Discrimination of ice types using SAR has been conventionally achieved by utilizing different combinations of linearly polarized backscattering coefficients [7–9]. Multiyear ice, smooth FYI, rough FYI, and new ice/open water in the Beaufort and Chukchi Seas during March have been identified using a single-polarization SAR image intensity-based classification scheme [9] while others used single-polarization SAR image texture analysis to discriminate new ice, FYI, and multiyear ice during the month of March in the Beaufort Sea and the Mould Bay, respectively [10, 11]. Dual

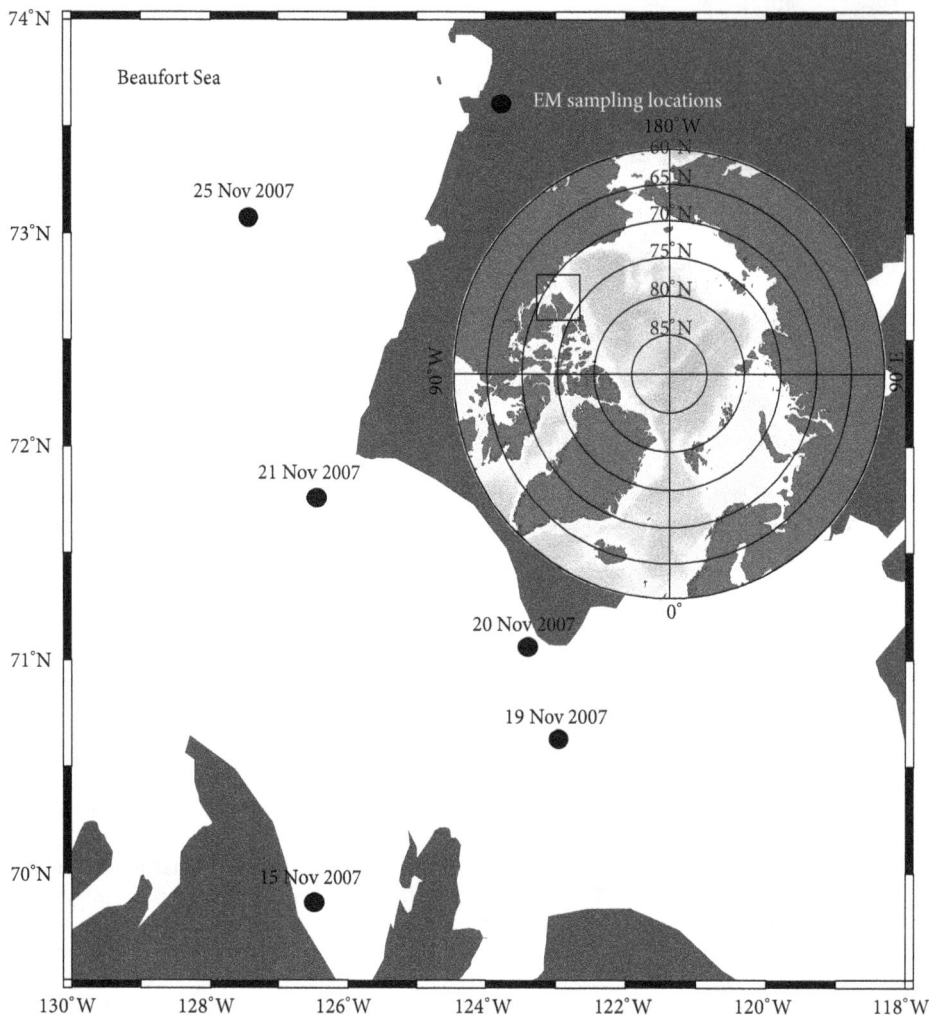

FIGURE 1: Geographic map of study area showing sampling locations.

copolarized backscattering coefficient differences in HH and VV have been used to discriminate FYI, multiyear ice, and lead areas in the Beaufort Sea during March [12]. However, the complexities in polarimetric signatures associated with the dynamic mixture of surface roughness and ice type conditions in an MIZ during fall freeze-up remain to be examined. Such an examination requires utilizing polarimetric radar backscatter, so that the material (dielectric) and geometrical properties of the surface, which influence backscatter, may be individually assessed.

In this study, ship-based observations of co- (linear) and crosspolarized backscatter, circular polarimetric coherences (ρ_{VVHH}, ρ_{HHVH}, and ρ_{RRLL}, resp.), and copolarized and crosspolarized polarization ratios (γ_{co} and γ_{cross}, resp.), are used to evaluate their utility for ice surface discrimination capabilities using a polarimetric radar operating in C-band (5.5 GHz). Characteristics of these polarimetric parameters for a variety of ice types in an MIZ during fall freeze-up are assessed with the following objectives:

(1) to investigate the performance of polarimetric ρ_{RRLL} for sea ice surface roughness discrimination by adapting the one-dimensional backscatter model of [13]

to two dimensions and introducing roughness as deviations in range and azimuth directions,

(2) to evaluate the utility of C-band polarimetric backscatter, coherences, and polarization ratios as a discriminator of surface roughness or ice type in a MIZ during fall freeze-up.

2. Methodology

2.1. Study Area. The study area is located in the southeastern Beaufort Sea and Amundsen Gulf regions in the western Canadian Arctic (Figure 1). The seasonal Cape Bathurst Polynya forms in the region and hosts a number of flaw leads during the winter [14]. During fall freeze-up, this area contains a variable mix of ice types under various stages of formation, for example, new ice, pancake ice, frost flowers, deformed ice, gray ice, and nilas (Figure 2). The photographs in Figures 2(a), 2(b), 2(c), and 2(e) were taken at an oblique angle from the port side of the Canadian Research Icebreaker *Amundsen* at approximately eight meters height using a handheld digital camera after a given scatterometer scan; Figure 2(d) was taken at nadir angle on the ice floe

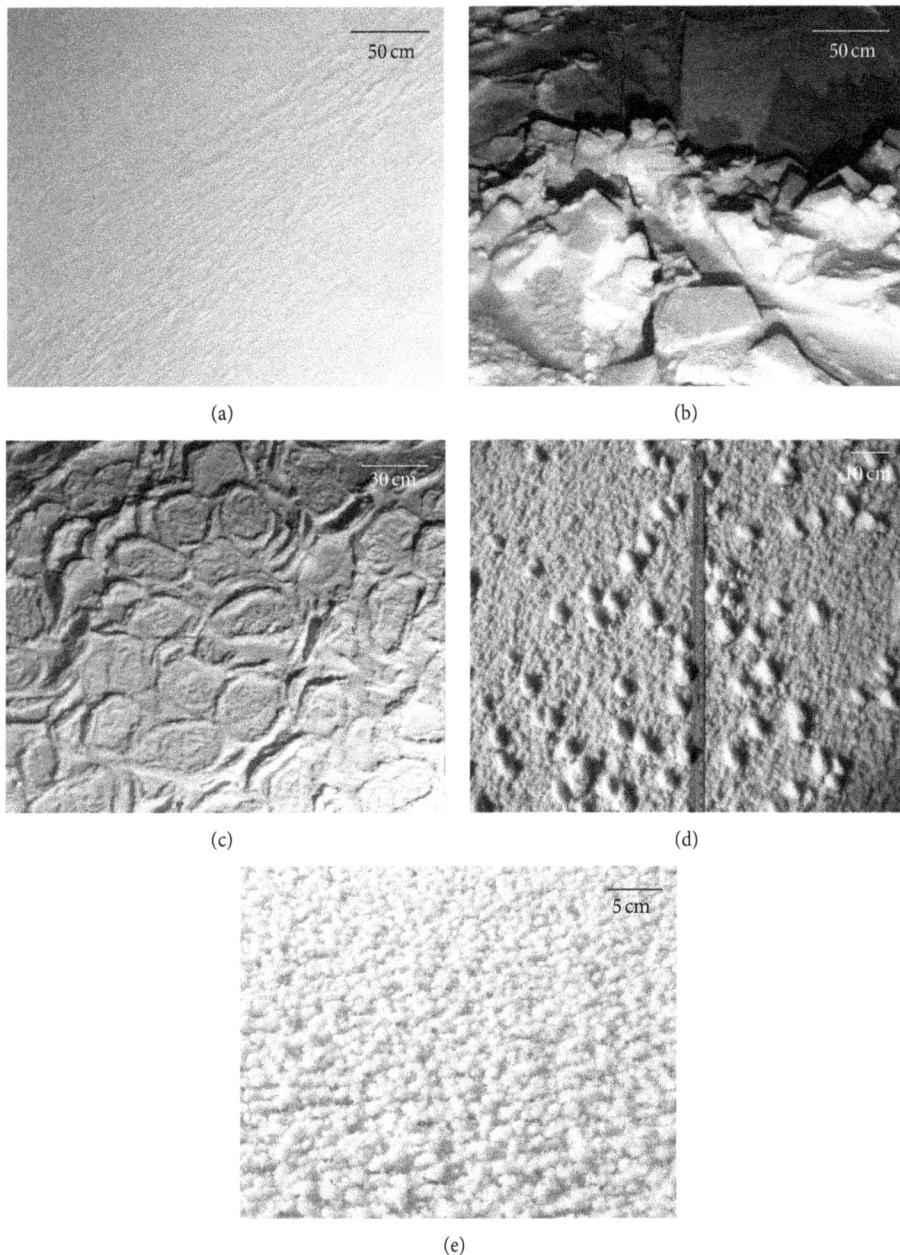

FIGURE 2: Photographs of ice types used in the study. (a) snow-covered first-year ice (SCFYI), (b) deformed first-year ice (DFYI), (c) consolidated pancake ice (PI), (d) snow-covered frost flower (SCFF), and (e) dense frost flower (DFF).

at about one meter height. In the present study, thin FYI types are considered (first stage: 30–50 cm—as per World Meteorological Organization nomenclature), which include snow-covered FYI, pancake ice, frost flowers, and deformed FYI located within the MIZ. Data described in the following sections were acquired as part of the Circumpolar Flaw Lead (CFL) System Study project of the International Polar Year (IPY) 2007-08 over the period October 2007–August 2008 [15]. Ancillary meteorological data were collected through a ship-based AXYS Technologies Inc. (Sydney, BC, Canada) Automatic Voluntary Observing Ships (AVOS) system. This system was mounted approximately 20 m above sea level on the wheelhouse to minimize the ship's influence and could measure air temperature and wind speed.

2.2. Theoretical Formulation. Sea ice is a distributed radar target, and the conditions of stationarity and homogeneity seldom hold for dynamically changing ice in a MIZ. The radar backscattering is therefore analyzed using temporally and spatially varying stochastic processes. Backscatter from sea ice is incoherent and either partially or completely polarized, as described by the polarimetric covariance matrix. The electric field vector of an incident (i) and scattered (s) EM wave can be given by

$$\mathbf{E}^i = E_{Hi}\widehat{\mathbf{h}} + E_{Vi}\widehat{\mathbf{v}},$$
$$\mathbf{E}^s = E_{Hs}\widehat{\mathbf{h}} + E_{Vs}\widehat{\mathbf{v}}, \tag{1}$$

TABLE 1: Technical properties and specifications of C-band scatterometer.

RF output frequency	5.25–5.75 GHz
Transmit power at bulkhead connector	12 dBm
Antenna diameter	0.61 m
Transmit bandwidth	500 MHz
Antenna beamwidth	5.5°
Antenna gain	28 dB, nominal
Crosspolarization isolation	>30 dB, measured at the peak of the beam
Transmit-receive polarizations	Linear, vertical, and horizontal
Sensitivity, minimum NRCS at 15 m range	−40 dB m^2/m^2

where H and V represent horizontal and vertical polarizations respectively. $\hat{\mathbf{h}}$ and $\hat{\mathbf{v}}$ are the unit vectors in the horizontal and vertical directions of polarization, respectively. The incident (i) and scattered field (s) can be either H or V. The scattered electric field is related to the incident electric field by the scattering matrix, S defined as

$$\begin{bmatrix} E_{Hs} \\ E_{Vs} \end{bmatrix} = \frac{e^{-jkr}}{r} \begin{bmatrix} S_{HH} & S_{HV} \\ S_{VH} & S_{VV} \end{bmatrix} \begin{bmatrix} E_{Hi} \\ E_{Vi} \end{bmatrix},$$

$$\begin{bmatrix} S_{HH} & S_{HV} \\ S_{VH} & S_{VV} \end{bmatrix} = S, \qquad (2)$$

where e^{-jkr}/r term accounts for wave propagation effects in amplitude and phase. If the orientation of a surface such as sea ice in azimuth direction is rotated by an angle, the corresponding new backscatter matrix can be constructed as provided by Lee et al. [16].

The coherency matrices can be derived as copolarized (3), crosspolarized (4) and circular (RRLL: right-right left-left rotation of the electric field vector about the line of sight) (5) coherences in magnitude form [13, 16] as (for derivation of ρ_{RRLL}, see Appendix A),

$$\rho_{VVHH} = \frac{\langle |S_{VV} S_{HH}^*| \rangle}{\sqrt{\langle |S_{VV}|^2 \rangle \langle |S_{HH}|^2 \rangle}}, \qquad (3)$$

$$\rho_{HHVH} = \frac{\langle |S_{HH} S_{VH}^*| \rangle}{\sqrt{\langle |S_{HH}|^2 \rangle \langle |S_{VH}|^2 \rangle}}, \qquad (4)$$

$$\rho_{RRLL} = \frac{\langle |S_{HH} - S_{VV}|^2 \rangle - 4 \langle |S_{HV}|^2 \rangle}{\langle |S_{HH} - S_{VV}|^2 \rangle + 4 \langle |S_{HV}|^2 \rangle}, \qquad (5)$$

$$\gamma_{co} = \frac{S_{VV}}{S_{HH}}, \qquad (6)$$

$$\gamma_{cross} = \frac{S_{HV}}{S_{HH}}, \qquad (7)$$

where S is the complex scattering matrix; an asterisk ($*$) represents the complex conjugate. The brackets $\langle \cdot \rangle$ represent ensemble averages of the observed data. There were approximately 34 pulses sent per incidence angle. An ensemble average was performed on those 34 pulses. Raw data were processed into range profiles and were averaged in the azimuth for each measured incidence angle. Polarimetric ratios γ_{co} and γ_{cross} are simply power ratios of backscattered energy. Polarimetric coherences and polarization ratios have demonstrated utility in reducing the ambiguities caused by the nonlinearity between system response and target properties. Regarding Arctic sea ice, some literature is available on the use of ρ_{VVHH}, ρ_{RRLL}, and γ_{co} at different EM frequencies. C-band backscatter coefficients (HH, HV, and VV) and ρ_{VVHH} have been used to characterize various FYI types (compressed, rubble and ridge, and smooth) and multiyear ice [17]. Thin sea ice has been effectively discriminated from FYI using C-band γ_{co} ratio [18]. ρ_{VVHH} and γ_{co} have been used to discriminate Arctic leads using L-band radar signatures [19]. In a similar study, Wakabayashi et al. [20] described polarimetric characteristics of different FYI types (thin ice, smooth, and rough) using L-band ρ_{RRLL} and γ_{co} and showed the utility of coherences and ratios in discriminating ice types. Nakamura et al. [21] discriminated ice surface using γ_{co} ratio in an observational study of lake ice using airborne L- and X-band SAR. These studies lack a holistic overview of the utility of different polarimetric coherences and ratios to discriminate thin FYI types in a MIZ.

2.3. Active Microwave Backscattering Data. C-band polarimetric backscattering data were collected using a completely stationary ship-mounted scatterometer system developed by ProSensing Inc., (Amherst, MA, USA) and mounted 7.56 m above the mean sea level on the port side of the *Amundsen* (Table 1). The system acquires backscatter and phase data in terms of the combinations of linear transmit-receive polarization combinations, HH, HV, VH, and VV at incidence angles 20–60° (5° increments) over a 60° azimuth range. The calibration of the instrument was performed through the methods given elsewhere [22, 23]. Polarimetric backscattering data were collected from homogeneous samples of snow-covered (dry and fresh) first-year ice (SCFYI), deformed FYI (DFYI), consolidated pancake ice (PI), snow-covered frost flowers (SCFF), and dense frost flowers (DFF) on different dates during November 2007. Data from each ice type sample comprised three to four contiguous scatterometer scans, which took up to 35 minutes to complete. The scatterometer had a footprint of 1.1 m^2 in the range direction at a 45° incidence angle [23] with the footprint increasing in size with incidence angle [22].

Radar

Range direction

$\hat{z}$

$\hat{y}$

ϕ

$\hat{n}$

γ

$\hat{x}$

Azimuth direction

ω

θ_1

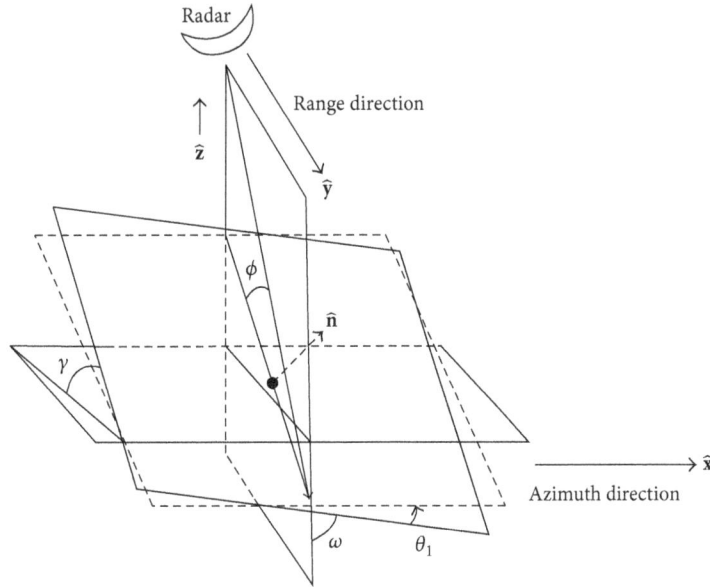

FIGURE 3: Illustration of scattering plane geometry with slight deviations in the orientation angles in azimuth (θ_1) and range directions (θ_2: not shown), respectively, as means of two-dimensional surface roughness.

Towards objective 2, scan data for each ice type were grouped by incidence angle representing near (20–25°), mid (35–40°), and far (55–60°) range groupings. These groupings best represent the diversity of scattering mechanisms available across the acquired incidence angle range. In the near range, surface scattering is expected to dominate the measured C-band backscatter, while surface-volume scattering is increasingly expected to influence C-band backscatter beyond approximately 30°, that is, mid to far ranges [24]. Furthermore, combining data from adjacent incidence angles doubled the number of samples from 8 to 18 depending on ice type, although at the expense of range resolution. Scatterometer data had unequal number of data points in each range group, which does not fulfil parametric ANOVA requirements for statistical significance testing. Polarimetric coherences and ratios of ice types were tested for independence from each other for each incidence angle grouping. Testing was done using the nonparametric Kruskal-Wallis statistic, with $\alpha = 0.01$ significance level (one tailed) used as the threshold for statistical independence.

2.4. Surface Roughness and Circular Coherence. In pursuit of objective 1, a polarimetric backscattering model is used which is mainly a Bragg backscattering (coherent scattering) model modified for surface roughness considering the surface slope by slightly changing the tilt of the surface from the horizontal. Microwave measurements of surface roughness using co- or crosspolarization backscattered power are most successful in flat areas. In sea ice microwave remote sensing, the dielectric constant and topography (slope in range and azimuth) are important. According to (22) in the one-dimensional scattering model of [13], the circular coherence is only sensitive to surface roughness. Surface roughness has been considered as a change in the slope of ice in azimuth and ground range directions [13, 16, 25]. This is implemented mathematically in

the Bragg backscattering model by considering roughness as a depolarizer which conforms to reflection symmetry; that is, the backscattering properties are identical on either side of the plane of incidence and HV = VH [25, 26]. The distribution of azimuth slope angles θ_1 is considered as one-dimensional Gaussian distributed [13].

The rotation matrix [16] and the coherency matrix [13] are calculated after introducing the rotation in azimuth anticlockwise about range direction. In this case, ρ_{RRLL} is derived as [13]

$$\rho_{\text{RRLL}} = e^{-8\upsilon_{\theta_1}^2},\tag{8}$$

where σ_{θ_1} is the standard deviation of the orientation angle distribution in azimuth direction and θ_1 is slope angle in azimuth direction. From (8), the ρ_{RRLL} is only dependent on the orientation of ice surface in the range direction, or the standard deviation of the orientation angle distribution (i.e., surface roughness). Here, the surface roughness is introduced through rotation by angle θ_2 in the range direction anticlockwise about azimuth direction (Figure 3). Angle θ_2 is not shown in Figure 3 due to complexity of the geometry. In this case also, the corresponding distribution of shift in orientation angle is Gaussian distributed.

The new rotation matrix U_2 is given by

$$U_2 = \begin{bmatrix} \cos 2\theta_2 & 0 & -\sin 2\theta_2 \\ 0 & 1 & 0 \\ \sin 2\theta_2 & 0 & \cos 2\theta_2 \end{bmatrix}.\tag{9}$$

The new averaged coherency matrix over the Gaussian distribution $p(\theta_2)$ can be calculated as

$$\langle T \rangle_{\theta_2} = \begin{bmatrix} \zeta A & \mu B & 0 \\ \mu B^* & 2C & 0 \\ 0 & 0 & (1-\zeta)A \end{bmatrix},\tag{10}$$

where $\mu(\theta_2) = \int \cos 2\theta_2 p(\theta_2)d\theta_2$ and $\zeta(\theta_2) = \int \cos^2 2\theta_2$ $p(\theta_2)d\theta_2$. B, a part of an element of coherency matrix, is defined according to scattering matrix, S [13]. B^* is the conjugate of B. Both B^* and B are not used in the computation of ρ_{RRLL}.

The ρ_{RRLL} can be computed as

$$\rho_{\text{RRLL}} = \frac{T_{22} - T_{33}}{T_{22} + T_{33}} = \frac{2C - (1 - \zeta(\theta_2))A}{2C + (1 - \zeta(\theta_2))A}, \quad (11)$$

$$\rho_{\text{RRLL}} = \frac{4C - \left(1 - e^{-8\sigma_{\theta_2}^2}\right)A}{4C + \left(1 - e^{-8\sigma_{\theta_2}^2}\right)A}, \quad (12)$$

where $A = |S_{\text{HH}} + S_{\text{VV}}|^2$, $C = (1/2)|S_{\text{HH}} - S_{\text{VV}}|^2$, and $\zeta(\theta_2) = (1/2)(1 + e^{-8\sigma_{\theta_2}^2})$. T_{ij} represents (i, j)th element of the matrix, $\langle T \rangle_{\theta_2}$ given in (10).

Given the above, the ρ_{RRLL} is dependent on the standard deviation of the orientation angle distribution in range and the dielectric constant of the surface. Thus, it is shown that the new ρ_{RRLL} is exponentially changing with the change in orientation angle in the azimuth direction, but it behaves in a way given by (12) and is dependent on both surface roughness (standard deviation) and the dielectric constant (scattering matrix) of the surface when roughness in two directions is considered. In our model, when two-dimensional roughness is considered, circular coherence is observed to be sensitive to both surface roughness and dielectric constant, thus, making it difficult to differentiate roughness. 2D model being more realistic requires further considerations of separating dielectrics from roughness.

Now, the slope-induced roughness is examined in the range direction only. Lee et al. [16] gave a relationship between slope in azimuth, slope in ground range, radar look angle (ϕ), and rotation in azimuth. Schuler et al. [13] expressed this relationship in terms of root mean square (rms) surface height (s) and correlation length (l), assuming that the range slope and orientation in azimuth are small perturbations around their means,

$$\rho_{\text{RRLL}} = e^{-16(s^2/l^2 \sin^2 \phi)}. \quad (13)$$

Figure 4 shows the incidence angle dependence of ρ_{RRLL} by varying the s^2/l^2 ratio. As the roughness increases, ρ_{RRLL} decreases. For $l \gg s$, that is, the surface is very smooth, the maximum value of ρ_{RRLL} approaches unity. ρ_{RRLL} decreases exponentially from unity to a fixed value of s^2/l^2 ratio at a particular incidence angle. A rough surface yields a smaller ρ_{RRLL}, which increases with increasing radar look angle. The range of s^2/l^2 for the presented ice classes is expected to lie between 0.001 and 0.1 [27].

The relationship between slopes in azimuth and range direction is further demonstrated. Corresponding shifts and radar incidence angle are given by (see Appendix B),

$$\frac{\tan \omega}{\tan \theta_1} = \sin \theta_2 (\tan \gamma \cdot \sin \phi + \cos \phi)$$

$$+ \cos \theta_2 (-\tan \gamma \cdot \cos \phi + \sin \phi), \quad (14)$$

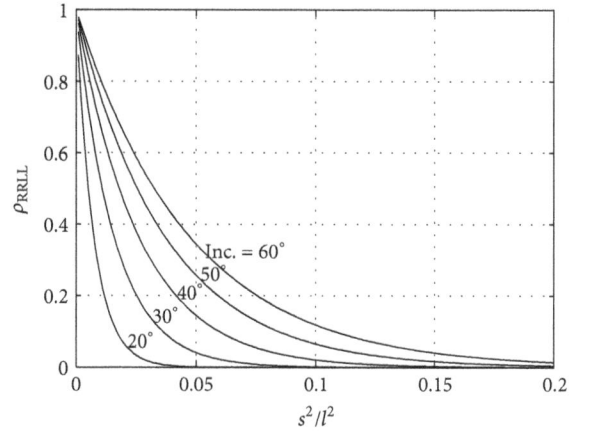

FIGURE 4: ρ_{RRLL} varying with squared ratio of rms surface height and surface correlation length; ρ_{RRLL} decays exponentially; however, it decays faster at steep incidence angles.

where $\tan \omega$ is azimuth slope, $\tan \gamma$ is range slope, θ_1 and θ_2 are the perturbations in orientation in azimuth and range directions, respectively, and ϕ is radar look angle. Figure 4 represents the case when orientation shift in the range direction is observed. In a sea ice remote sensing context, both surface roughness and the dielectric constant of ice affect ρ_{RRLL} when slope is changed in azimuth direction, whereas only surface roughness affects ρ_{RRLL} when slope is changed in range direction.

3. Field Results

3.1. Sea Ice Type Discrimination (Coherences and Ratios). The date and hour of scatterometer data acquisitions corresponding to each sea ice type, as well as coincident meteorological parameters, namely, wind speed, air temperature, and relative humidity, are provided in Table 2. The photographs of the selected ice samples are shown in Figure 2. With the exception of wind speed, there is negligible variation in meteorological conditions between ice type scans. As such, it is expected that between-scan, temperature-induced effects on the dielectric properties, and backscattering intensities from the different ice types are negligible.

Figure 5 shows backscattering coefficients for co- (HH and VV) and crosspolarization (HV) configurations of each ice type. The two frost flower cases (DFF and SCFF) are plotted separately to exemplify differences in backscattering behavior on the basis of their different frost flower concentrations. The DFF and SCFF have a visually measured concentration of approximately >95% and 20%, respectively. While SCFYI is visually separable using HH, HV, and VV polarizations at all incidence angles (low backscatter), PI and DFYI signatures overlap and are difficult to separate from each other. This may be indicative of PI geometry within the scatterometer footprint, as PI comprises a series of upturned edges and flat areas of ice (see Figure 2). The curvature of upturned PI edges causes a backscatter response similar to that caused by the deformations (upturned ice) in the DFYI.

TABLE 2: Meteorological parameters associated with each ice type on different dates.

	Sea ice type	Wind speed (m/s)	Air temperature (°C)	Relative humidity %
Nov. 15, 2007 (Stn. 1117, 1400 hrs)	SCFYI	14.4	−16.0	85
Nov. 19, 2007 (Stn. 1100, 0030 hrs)	DFYI	13.9	−16.2	73
Nov. 20, 2007 (Stn. 1910, 0300 hrs)	PI	2.6	−13.5	79
Nov. 21, 2007 (Stn. 437, 1630 hrs)	SCFF	5.1	−16.2	82
Nov. 25, 2007 (Stn. 1812, 2100 hrs)	DFF	3.6	−16.6	86

TABLE 3: Mean C-band polarimetric coherences and ratios of selected ice types, for near (N), middle (M), and far (F) range incidence angle groupings (also shown graphically in Figure 6). The number of data samples is: SCFYI, $N = 14$; DFYI, $N = 8$; PI, $N = 14$; SCFF, $N = 18$; DFF, $N = 10$.

	ρ_{VVHH}	ρ_{HHVH}	ρ_{RRLL}	γ_{co} (dB)	γ_{cross} (dB)
SCFYI					
N	0.95	0.09	0.47	0.40	−16.44
M	0.81	0.09	0.65	3.64	−13.72
F	0.55	0.08	0.81	4.54	−12.25
DFYI					
N	0.88	0.10	0.42	1.36	−12.38
M	0.91	0.08	0.52	2.47	−12.94
F	0.58	0.05	0.58	1.86	−11.40
PI					
N	0.96	0.04	0.61	1.78	−16.89
M	0.84	0.04	0.61	1.89	−13.73
F	0.80	0.05	0.59	2.10	−12.70
SCFF					
N	0.96	0.07	0.73	0.89	−17.88
M	0.89	0.07	0.70	−0.59	−15.83
F	0.66	0.05	0.77	0.46	−14.08
DFF					
N	0.84	0.15	0.35	2.41	−09.71
M	0.91	0.15	0.59	3.65	−12.06
F	0.79	0.12	0.72	3.23	−12.23

DFF and SCFF are differentiable at HV and VV polarization at mid to far incidence angles.

Mean coherences and polarization ratios for each ice type as a function of incidence angle grouping are documented in Table 3. All sea ice types show high ρ_{VVHH}, indicating low depolarisation and primarily single (surface) backscattering. The ρ_{HHVH} for DFF is notably higher than that from the other ice types, which points to strong depolarisation caused by the frost flower structures. As shown in the previous section, a low value of ρ_{RRLL} indicates a rougher surface. At mid to far ranges in Table 3, the ρ_{RRLL} for DFYI is the lowest while for SCFYI it is the highest, which is consistent with the roughest and smoothest ice types, respectively. Furthermore, for frost flower-covered surfaces, that is, SCFF and DFF, the lower magnitude of ρ_{RRLL} is consistent with the higher concentration of frost flowers. At near-incidence angle range, the SCFYI shows higher roughness (i.e., lower $\rho_{RRLL} = 0.47$, Table 3) compared to that of PI (0.61). This may be due to the fact that the snow is dry and has low salinity, which allows EM waves to penetrate through the snow. This is likely to provide roughness of snow-ice interface rather than air-snow interface. At mid incidence angle range, as expected, SCFYI shows lower roughness (i.e., higher $\rho_{RRLL} = 0.65$, Table 3) compared to that of PI (0.61). Mid incidence angles are well suited for differentiating ice roughness/types using ρ_{RRLL}.

Looking at polarization ratios in Table 3, the γ_{co} increases rapidly with incidence angle and is the highest at the far range for SCFYI. The γ_{co} ratio is also high for DFF, but it remains fairly constant across all incidence angles. High γ_{co} is also representative of saline ice surface (FYI in this case) or surface scattering. The presence of dry snow (~1-2 cm) allows the EM waves to penetrate through snow, which causes reflection from the ice-snow interface. The γ_{co} behavior of SCFYI is consistent with that of a surface, which is very smooth (i.e., a Bragg surface), where the ratio between backscattered H and V is only dependent on incidence angle and dielectric constant [28]. On the other hand, the γ_{co} behavior for DFF is consistent with that of a rough surface exhibiting backscatter from features with preferential vertical orientation [22]. Including the γ_{cross} ratio in this comparison further supports

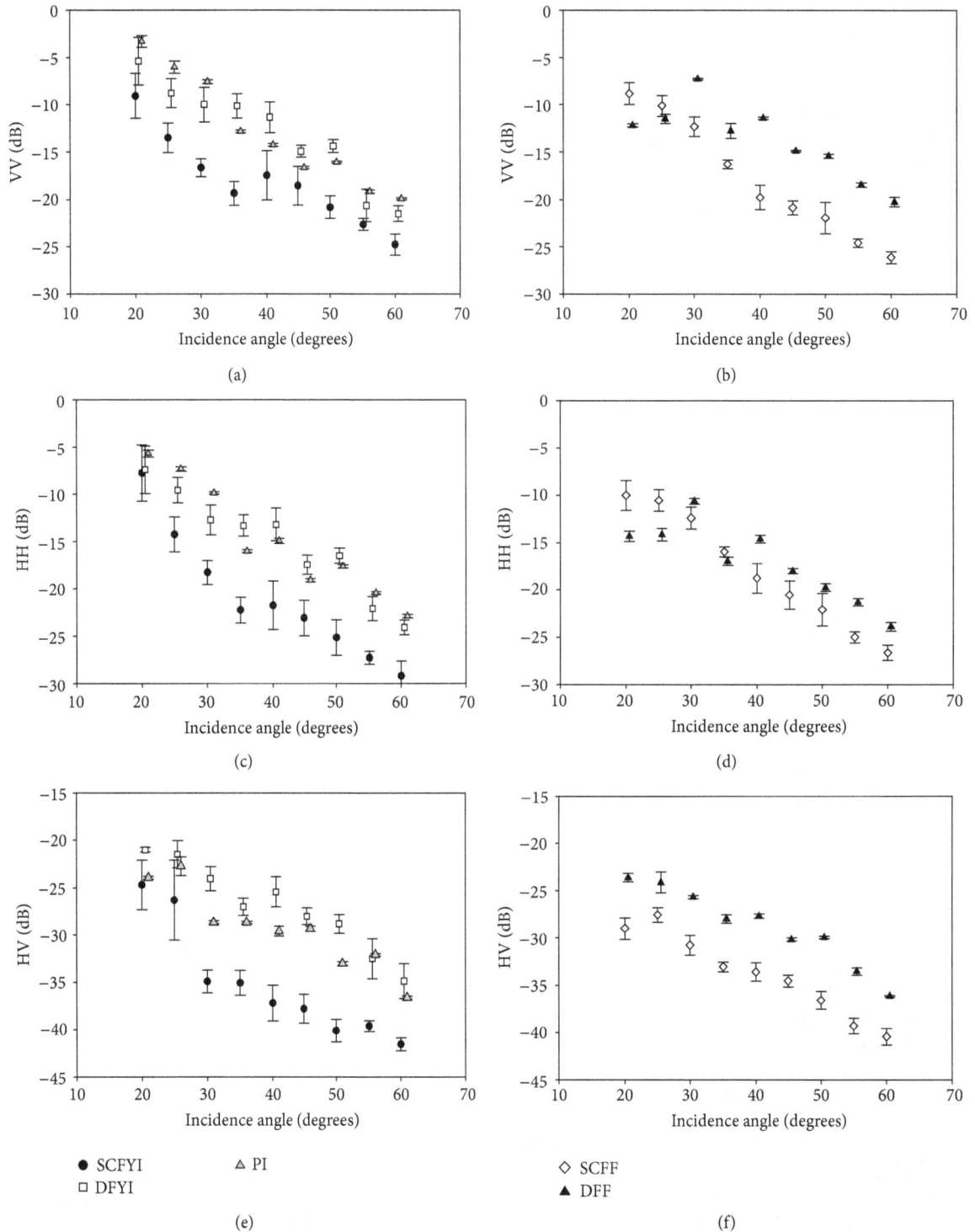

FIGURE 5: Co- (HH and VV) and cross- (HV) polarization backscatter intensities of snow-covered first-year sea ice (SCFYI), deformed first-year sea ice (DFYI), consolidated pancake ice (PI), snow-covered frost flowers (SCFF), and dense frost flowers (DFF).

the distinction in backscattering mechanisms. The near range γ_{cross} ratio is much smaller for SCFYI than DFF, indicating it to be much smoother. The DFYI and DFF show the highest overall γ_{cross}, due to multiple scattering within deformities for DFYI and depolarisation caused by frost flowers for DFF.

Figure 6 shows box plots of coherences and polarization ratios of each ice type. Table 4 provides the significance values

resulting from statistical tests for independence between each ice type based on a given coherence or ratio. All data in Figure 6 and Table 4 are based on the aforementioned incidence angle groupings from near to far range and, together, facilitate a conceptual approach to assessing the utility of each parameter for distinguishing ice types within an MIZ. Summarizing Figure 6 and Table 4, the near range ρ_{HHVH} and

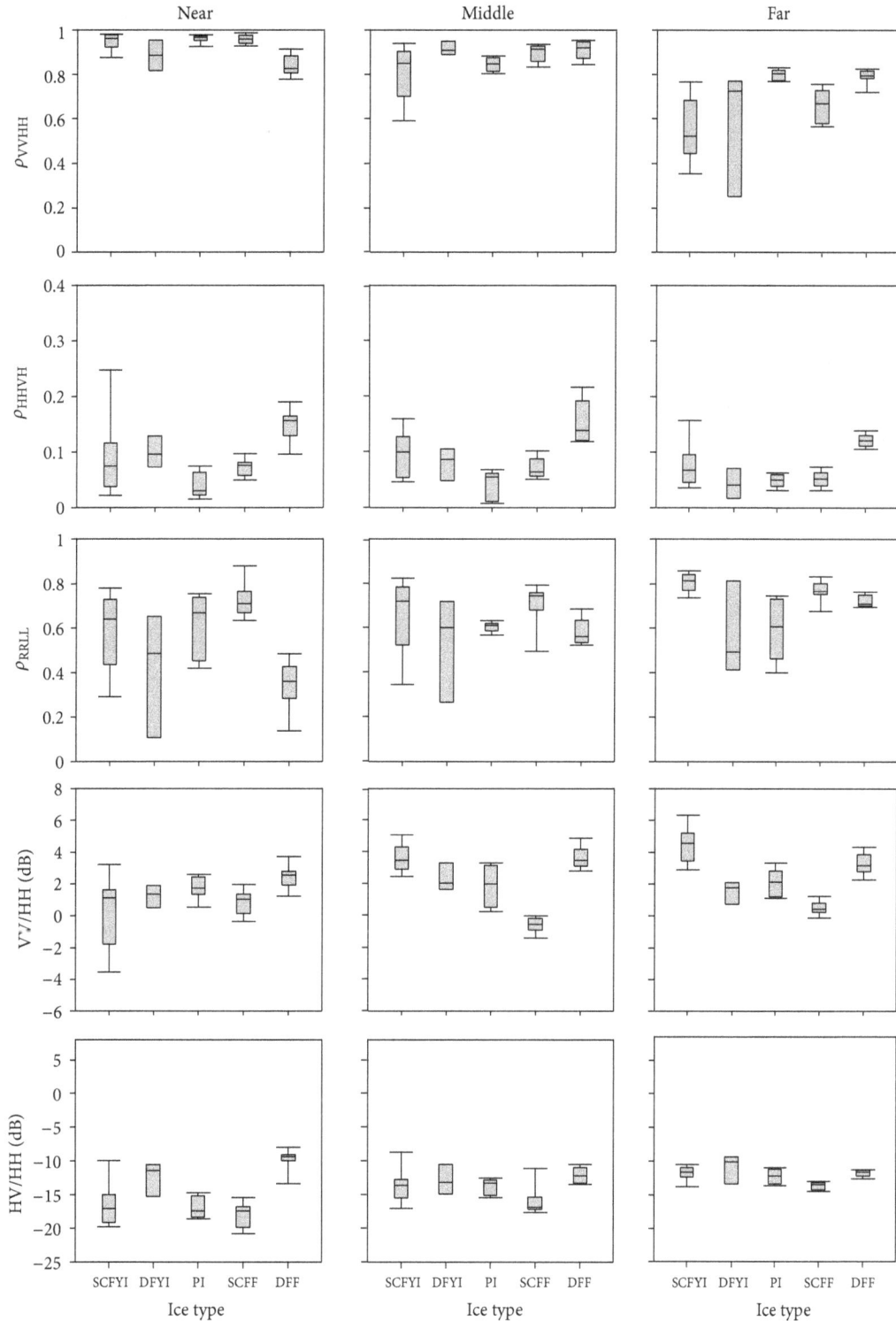

FIGURE 6: Box plots of coherences and polarization ratios of ice types based on near, middle, and far range incidence angle groupings. Significance values are provided in Table 4.

γ_{cross} provide the greatest separation between classes, while the far range ρ_{VVHH} and γ_{co} provide the greatest separation. By combining γ_{co} (far) with either of ρ_{HHVH} or γ_{cross} (near), all ice types are independent of each other. From Figure 4 it

is known that a lower ρ_{RRLL} is associated with a rougher ice surface.

It is demonstrated using theory that lower values of ρ_{RRLL} indicate a rougher ice surface. Referring to Figures 4 and 6,

TABLE 4: Matrix of significance values from non-parametric Kruskal-Wallis tests for independence between ice types based on polarimetric parameters and near (N), middle (M), and far (F) range groupings. The number of data samples is: SCFYI, $N = 14$; DFYI, $N = 8$; PI, $N = 14$; SCFF, $N = 18$; DFF, $N = 10$.

	Near range					Middle range					Far range				
	SCFYI	DFYI	PI	SCFF	DFF	SCFYI	DFYI	PI	SCFF	DFF	SCFYI	DFYI	PI	SCFF	DFF
ρ_{VVHH}															
SCFYI															
DFYI	.285					**.015**					.094				
PI	.028	.094				.509	**.000**				**.000**	**.001**			
SCFF	.463	.119	**.002**			.011	.322	**.001**			.014	.199	**.000**		
DFF	**.004**	**.013**	.028	**.000**		**.005**	.483	**.000**	**.000**		**.000**	**.004**	.420	**.000**	
ρ_{HHVH}															
SCFYI															
DFYI	.201					.201					.035				
PI	**.006**	**.000**				**.005**	**.005**				.018	.221			
SCFF	.429	**.008**	**.000**			.125	.156	**.001**			.046	.184	.277		
DFF	**.002**	**.002**	**.000**	**.000**		**.001**	**.000**	**.000**	**.000**		**.001**	**.000**	**.000**	**.000**	
ρ_{RRLL}															
SCFYI															
DFYI	.308					.048					.030				
PI	.164	.106				.082	.513				**.000**	.357			
SCFF	**.006**	**.001**	.039			.418	**.005**	**.002**			.015	.023	**.000**		
DFF	.023	.197	**.000**	**.000**		.069	.483	.170	**.003**		**.000**	.042	.117	**.002**	
γ_{co}															
SCFYI															
DFYI	.285					**.018**					**.001**				
PI	.028	.094				**.001**	.183				**.000**	.308			
SCFF	.463	.119	**.002**			**.000**	**.000**	**.000**			**.000**	**.001**	**.000**		
DFF	**.004**	.013	.028	**.000**		.466	.017	**.001**	**.000**		**.000**	**.003**	**.005**	**.000**	
γ_{cross}															
SCFYI															
DFYI	**.005**					.149					.041				
PI	.437	**.001**				.214	.357				.164	.041			
SCFF	.118	**.000**	.179			**.006**	**.005**	**.001**			**.000**	.013	**.001**		
DFF	**.000**	.010	**.000**	**.000**		**.004**	.204	**.006**	**.000**		.420	.042	.354	**.000**[#]	

[#]Bold numbers indicate important significant values.

ρ_{RRLL} is high for increasing incidence angles and for low surface roughness. This is only true for SCFYI and SCFF. In the presence of dry and fresh snow the volume contribution from FYI can be ignored, in which case ρ_{RRLL} dictates surface roughness of the snow-ice interface rather than air-snow interface. The coherence estimates are negligibly affected by the signal-to-noise ratio (typically >10 dB) during the processing of scatterometer data. These coherences can also be computed using polarimetric observations from space-based platforms.

4. Summary and Conclusions

The one-dimensional backscatter model of Schuler et al. [13] was modified to two dimensions of surface roughness by considering deviation in the orientation angles (i.e., the slopes) in azimuth and range direction simultaneously as an improvement in the model. Parameters derived from the fully polarimetric C-band microwave backscatter response from sea ice targets were demonstrated to have utility for small-scale (cm level) sea ice roughness identification. Circular coherence has been investigated for its usefulness in discriminating surface roughness among other polarimetric parameters. Circular coherence is theoretically shown to detect measurement sensitivity to surface roughness.

The conclusions with reference to objective 1 are as follows. It was shown theoretically that the ρ_{RRLL} decreases exponentially with increasing surface roughness. However, ρ_{RRLL} responds to both roughness (standard deviation) and dielectric constant (scattering matrix) of the surface in the case when the orientations of the ice target in azimuth direction are changed. It remains challenging to separate roughness effects from the dielectric effects using C-band backscatter measurements. ρ_{RRLL} independently does not provide a robust sea ice roughness discrimination scheme. However, ρ_{RRLL} provides an improved insight of sea ice surface roughness combined with other polarimetric coherences and channel ratios in the chosen samples. The experimental data also show that rougher ice surface exhibits lower mean value of ρ_{RRLL} (Table 3, Figure 4), though a complete validation of the effect of changing orientations of ice floe on ρ_{RRLL} is required. This would require polarimetric backscattering data and surface roughness information to be acquired at different lines of sight (i.e., orientation of ice floes). Unfortunately, difficulties associated with extreme weather conditions and limitations to navigation in the Arctic restrict such detailed data acquisition; however, a tank experiment could be a useful alternative.

The utility of C-band polarimetric coherences and ratios is addressed in the light of objective 2 as follows: for coherences, ρ_{VVHH} is smaller at far range incidence angles for all ice types. ρ_{HHVH} is less sensitive to roughness and is not a good discriminator of roughness. Regarding channel ratios, based on Kruskal-Wallis test, γ_{co} is more sensitive to increasing surface roughness compared to γ_{cross} and demonstrates utility for separating ice types compared to the other observed parameters.

The knowledge obtained through surface-based polarimetric coherences and ratios can readily be extended to discriminate sea ice roughness on small scales using C-band microwave satellites (currently in orbit RADARSAT-2, RISAT-1). Future work will be to develop an algorithm combining all polarimetric coherences and ratios to discriminate individual ice type in a MIZ. These observations may become particularly useful for satellite measurements once planned SAR constellations (Sentinel series) systems are available, as currently planned by National Aeronautics and Space Administration and European Space Agency.

Appendices

A. Derivation of ρ_{RRLL}

To understand how to extract best information from the scattering matrix $\mathbf{S}$, it is represented by the vector, $\mathbf{V}$, built as follows:

$$\mathbf{S} = \begin{bmatrix} HH & HV \\ VH & VV \end{bmatrix}; \qquad \mathbf{k} = \mathbf{V}(\mathbf{S}) = \frac{1}{2}\,\mathrm{Tr}\left([\mathbf{S}]\,\boldsymbol{\psi}\right),$$
$$\mathbf{k} = (k_0 \ \ k_1 \ \ k_2 \ \ k_3)^{\mathrm{T}}, \tag{A.1}$$

where $\boldsymbol{\psi}$ is a basis matrix which is constructed as an orthonormal set under the Hermitian inner product. The polarimetric coherency matrix is based on linear combinations arising from Pauli matrices [28] given as

$$\boldsymbol{\psi} \equiv \sqrt{2}\begin{bmatrix} 1 & 0 \\ 0 & 1 \end{bmatrix} \sqrt{2}\begin{bmatrix} 1 & 0 \\ 0 & -1 \end{bmatrix} \sqrt{2}\begin{bmatrix} 0 & 1 \\ 1 & 0 \end{bmatrix} \sqrt{2}\begin{bmatrix} 0 & -i \\ i & 0 \end{bmatrix}. \tag{A.2}$$

The factor of $\sqrt{2}$ arises from the requirement to keep $\mathrm{Tr}([\mathbf{S}])$, the total power scattered, an invariant. The target vector in above base is constructed as

$$\mathbf{k} = \frac{1}{\sqrt{2}}$$
$$\times \left[(S_{HH}+S_{VV})\,(S_{HH}-S_{VV})\,(S_{HV}+S_{VH})\,i\,(S_{HV}-S_{VH})\right]^{T}. \tag{A.3}$$

From the vector form of scattering matrix, the Pauli coherency matrix is generated from the outer product of the vector with its conjugate transpose as

$$\mathbf{T_4} = \left\langle \mathbf{k}\cdot\mathbf{k}^{*\mathrm{T}} \right\rangle. \tag{A.4}$$

For reciprocal target matrix (as in monostatic backscattering), $S_{HV} = S_{VH}$; the four-dimensional polarimetric coherency matrix reduced to three-dimensional polarimetric coherency matrix is obtained as

$$\mathbf{T_3} = \begin{bmatrix} \left\langle |S_{HH}+S_{VV}|^2 \right\rangle & \left\langle (S_{HH}+S_{VV})(S_{HH}-S_{VV})^* \right\rangle & 2\left\langle (S_{HH}+S_{VV})S_{HV}^* \right\rangle \\ \left\langle (S_{HH}-S_{VV})(S_{HH}+S_{VV})^* \right\rangle & \left\langle |S_{HH}-S_{VV}|^2 \right\rangle & 2\left\langle (S_{HH}-S_{VV})S_{HV}^* \right\rangle \\ 2\left\langle S_{HV}(S_{HH}+S_{VV})^* \right\rangle & 2\left\langle S_{HV}(S_{HH}-S_{VV})^* \right\rangle & 4\left\langle |S_{HV}|^2 \right\rangle \end{bmatrix}. \tag{A.5}$$

The ρ_{RRLL} is computed as [28, 29]

$$\rho_{RRLL} = \frac{T_{22}-T_{33}}{T_{22}+T_{33}} = \frac{\left\langle |S_{HH}-S_{VV}|^2 \right\rangle - 4\left\langle |S_{HV}|^2 \right\rangle}{\left\langle |S_{HH}-S_{VV}|^2 \right\rangle + 4\left\langle |S_{HV}|^2 \right\rangle}, \tag{A.6}$$

where T_{ij} represents the (i, j) element of the matrix, $\mathbf{T_3}$, given in (A.5). For the one-dimensional polarimetric scattering model described elsewhere [13, 30], ρ_{RRLL} is expressed as independent of dielectrics, thus depending only on surface roughness. Fore more detailed, step by step derivation of ρ_{RRLL}, the reader is directed to [28–30].

B.

Here, the relationship between slope in azimuth and ground range, radar look angle, shift in azimuth, and shift in ground range is derived. The slope equation given by Lee et al. [16] does not include shift in range direction. Figure 3 shows the geometry of backscattering plane. Suppose that the backscattering plane is shifted in azimuth direction by angle

θ_1 and in range direction by angle θ_2. $\hat{\mathbf{n}}$ represents the surface normal on the backscattering plane before rotating:

$$\hat{\mathbf{n}} = n_1\hat{\mathbf{x}} + n_2\hat{\mathbf{y}} + n_3\hat{\mathbf{z}}, \tag{B.1}$$

$$\widehat{\mathbf{N}} = \begin{bmatrix} 0 & -\sin\phi & \cos\phi \\ 1 & 0 & 0 \\ 0 & \cos\phi & \sin\phi \end{bmatrix} \begin{bmatrix} n_1 \\ n_2 \\ n_3 \end{bmatrix}$$

$$= \begin{bmatrix} -n_2\sin\phi + n_3\cos\phi \\ n_1 \\ n_2\cos\phi + n_3\sin\phi \end{bmatrix}. \tag{B.2}$$

The surface normal changes after a shift in azimuth and range directions. The transformed normal is

$$\hat{\mathbf{n}}\left(\theta_1,\theta_2\right) = \begin{bmatrix} 1 & 0 & 0 \\ 0 & \cos\theta_1 & \sin\theta_1 \\ 0 & -\sin\theta_1 & \cos\theta_1 \end{bmatrix} \begin{bmatrix} \cos\theta_2 & 0 & -\sin\theta_2 \\ 0 & 1 & 0 \\ \sin\theta_2 & 0 & \cos\theta_2 \end{bmatrix} \widehat{\mathbf{N}}. \tag{B.3}$$

If $\cos\theta_1 = c_1$ and $\sin\theta_2 = s_2$, and substitute $\widehat{\mathbf{N}}$ from (B.2), then (B.3) is

$$= \begin{bmatrix} -c_2\left(n_2\sin\phi + n_3\cos\phi\right) - s_2\left(n_2\cos\phi + n_3\sin\phi\right) \\ s_1s_2\left(-n_2\sin\phi + n_3\cos\phi\right) + n_1c_1 + s_1c_2\left(n_2\cos\phi + n_3\sin\phi\right) \\ c_1s_2\left(-n_2\sin\phi + n_3\cos\phi\right) - n_1s_1 + c_1c_2\left(n_2\cos\phi + n_3\sin\phi\right) \end{bmatrix}. \tag{B.4}$$

After the rotation the surface normal is in new plane where the second component must be zero:

$$s_1s_2\left(-n_2\sin\phi + n_3\cos\phi\right) + n_1c_1$$
$$+ s_1c_2\left(n_2\cos\phi + n_3\sin\phi\right) = 0. \tag{B.5}$$

With range slope, $\tan\gamma = -(n_2/n_3)$ and azimuth slope $\tan\omega = -(n_1/n_3)$,

$$\frac{\tan\omega}{\tan\theta_1} = \sin\theta_2\left(\tan\gamma\cdot\sin\phi + \cos\phi\right)$$
$$+ \cos\theta_2\left(-\tan\gamma\cdot\cos\phi + \sin\phi\right). \tag{B.6}$$

In (B.6), if the perturbation in orientation in range direction is zero, that is, $\theta_2 = 0$, it reduces to equation given by Lee et al. [16].

Acknowledgments

The main funding for the project was provided by the IPY-Canada, the Natural Sciences and Engineering Research Council (NSERC), and the Canada Research Chairs (CRC) Program, each to DGB. Authors gratefully thank icebreaker *Amundsen* crew for their exceptional support in acquiring data and excellent navigation during overwintering in the Arctic. They thank their colleagues from various Canadian and international organizations who helped in deployment and recuperation of instruments in the field. Authors thank Dr. Dustin Isleifson for data preprocessing and providing multiple and very useful reviews in improving the paper.

References

[1] R. Kwok and G. F. Cunningham, "Contribution of melt in the Beaufort Sea to the decline in Arctic multiyear sea ice coverage: 1993–2009," *Geophysical Research Letters*, vol. 37, Article ID L20501, 2010.

[2] R. Kwok and D. A. Rothrock, "Decline in Arctic sea ice thickness from submarine and ICESat records: 1958–2008," *Geophysical Research Letters*, vol. 36, Article ID L15501, 2009.

[3] S. V. Nghiem, I. G. Rigor, D. K. Perovich, P. Clemente-Colon, J. W. Weatherly, and G. Neumann, "Rapid reduction of Arctic perennial sea ice," *Geophysical Research Letters*, vol. 34, Article ID L19504, 2007.

[4] M. G. McPhee, J. H. Morison, and F. Nilsen, "Revisiting heat and salt exchange at the ice-ocean interface: ocean flux and modelling considerations," *Journal of Geophysical Research*, vol. 113, Article ID C06014, 2008.

[5] E. L. Andreas, T. W. Horst, A. A. Grachev et al., "Parameterizing turbulent exchange over summer sea ice and the marginal ice zone," *Quarterly Journal of the Royal Meteorological Society*, vol. 136, pp. 927–943, 2010.

[6] D. Lavoie, R. W. Macdonald, and K. L. Denman, "Primary productivity and export fluxes on the Canadian shelf of the Beaufort Sea: a modelling study," *Journal of Marine Systems*, vol. 75, no. 1-2, pp. 17–32, 2009.

[7] R. Kwok, E. Rignot, B. Holt, and R. Onstott, "Identification of sea ice types in spaceborne synthetic aperture radar data," *Journal of Geophysical Research*, vol. 97, no. 2, pp. 2391–2402, 1992.

[8] H. Melling, "Detection of features in first-year pack ice by synthetic aperture radar (SAR)," *International Journal of Remote Sensing*, vol. 19, no. 6, pp. 1223–1249, 1998.

[9] G. M. Wohl, "Operational sea ice classification from synthetic aperture radar imagery," *Photogrammetric Engineering & Remote Sensing*, vol. 61, no. 12, pp. 1455–1462, 1995.

[10] D. G. Barber and E. F. LeDrew, "SAR sea ice discrimination using texture statistics: a multivariate approach," *Photogrammetric Engineering & Remote Sensing*, vol. 57, no. 4, pp. 385–395, 1991.

[11] Q. A. Holmes, D. R. Nüesch, and R. A. Shuchman, "Textural analysis and real-time classification of sea-ice types using digital SAR data," *IEEE Transactions on Geoscience and Remote Sensing*, vol. 22, no. 2, pp. 113–120, 1984.

[12] S. V. Nghiem and C. Bertoia, "Study of multi-polarization C-band backscatter signatures for arctic sea ice mapping with future satellite SAR," *Canadian Journal of Remote Sensing*, vol. 27, no. 5, pp. 387–402, 2001.

[13] D. L. Schuler, J. S. Lee, D. Kasilingam, and G. Nesti, "Surface roughness and slope measurements using polarimetric SAR data," *IEEE Transactions on Geoscience and Remote Sensing*, vol. 40, no. 3, pp. 687–698, 2002.

[14] D. G. Barber and J. M. Hanesiak, "Meteorological forcing of sea ice concentrations in the southern Beaufort Sea over the period 1979 to 2000," *Journal of Geophysical Research C*, vol. 109, no. 6, pp. C06014–16, 2004.

[15] D. G. Barber, M. G. Asplin, Y. Gratton et al., "The International Polar Year (IPY) Circumpolar Flaw Lead (CFL) system study: overview and the physical system," *Atmosphere*, vol. 48, no. 4, pp. 225–243, 2010.

[16] J. S. Lee, D. L. Schuler, and T. L. Ainsworth, "Polarimetric SAR data compensation for terrain azimuth slope variation," *IEEE*

Transactions on Geoscience and Remote Sensing, vol. 38, no. 5 I, pp. 2153–2163, 2000.

[17] R. Rignot and M. R. Drinkwater, "Winter sea-ice mapping from multi-parameter synthetic-aperture radar data," *Journal of Glaciology*, vol. 40, no. 134, pp. 31–45, 1994.

[18] T. Geldsetzer and J. J. Yackel, "Sea ice type and open water discrimination using dual co-polarized C-band SAR," *Canadian Journal of Remote Sensing*, vol. 35, no. 1, pp. 73–84, 2009.

[19] D. P. Winebrenner, L. D. Farmer, and I. R. Joughin, "On the response of polarimetric synthetic aperture radar signatures at 24-cm wavelength to sea ice thickness in Arctic leads," *Radio Science*, vol. 30, no. 2, pp. 373–402, 1995.

[20] H. Wakabayashi, T. Matsuoka, K. Nakamura, and F. Nishio, "Polarimetric characteristics of sea ice in the sea of okhotsk observed by airborne L-band SAR," *IEEE Transactions on Geoscience and Remote Sensing*, vol. 42, no. 11, pp. 2412–2425, 2004.

[21] K. Nakamura, H. Wakabayashi, K. Naoki, F. Nishio, T. Moriyama, and S. Uratsuka, "Observation of sea-ice thickness in the sea of okhotsk by using dual-frequency and fully polarimetric airborne SAR (Pi-SAR) data," *IEEE Transactions on Geoscience and Remote Sensing*, vol. 43, no. 11, pp. 2460–2468, 2005.

[22] T. Geldsetzer, J. B. Mead, J. J. Yackel, R. K. Scharien, and S. E. L. Howell, "Surface-based polarimetric C-band scatterometer for field measurements of sea ice," *IEEE Transactions on Geoscience and Remote Sensing*, vol. 45, no. 11, pp. 3405–3416, 2007.

[23] D. Isleifson, B. Hwang, D. G. Barber, R. K. Scharien, and L. Shafai, "C-band polarimetric backscattering signatures of newly formed sea ice during fall freeze-up," *IEEE Transactions on Geoscience and Remote Sensing*, vol. 48, no. 8, pp. 3256–3267, 2010.

[24] S. V. Nghiem, R. Kwok, S. H. Yueh, and M. R. Drinkwater, "Polarimetric signatures of sea ice 2. Experimental observations," *Journal of Geophysical Research*, vol. 100, no. C7, pp. 13681–13698, 1995.

[25] S. V. Nghiem, S. H. Yueh, R. Kwok, and F. K. Li, "Symmetry properties in polarimetric remote sensing," *Radio Science*, vol. 27, no. 5, pp. 693–711, 1992.

[26] S. H. Yueh, R. Kwok, and S. V. Nghiem, "Polarimetric backscattering and emission properties of targets with reflection symmetry," *Radio Science*, vol. 29, no. 6, pp. 1409–1420, 1994.

[27] F. D. Carsey, Ed., *Microwave Remote Sensing of Sea Ice*, American Geophysical Union, Washington, DC, USA, 1992.

[28] S. R. Cloude and E. Pottier, "A review of target decomposition theorems in radar polarimetry," *IEEE Transactions on Geoscience and Remote Sensing*, vol. 34, no. 2, pp. 498–518, 1996.

[29] J. S. Lee, D. L. Schuler, T. L. Ainsworth, E. Krogager, D. Kasilingam, and W. M. Boerner, "On the estimation of radar polarization orientation shifts induced by terrain slopes," *IEEE Transactions on Geoscience and Remote Sensing*, vol. 40, no. 1, pp. 30–41, 2002.

[30] I. Hajnsek, E. Pottier, and S. R. Cloude, "Inversion of surface parameters from polarimetric SAR," *IEEE Transactions on Geoscience and Remote Sensing*, vol. 41, no. 4, pp. 727–744, 2003.

15

Elastic Properties of Natural Sea Surface Films Incorporated with Solid Dust Particles: Model Baltic Sea Studies

Adriana Z. Mazurek[1] and Stanisław J. Pogorzelski[2]

[1] Department of Physics, Warsaw University of Life Sciences, Nowoursynowska 159, 02-776 Warsaw, Poland
[2] Institute of Experimental Physics, University of Gdańsk, Wita Stwosza 57, 80-952 Gdańsk, Poland

Correspondence should be addressed to Stanisław J. Pogorzelski, fizsp@ug.edu.pl

Academic Editor: Robert Frouin

Floating dust-originated solid particles at air-water interfaces will interact with one another and disturb the smoothness of such a composite surface affecting its dilational elasticity. To quantify the effect, surface pressure (Π) versus film area (A) isotherm, and stress-relaxation (Π-time) measurements were performed for monoparticulate layers of the model hydrophobic material (of μm-diameter and differentiated hydrophobicity corresponding to the water contact angles (CA) ranging from 60 to 140°) deposited at surfaces of surfactant-containing original seawater and were studied with a Langmuir trough system. The composite surface dilational modulus predicted from the theoretical approach, in which natural dust load signatures (particle number flux, daily deposition rate, and diameter spectra) originated from *in situ* field studies performed along Baltic Sea near-shore line stations, agreed well with the direct experimentally derived data. The presence of seawater surfactants affected wettability of the solid material which was evaluated with different CA techniques applicable to powdered samples. Surface energetics of the particle-subphase interactions was expressed in terms of the particle removal energy, contact cross-sectional areas, collapse energies, and so forth. The hydrophobic particles incorporation at a sea surface film structure increased the elasticity modulus by a factor K (1.29–1.58). The particle-covered seawater revealed a viscoelastic behavior with the characteristic relaxation times ranging from 2.6 to 68.5 sec.

1. Introduction

The atmospheric transport and deposition of mineral particles strongly influence the physics and chemistry of the marine atmosphere, and the biogeochemical cycles in seawater. So far, our emphasis has been closed on the multicomponent character of natural surfactant films, and the consequent complexities involved in any attempt to predict the interfacial viscoelastic properties playing a crucial role in modeling of physical systems with surface film-mediated interfaces. A variation in the surface rheological parameters of the natural surfactant seawater films has been conceived as a different in source of surfactant materials and in physical dynamics reflecting organic matter migration, degradation, and spatial-temporal dynamics in natural waters [1]. In nearly all cases, uniform, homogeneous surfaces have been studied. However, in "real" systems, in technology, biology, and oceanography, surfaces are very often non-uniform. For instance, a flat surface containing a surfactant monolayer which has undergone a two-dimensional phase separation falls under this definition, as well as air-water and oil-water interfaces with droplets, solid particles, or even thin layers of a microemulsion, foam, or a bicontinuous phase. The composite surfaces seem to be homogeneous by macroscopic observation but heterogeneous at a microscopic level. Monoparticulate layers of fine solid particles at air-water interfaces, like the monomolecular films of insoluble surfactants, can be formed by spreading from volatile organic dispersions or scattering onto the water surface. In such a definition, a composite surface consists of a mosaic of different surfaces leading to physicochemical or geometrical roughness. The exhibited film parameters variability with the environmental factors (film temperature, ionic strength, and pH of the aqueous subphase, wind speed, and time scale of relaxation processes taking place in a multicomponent natural film) has been already discussed in detail elsewhere [1, 2].

In this paper, for the first time, the influence of mineral dust particles (of micrometer diameters) deposition at the natural seawater interfaces on the surface dilational viscoelasticity was addressed. In the subsequent sections a general approach for the treatment of the dynamic properties of the composite surfaces is proposed and further quantified on several natural environment interfacial systems.

The composite surface has the particular surface rheological properties dependent not only on the particle number flux, particle shape, and its dimension but also on the wettability of solid material in contact with seawater [3]. The contact angle (CA) is a common measure of the hydrophobicity of a solid surface. Such particulate monolayers have been shown to possess the compression characteristics that are similar to insoluble monomolecular monolayers [4] and have been found to be capable of stabilizing foams and emulsions [5]. However, the interpretation of the observed contact angle is complicated by many factors such as the physical and chemical heterogeneity of the system, smoothness of the surface, presence of surfactants (autophobing affect), which all can affect the measured value. The range of contact angles of natural solid particles in seawater is not known, although more typically natural particles are characterized by contact angles of less than 90° (being of hydrophilic solid material). The input data for the model studies were derived from the supplementary field measurements of several natural environmental systems quantified in terms of the natural dust load signatures (the particle number flux, daily deposition rate, particle shape, and its diameter spectra) registered in several sampling stations located nearby sea shore (Gulf of Gdańsk, Baltic Sea, Poland). The aim of the paper was to quantify the natural dry aerosol particles effect on the dilational viscoelasticity of the apparent seawater surface.

2. Theoretical Background

2.1. Surface Dilational Viscoelasticity. The dilational elasticity modulus E_{isoth} (or Gibb's modulus), expressing the static, compressional response of a film to the surface compression or dilation corresponding to isotherm registration in its thermodynamic equilibrium, is defined as [8]

$$E_{\text{isoth}} = -A\left(\frac{d\Pi}{dA}\right), \qquad (1)$$

where Π is the surface pressure of a film, and A is the film area.

The establishment of thermodynamic equilibrium in the film is not trivial since the most real surfactant interfacial systems are viscoelastic. The effect depends on the dimensionless parameter Deborah number (De) defined as the ratio of the film relaxation time τ to the "time of observation" (a reciprocal of the strain rate of a film: $t_{\text{obs}} = [(dA/A)/dt]^{-1}$, as argued in [9]). The interfacial system appears to be at the quasi-equilibrium thermodynamic state if De number is less than unity.

Any relaxation process in films leads to dilational viscoelasticity, and the surface dilational viscoelastic modulus E is a complex quantity composed of real E_d (dilational elasticity) and imaginary E_i (= $\omega\eta_d$, where η_d is the dilational viscosity) parts $E = E_d + iE_i = E_o\cos\varphi + iE_o\sin\varphi$, where ω is the angular frequency of periodic oscillations, $E_o = -\Delta\Pi/(\Delta A/A)$ represents the amplitude ratio between the surface stress and strain, and φ is the loss angle of the modulus [10]. For surface layers exhibiting a pure elastic behavior the linear relation between $\Delta\Pi$ and ΔA appears, as shown by (1). In the case of the viscoelastic film, the relation contains an additional term depending on the surface deformation rate:

$$\Delta\Pi = E_{\text{isoth}}\frac{\Delta A}{A} + \eta_d\frac{d(\Delta A/A)}{dt}. \qquad (1a)$$

The time scale of the relaxation processes taking place in surface films, and the viscoelasticity modulus parts can be derived from the stress-relaxation studies [11]. The surface pressure-time ($\Pi - t$) response of a film to a rapid step ($\Delta t = 0.2$–1.5 s) relative surface area deformation $\Delta A/A_o$ (= 0.07–0.23) is registered and presented in the following form [12]:

$$\ln\left[\frac{(\Pi_\infty - \Pi_t)}{(\Pi_\infty - \Pi_0)}\right] = -\lambda_i t, \qquad (2)$$

where Π_∞, Π_o, and Π_t are the surface pressures at steady-state condition ($t \to \infty$), at time $t = 0$, and at any time t; λ_i is the fist-order rate constant related to the relaxation time τ_i (= $1/\lambda_i$).

In the framework of a model for dilational viscoelasticity, adapted to the stepwise deformation mode, the real and imaginary parts of E can be obtained from the following relations [13]:

$$E_d = E_o\left[\frac{(1+\Omega)}{(1+2\Omega+2\Omega^2)}\right],$$

$$E_i = E_o\left[\frac{\Omega}{(1+2\Omega+2\Omega^2)}\right],$$

$$E_o = \frac{(\Pi_o - \Pi_\infty)}{(\Delta A/A_o)} \text{ is an amplitude of the modulus } E,$$

$$\text{where } \Omega = \left(\frac{\Delta t}{\tau_i}\right)^{1/2}, \qquad \tan\varphi = \frac{\Omega}{(1+\Omega)}.$$

$$|E| = \left(E_d^2 + E_i^2\right)^{1/2} \text{ is the modulus of the complex quantity } E;$$

$$\qquad (3)$$

Δt is the applied step film area deformation time.

At sufficiently low film area compression rates (De $\ll$ 1), the dilational viscoelasticity modulus can be approximated by E_{isoth}.

2.2. Structural Parameters of Solid Particle Layers at the Air-Water Interface. The data collected in the surface film elasticity measurements performed on the model particle-incorporated inhomogeneous films allowed the particle removal energy, contact cross-sectional areas, and collapse film energies to be determined, that is, the key parameters in surface rheology.

Surface pressure Π versus surface area A isotherms can be determined for monoparticulate layers by means of Langmuir trough film, in order to obtain information on particle sizes, particle-particle repulsive interactions, and wettabilities. Upon compressing the layers, collapse takes place at a certain pressure Π_c, at which the particles are forced out of the air-water interface. Beyond this point, it may be presumed that the compression work done on the system is entirely channeled into the removal of particles from the interface [14]. The removal energy E_r for one particle can be written as follows:

$$E_r = \Pi_c A_c, \qquad (4)$$

where A_c is the area per particle at collapse ($= A/N$, A is the actual film balance area at Π_c, $N = (m/\rho)/(4R^3\pi/3)$ is the number of all particles spread onto the water surface, R is the particle radius, and ρ is the particle density). Information on the particle wettability can be obtained assuming equality of E_r to the work of adhesion W_r [15].

The collapse pressure can be related to the air-water contact angle θ, for monodisperse and spherical particles, according to:

$$\Pi_c A_c = \gamma_{LV} R^2 \pi [1 \pm \cos\theta]^2, \qquad (5)$$

where γ_{LV} is the air-water surface tension. If a particle is removed from the interface into the upper phase, then the cosine in the brackets is taken positive, and θ signifies the receding contact angle θ_R; if it moves into the lower phase, the cosine is taken negative (advancing contact angle θ_A). A position of a solid spherical particle at the air-water interface, for hydrophilic (CA < 90°) and hydrophobic (CA > 90°) spherical particles is shown in Figure 1.

For the closest packed (hexagonal) arrangement of the monodisperse and spherical particles at collapse (i.e., if the maximum surface coverage SC = $(R^2\pi/A_c)$ = 0.91 can be reached during the compression), the contact angle can be calculated from the following expression [14]:

$$\cos\theta = \pm \left[\left\{ \frac{\left(2(3)^{1/2}\Pi_c \right)}{(\gamma_{LV}\pi)} \right\}^{1/2} - 1 \right]. \qquad (6)$$

The number of particles per unit area for hexagonal close packing is $N_p = 1/[2(3)^{1/2}R^2]$ and is dependent only on R and corresponds to the contact cross-sectional area $A_{cc} = 1/N_p$. In this case, the contact angle can be determined without knowing the particle size and density. The collapse energy E_c can be calculated by integration of the isotherm plot $\Pi(A)$ from A_∞—the area for a particle at which the surface pressure does not exceed the zero value to the collapse area A_c [4].

2.3. Dilational Modulus of a Composite Surface-Solid Particles Covering a Variable Area. The difficult task is to derive the measured stress (surface tension change) resulting from an externally applied two-dimensional (dilational) strain (surface area change). The complication for a composite

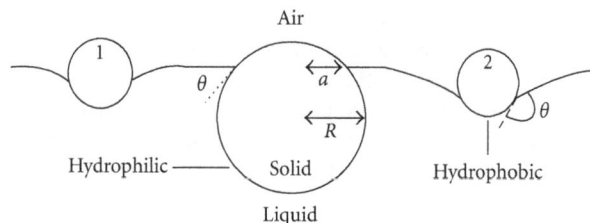

FIGURE 1: Schematic drawing of a spherical particle composed of hydrophilic (CA < 90°) and hydrophobic (CA > 90°) material of radius R showing its position at an air-water interface determined by the contact angle θ where gravitational effects are negligible, for spheres of μm-diameter.

surface is its nonuniformity which leads to an uneven distribution of the applied strain and the resulting effect on the measured stress.

A solid particle which is spherical and so small (in reference to the capillary length, i.e., $R \ll (2\gamma_{LV}/\rho_L g)^{1/2}$, where γ_{LV} and ρ_L are the surface tension and density of the liquid phase, resp.) that gravity can be ignored if compared to the surface forces will adopt a position in the air-fluid interface which is fully determined by the wetting angle (Figure 1). The area effectively occupied by such particles when there are n of them per unit surface area A is given by

$$A_p = n\pi R^2 \sin^2\theta. \qquad (7)$$

Considering the total surface covered with a collection of the solid particles and remaining particle-free liquid surface (having E_{free} modulus), for the modulus of the composite surface E_{com} we have [3]

$$E_{\text{com}} = \frac{E_{\text{free}}}{\left\{ 1 + n\pi R^2 \left[\left(2 E_{\text{free}}\cos^2\theta/\gamma_{LV} \right) - \sin^2\theta \right] \right\}}. \qquad (8)$$

As can be noticed, the presence of partially wetted spherical particles can either increase or decrease the apparent dilational modulus of the whole surface depending on the sign of the term within brackets in the denominator of (8). Thus, when θ is close to 90°, or when $E_{\text{free}}/\gamma_{LV}$ (values ~0.1 are found in practice) is very small, there will be increase. In contrast, when θ is close to 0° or when $E_{\text{free}}/\gamma_{LV}$ is large, the surface dilational modulus will decrease due to the presence of particles.

3. Materials and Methods

3.1. Materials. Distilled water, used for Langmuir trough isotherm studies and contact angle measurements, was taken from a water deionization apparatus (Millipore, conductivity 0.05 μS cm^{-1}) with pH 5.8 ± 0.1, and surface tension γ_{LV} = 72.5 ± 0.1 mN m^{-1}. Original seawater was collected at Jelitkowo (Gulf of Gdańsk, Baltic Sea, Poland) on December 7–9, 2006 having γ_{LV} = 60.5 ± 0.1 and pH 8.2 ± 0.1. As the model solid material, in Langmuir trough measurements, the following particles of differentiated wettability were chosen: silica microsphere (mean diameter ± standard deviation;

88.00 ± 10.56 μm), talc (22.94 ± 2.78 μm), and combustion dust (3.79 ± 0.48 μm).

3.2. Methods

3.2.1. Langmuir Trough and Stress-Relaxation Measurements.
For the determination of surface pressure versus surface area isotherms, a laboratory built Langmuir trough-Wilhelmy plate system was used, described in detail elsewhere [1].

The surface pressure was measured with a Wilhelmy plate technique using a piece of filter paper (Whatman No. 1, Madstone, England; 5 cm wide) attached to the arm of a force sensor (GM 2 + UL 5, Scaime, France); they were accurate to within 0.1 mN m^{-1}.

In order to register force-area isotherms, the surface area in the trough was continuously decreased during the compression from 850 to 25 cm^{-2} at a constant rate of 3 cm s^{-1}. It corresponds to the surface deformation velocity $u = \Delta A/\Delta t$ equal to 0.95 cm^2 s^{-1}. Any relaxation process in films leads to the surface viscoelasticity and may affect the shape of isotherms, and consequently the recovered film parameters. As shown in [9], the effect depends on the dimensional parameter Deborah number (De) defined as the ratio of the film relaxation time τ to the time of observation t_{obs} (= $[(dA/A)/dt]^{-1}$). For several marine film samples already studied in the Baltic Sea [2], the relaxation times were found to lie in rather narrow ranges [2]: $\tau_1 = (1.3$–2.9 s) and $\tau_2 = (10.1$–25.6 s). The film deformation rate adapted in these isotherm studies leads to De number values as low as 0.07, that is, much less than unity, and the interfacial system appears to be in its quasi-equilibrium thermodynamic state.

A new type of Langmuir trough has been used in this study to perform stress-relaxation studies, adapted after [16]. A solid PTFE frame, or barrier, is placed in a rectangular plastic tray of dimensions 90 cm × 60 cm × 6 cm. The barrier has four, rigid solid sides 60 cm long, and 5 cm high. The four sides are hinged at the corners so that they provide a continuous, leak-free enclosure. The walls have PTFE pegs underneath so that liquid can flow under the walls to create a well-defined interfacial area inside the barrier. The barrier contains the film of interest, which is omnidirectionally expanded or compressed by flexing the corners, thus changing the shape of the barrier. Two opposite corners of the barrier are connected to a geared stepper motor. The steeper motor system is driven and controlled via a PC-class computer. The drive movement is accurate to 0.1 mm. On driving the corners together or apart the interfacial area inside the barrier is changed stepwise from its initial value $A_0 = 1600$ cm^2. The surface pressure is measured with a Wilhelmy plate dipping into the interface, situated at the centre of the film, suspended from a force transducer. The surface pressure-time response to a rapid, step ($\Delta t = 0.2$–1.5 s) relative area deformation $\Delta A/A_0$ (= 0.07–0.23) is registered to derive the dynamic surface rheology parameters.

Solid samples (~600 mg) were carefully sprinkled onto the trough water surface by using a spatula. A particle dust cloud generator was also used in this study. A vibration feeder supplies the test particles to the fun-generated air

FIGURE 2: Natural dust deposition sampler system with a horizontal deposition plate to collect particles (1), self-oriented to the wind direction by wing (adapted after [6]).

stream. The model particles are mixed with an air stream and blown upward. The air velocity, at the middle of the outlet stack oriented over the water surface, was 3-4 m s^{-1}. The particle surface density (number cm^{-2}) and particle diameter distribution were derived from the microscope picture analyses of the greased cover glass palates. The dry deposition collector was placed nearby the studied surface. Detailed information on the applied optical method and the sampling procedure can be found elsewhere [6].

3.2.2. Natural Dust Solid Particles Collection.
Figure 2 shows a top of the dry deposition sampler, and the deposition plate is used in this study to collect the airborne particles. The sampler was designed to provide minimum airflow disruption. A pivoting support system pointed the horizontally oriented deposition plate into the wind. The deposition film consisted of a greased (with Apiezon grease type L as a deposition sampling substrate not absorbing water vapor) cover glass supported on a microscope slide glass (24 × 40 mm) placed in a groove on the deposition plate. More details on dry deposition methodology with the mentioned sampler can be found in [6].

(1) Particle Water Contact Angle Characterization. Since the model materials are in a powdered form, there is no universal method for solid surface/surfactant-containing water contact angle determination. Here we used three most widely used methods to select the most practical one giving the most accurate and reproducible CA values.

(2) Sessile Drop Technique. This technique is among the most popular methods used in surface chemistry laboratories, especially due to their simplicity. In the sessile drop technique a small liquid drop (2–5 mm diameter) was placed on the solid surface (microscope glass slide covered with a thin layer of the studied solid material) using a microliter syringe. Horizontal sessile drop profiles, taken with a digital camera, were analyzed with the Axisymmetric Drop Shape Analysis-Profile (ADSA-P) technique, from which all the sessile drop

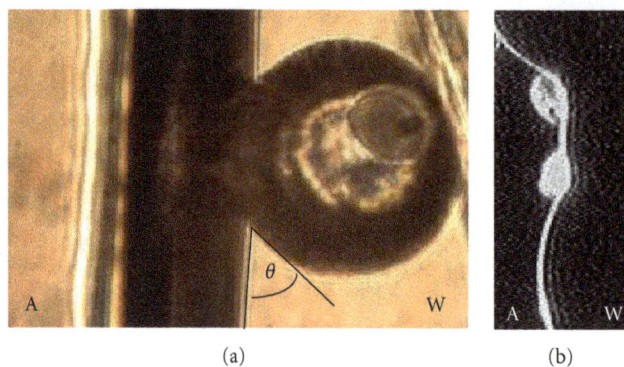

FIGURE 3: Exemplary photographic records of model silica microsphere (a) and talc particle (b) floating at the air (A)—original seawater (W) interface taken for evaluation of CA in direct microscopic method.

parameters such as CA, contact radius, and volume can be extracted. The adapted apparatus is described in [17].

(3) Microscope Technique. The surface of solid particles was characterized by measuring the static contact angles from the microscopic images of the particles placed at the air-water interface formed between two plane-parallel microscope glass slides. The details of the applied optical method can be found elsewhere [4]. An example of photographic records of the model dust material contained in seawater is presented in Figures 3(a) and 3(b).

(4) Thin Layer Wicking Technique. The most reliable technique for the measurement of the contact angle on the powered samples is "the thin layer wicking technique" based on the Washburn equation [18, 19]. In this method, a powdered sample is deposited on a microscopic glass slide in the form of aqueous slurry on which a thin layer of the powdered mineral has been formed. An aqueous suspension of the model material was prepared by dispersing a known amount of sample material in distilled water (5% weight by volume). The appropriate volume of the suspension was withdrawn with a pipette and sprayed over a glass slide (24×60 mm). After drying the sample at $110°$C, one end of the glass slide was contacted with a vial containing cotton wool filled with a spreading liquid (seawater, distilled water, and pentadecane) with the marked appropriate scale (1–15 cm), as shown in Figure 4. The liquid will start to move along the horizontally-oriented slide through the capillaries formed between the particles deposited on the glass surface. The velocity of moving liquid line is measured and then converted to the contact angle using the Washburn equation. The calibration procedure, as required, was performed with pentadecane ($C_{15}H_{32}$) as a reference liquid (giving the complete spreading condition i.e., CA = 0). It should be borne in mind that such a method gives the so-called advancing CA that is usually higher than the equilibrium Young CA. The applied apparatus and methodology is described in details in [7].

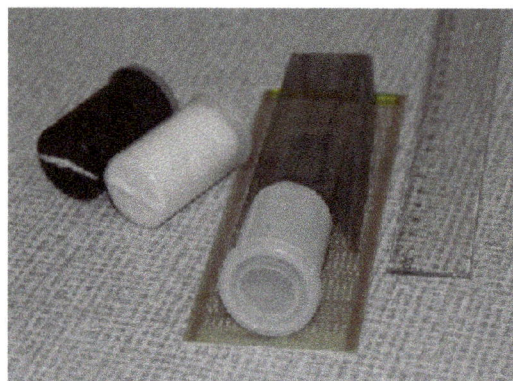

FIGURE 4: Cylindrical cotton wool filled plastic containers saturated with probe liquids attached to a horizontal glass slide covered with a powdered material used in "the thin layer wicking technique" to determine CA of the model materials (adapted after [7]).

4. Results and Discussion

4.1. Monoparticulate Solid Layer Interfacial Parameters. The surface pressure (Π) versus film area (A) measurements of composite layers provided fruitful information about the particle size distribution, particle-particle and particle subphase interactions, and surface wettability signatures [20]. The surface energetics of the composite surface can be quantified by means of the surface parameters evaluated according to (4)–(6). Typical surface pressure (Π) versus surface area (A) isotherms obtained for the model hydrophobic particles are depicted in Figures 5(a)–5(c). Compression led to a collapse phenomenon of the structured monoparticle layers. The extrapolation procedure for determination of Π_c and A_c for silica monoparticles is shown in Figure 6. The collapse of solid monoparticulate layers is not a real collapse phenomenon (e.g., surface pressure does not drop to zero at this point). The "knee" in the curve, taking place in a surface pressure range about 15 mN m^{-1} for microsphere, is followed to area in which the particles become close packed. Above the particular surface pressure, different for the model materials, depending on their hydrophobicity,

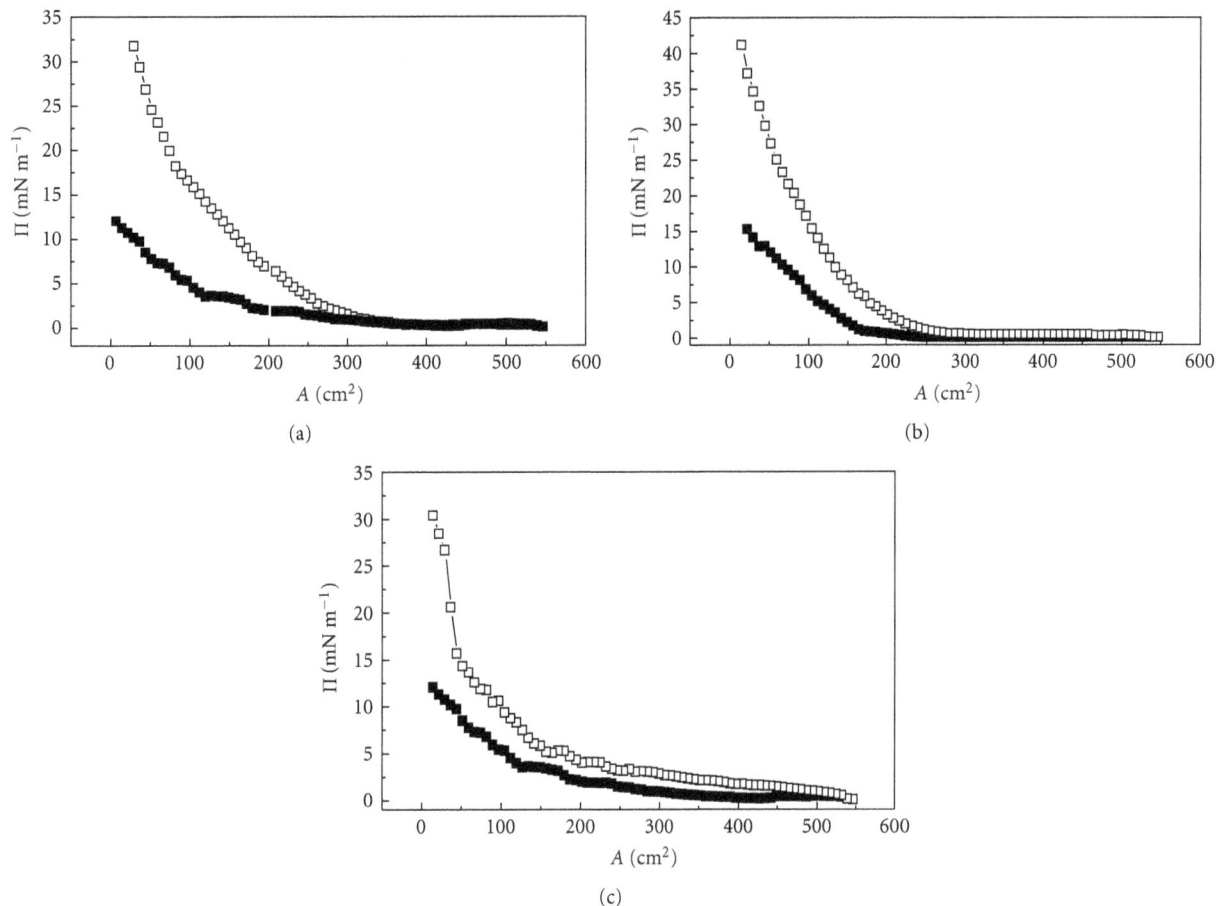

FIGURE 5: Surface pressure (Π) versus surface area (A) curves for (a) silica microsphere $\square$, (b) talc $\square$, and (c) combustion dust $\square$ particles spread on an original seawater subphase collected at Jelitkowo (on 08.12.2006). Reference particle-free isotherm $\blacksquare$. Water sample at $T = 22.4°C$ and pH = 8.4.

monolayer pleated that leads to gradually create a three-dimensional monoparticulate layer. The monolayer collapse effect is dependent on the size and nature of the particles. Consequently, continued compression above this point causes a slow further increase in surface pressure, and the particles are gradually squeezed out of the interface. The steepness of the surface pressure—surface area isotherms facilitated the determination of the packing area, which, by analogy to monolayers formed by surfactants—is referred to as the contact cross-sectional areas CCSA. CCSA can be determined from the Π-A isotherms by fitting a straight line to the steepest, nearly linear (solid state) part of the isotherm (the limiting area for insoluble films is determined in the same way). CCSA and collapse energies E_c (E_c results from the integration of the isotherm plot between $\Pi = 0$ and $\Pi = \Pi_c$) provided semiquantitative information on the strength of structural interfacial forces. That is, the higher the particle hydrophobicity, the greater the structural strength of the monoparticulate layer. So, the value of CCSA refers to the particle hydrophobicity; greater values of CCSA stand for higher hydrophobicity of particles [20]. Monoparticulate layers which were formed from the most hydrophobic spheres had a structural strength greater than

that of those which were formed from the least hydrophobic material as revealed by E_c and CCSA values collected in Table 1. The collapse energy E_c, necessary to compress the most hydrophobic particles (combustion dust) into a close-packed hexagonal array, was almost two times greater than that required for the compression of the least hydrophobic particles (talc and microsphere). The total pair-interaction energy for interfacial particles floating at liquid-gas interface is expressed as $V_T = V_A + V_R + V_S + V_D$ (where V_A is the van der Waals attraction; V_R is the electric-double-layer repulsion; V_S is the structural interaction energy; V_D can be either attractive (hydrophobic attraction energy) or repulsive (salvation repulsion energy) depending on the hydrophobic-hydrophilic nature of the particle surface. If the water contact angle is lower than 15°, the surface will be hydrophilic, and above 64° it should be considered as hydrophobic [20].

The Langmuir trough method makes possible to study the wettability of mineral particles (with CA) by a surfactant solution. It is also accepted that the surface pressure measurements can be related to the work necessary for the compression of particulate layers, and for the particle removal W_r from the liquid-gas interface. One could obtain the contact angles in sea surface containing solid particles if E_r and W_r

TABLE 1: Characteristic parameters particle-particle and particle-subphase interaction parameters evaluated for monoparticulate layers of the model material deposited at distilled (W) and original sea water surfaces (S), collected at Jelitkowo on December 8, 2006; $T = 22.4°C$; pH = 8.4; $\gamma_{A/W} = 60.5$ mN m^{-1}. Standard deviation is given in brackets.

Sample	$\langle R \rangle$ (μm)	N (mm^{-2})	Π_c (mN m^{-1})	A_c (μm^2/particle) ×10^3	CCSA (μm^2/particle) ×10^3	E_r (J/particle) ×10^{-12}	W_r (J/particle) ×10^{-10}	E_c (J/particle) ×10^{-12}	θ (°)
Talc^W	11.47 (0.1)	16	24.00 (2.89)	0.234 (0.028)	790.0 (94.9)	5.61 (0.67)	0.12 (0.01)	15.10 (1.81)	110.9 (6.5)
Talc^S	11.47 (0.1)	16	24.84 (2.99)	0.364 (0.044)	1088.0 (131.6)	9.05 (1.07)	0.26 (0.03)	19.42 (2.34)	112.0 (6.7)
Silica Microsphere^W	44.00 (1.0)	24	30.84 (3.74)	0.082 (0.009)	260.0 (33.9)	2.52 (0.31)	2.07 (0.25)	8.24 (0.97)	88.3 (5.3)
Silica Microsphere^S	44.00 (1.0)	12	27.14 (3.27)	0.264 (0.032)	592.8 (72.1)	7.17 (0.88)	3.75 (0.44)	19.42 (2.37)	92.2 (5.5)
Combustion dust^W	1.89 (0.01)	4	18.74 (2.24)	0.300 (0.037)	1740.0 (211.4)	5.62 (0.68)	0.06 (0.01)	22.17 (2.71)	117.7 (6.9)
Combustion dust^S	1.89 (0.01)	9	37.26 (4.45)	0.458 (0.053)	1820.0 (219.5)	9.61 (1.19)	0.07 (0.01)	37.29 (4.52)	120.1 (7.2)

TABLE 2: Contact angles (CA) of studied mineral dust particles with distilled (W) and original sea water (S) derived by means of different techniques. r^*: pore radius.

Sample/phase	CA determination technique			
	Sessile drop	Isotherm derived from (8)	Microscope technique	Thin layer wicking technique
Talc^W	108.6 ± 6.8	110.9 ± 6.5	88.8 ± 5.4	80.3 ± 4.8 (r^* = 402.6 nm)
Talc^S		112.0 ± 6.7	93.9 ± 5.7	86.7 ± 5.2
Microsphere^W	99.4 ± 6.1	88.3 ± 5.3	62.0 ± 4.2	78.0 ± 4.6 (r^* = 1301.1 nm)
Microsphere^S		92.2 ± 5.5	68.1 ± 4.8	83.0 ± 4.9
Combustion dust^W	120.6 ± 7.1	117.7 ± 6.9	116.9 ± 6.0	124.2 ± 8.3 (r^* = 695.7 nm)
Combustion dust^S		120.1 ± 7.2	126.2 ± 6.8	128.7 ± 4.6

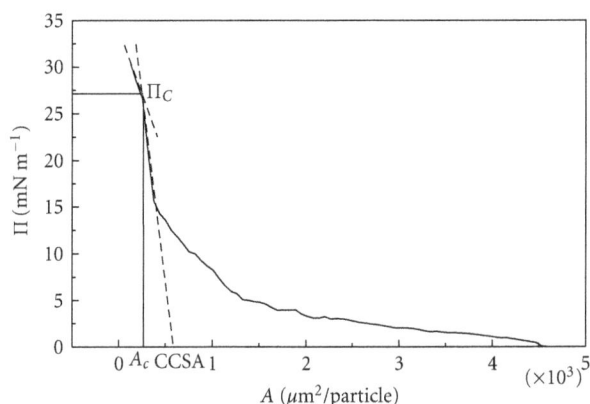

FIGURE 6: Surface pressure (Π) versus surface area (A) isotherm, for silica microsphere layers deposited at original seawater collected at Jelitkowo (08.12.06; T = 22.4°C; pH = 8.4) as an exemplary plot illustrating determination and extrapolation procedures to derive collapse pressure Π_c, surface area A_c at the collapse pressure, and contact cross-sectional area (CCSA).

noticed here, for the most hydrophobic solid material, that is, combustion dust particles. These observations are related to differences in the structural force strength between the model composite particle layers. Moreover, there may be a strong fluctuation of the liquid-air interface due to the trapping the test particles, which probably results in a lower value of the wetting angle (the capillary forces supply significant kinetic energy for the particles during their trapping) which, due to the elasticity of water-air interface, leads to an oscillation of the particles [4]. It was also found that the collapse pressure and contact angles of hydrophobic particles can be depended on the amount of spread particles to a certain extent [15]. The effect was attributed to a surface pressure gradient along the very cohesive particulate layer.

4.2. Composite Surface Elasticity.

The composite modulus E_{com} of the particle-incorporated seawater interface can be theoretically obtained from (8) with the following entering input data characteristics for the model materials: particle concentration, diameter distribution, and solid/water phase contact angles. E_{com} values were comparable to the result obtained from the isotherm studies, and the ones estimated for the clean sea surface (without particles), as collected in Table 3. Theoretically predicted values of E_{com} were in agreement with the experimental data within a range of 3% error, for microsphere and combustion dust, although apparently higher. A variety was considerably highest for talc particles ($\approx$21%), that has the unique surface properties. Particles of talc have the shape of platelets due to the layer structure of the mineral. It is well known that the basal surfaces are hydrophobic, while the edge surfaces are hydrophilic [22]. The hydrophobicity of the basal surfaces arises from the fact that the atoms exposed on the surface are linked together by siloxane (Si–O–Si) bonds and do not form strong hydrogen bonds with water. The edge surfaces are composed of hydroxyl ions, magnesium, silicon, and substituted cations, all of which undergo hydrolysis. As a result, the edges are hydrophilic and can form strong hydrogen bonds with water molecules and polar substances. The presence of the particles in sea surface film caused an increase of the apparent modulus by a factor K (E_{com}/E_{free} = 1.29–1.58), as summarized in Table 3.

are equal. Three different CA wettability measurement techniques were taken in for the determination of contact angle of the model powders. The particle-subphase interaction can be responsible for the observed difference between contact angles derived from the isotherms (larger by 8–10°) than these obtained from the direct contact angle measurements (the sessile drop, microscope, and thin layer wicking technique), as can be noticed from Table 2. The higher values (by 5–6°) of the contact angle were observed for the model particles in contact with the surfactant-containing seawater if referred to the distilled water. The effect is attributed to the autophobing phenomenon leading to the solid surface free energy decrease in the surfactant water solutions accompanied by the corresponding increase of θ [21].

To sum up, the effect of surfactant-containing seawater on the structural strength and particle-particle interaction was expressed in higher hydrophobicity of the model materials deposited at the seawater surface (A_c↑ and CCSA↑), higher twice E_c, larger is the energy required to remove the particle (E_r↑), and corresponds well with an increase of the contact angle (θ↑). The largest parameters values were

(a)   (b)

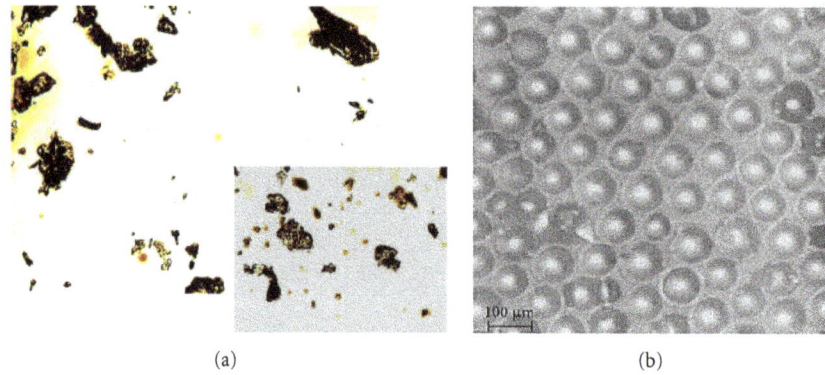

FIGURE 7: (a) Atmospheric irregularly shaped dust particles on the deposition film photographed in Baltic Sea coastal areas (at Gdynia, mean diameter = $5.86 \pm 0.85\,\mu$m). (b) Monoparticular layer of silica spheres at A/W interface.

TABLE 3: Elasticity modules of composite surfaces obtained from experimental isotherm measurements (a), and theoretical predictions (from (8)) (b). Original sea water collected at Jelitkowo on 08.12.06; $T = 22.4°$C, pH = 8.4, $\gamma_{A/W} = 60.5\,$mN m^{-1}. K is the ratio E_{isoth}/E_{free}, where E_{free} is the modulus for clean particle-free seawater surface (c). Standard deviation is given in brackets.

Sample	E_{isoth}[a] (mN m^{-1})	E_{com}[b] (mN/m)	K
Talc			
$\quad \bar{R} = 11.47\,\mu$m			
$\quad N = 16\,$mm^{-2}	10.60 (1.26)	12.90 (1.55)	1.29
$\quad \theta = 112.0°$			
Microsphere			
$\quad \bar{R} = 44.00\,\mu$m			
$\quad N = 12\,$mm^{-2}	12.96 (1.57)	13.39 (1.61)	1.58
$\quad \theta = 92.2°$			
Combustion dust			
$\quad \bar{R} = 1.89\,\mu$m			
$\quad N = 9\,$mm^{-2}	11.84 (1.43)	12.25 (1.48)	1.45
$\quad \theta = 120.1°$			
Jelitkowo	8.17[c] (1.03)	—	—

4.3. Morphology, Chemical Composition, Spatial Distribution, and Wettability of Natural Dry Deposition Events at Baltic Sea Coastal Areas. In order to validate the dilational composite modulus approach in reference to the natural particular dust deposition event conditions, that is, the seawater interface incorporated with natural mineral dust particles, dry dust characterization studies were performed in Baltic Sea shore regions (Gulf of Gdańsk, Poland) in September 2006. Mineral fly dust mainly consists (identified by X-ray diffraction) of a mixture of silicates (clay minerals, feldspars, quartz) and sometimes carbonates and sulfates [23]. It was observed that the mineralogical composition of the dust directly depends on its origin [23]. By order of abundance, the minerals identified in the dust were as follows: illite > quartz > smectite > palygorskite > kaolinite > calcite > dolomite > feldspars [24]. The mean density of the compounds was 2.65 g cm^{-3} [25]. Morphologically, mineral atmospheric particles have an irregular shape (Figure 7), which demands applying, in the aim of the particle diameter distribution to be characterized, the equivalent maximum diameter defined as the maximum diameter corresponding to an axisymmetric drop with the same volume as the actual drop ($\sqrt{\langle d_{max}^2 \rangle} \equiv 2\sqrt{A/\pi'}$; the drop mean quadratic diameter—$\sqrt{\langle d_{max}^2 \rangle}$; particle area—$A$). For comparison, a monoparticular layer of the silica microspheres at the air-water interface is shown in Figure 7(b). Histograms of the dust particle diameter distribution, for a 24-hour collection time, registered at Brzezno (a), Sopot (b), and Gdynia (c) are depicted in Figure 8. Mean values of the diameter were ranging from 5.8 to 7.6 μm depending apparently on the distance from the shore line. As evidenced from the dust diameter profile taken along the Sopot pier shown in Figure 9(a), the mean value steeply dropped down from 5.5 μm to around 3.5 μm within the first 70 meters from the shore line remaining almost the same later on. The particle number flux (mean value) at the shore line was equal to 47.5 per 1 mm^{-2} and continuously decreased with the distance passed toward the sea attaining 26.3 at the pier end (see Figure 9(b)). A deposit gauges are usually operated on a monthly basis with results being expressed as the mean daily deposition rate (mg m^{-2} day^{-1}) of undissolved solids. The ambient daily dry deposition rate spatial characteristics collected in the Sopot pier revealed an exponential decay character with the values ranging from 10.82 to 2.30 mg m^{-2} day^{-1}(Figure 9(c)). For example, dust gauges located within 100 m off heavily trafficked roads evidenced the deposition rates frequently exceeded 200 mg m^{-2} d^{-1}. The following approximate dust fluxes could be expected in the open country-village area (general deposits) ~50 mg m^{-2} d^{-1}, commercial centre of town ~100 mg m^{-2} d^{-1}, and purely industrial area ~150 mg m^{-2} d^{-1} [26].

Large quantities of the suspended solid particles of different size spectra are often found in estuaries, shallow water, and near ocean sewage outfalls of large cities [27]. A numerous part of suspended particulate matter in estuaries and coastal waters can exist as aggregates, or flocs, which are composed of inorganic material particles and biogenic debris as well as organic matter as cells, cellular exudates,

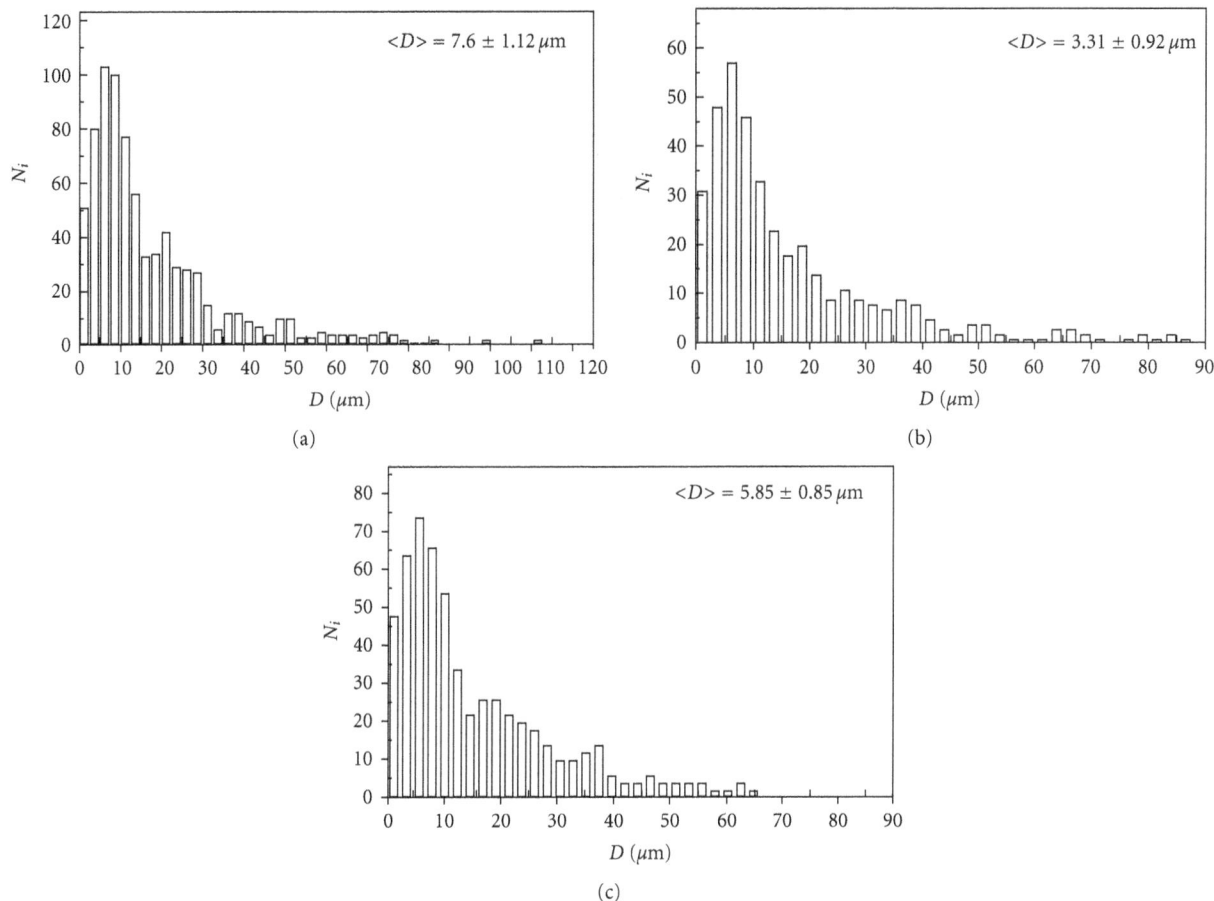

FIGURE 8: Histogram of the natural dust particle diameter distribution, for a 24-hour collection time, measured at (a) Brzezno (collection point 55-meter distant from the shore line), (b) Sopot (25-meter distant), and (c) Gdynia (5-meter distant).

and humic material [28]. Both the concentration and particle size distribution are primary information for the analysis of the suspended material, and their influence on the change of sea surface film elasticity. Temporal and spatial variabilities in the concentration and composition of the particle material depend on biological, chemical, geological, and physical factors [29]. For instance, the size distribution of suspended particles observed in the Humber Estuary exhibited a large range with the median diameters extending from $50\,\mu m$ in surface water to $500\,\mu m$ near the bed [30]. Whereas, in coastal waters near a sewage outfall in Sydney (Australia), it can be seen that 90 per cent of particles lie between diameters 0.4 and $2.4\,\mu m$. The mass-particle size distributions of mineral dust particles transported long distances are continuously reshaped by the dry removal of particles, and under some circumstances by mixing or aggregation [31]. The mass median diameters for the north Atlantic were ~2-3 μm (2.4 (Bermuda spring); 2.0 (Bermuda summer); 2.3 (Barbados); 1.2 (Izana); 3.0 (at-sea ship-collected). It has also been shown that the typical diameter range of the road (the highway) dust is about 5–30 μm [25].

If it occurs in the marine boundary layer over the remote ocean, the implications for the air-sea exchange of dust will be far different than if the aggregation takes place close to the source region [32, 33]. CA of the natural dust particles

performed by means of direct microscopic and thin layer wicking techniques were varying in a rather wide range from 47.3° to 106° depending on the collection site, and probably on a variable proportion of hydrophobic and hydrophilic material composing the particle. In reference to the obtained results, mineral dust consists of variety hydrophobicity properties particles. The sea surface is supposed to be covered by the most hydrophobic particles ($\theta > 90°$), whereas the hydrophilic ones ($\theta < 90°$) are submerged in the subphase layer. Both kinds of materials appear as the mixed, aggregated interfacial structures which possess significant dilational viscoelasticity. The real dust characteristics registered in the shore line of Gdańsk Gulf differed significantly from the model material properties (i.e., irregular particles of smaller diameters with higher surface concentrations), that should lead to the particular elastic properties of natural composed sea surface films.

4.4. Viscoelasticity of Solid Particles Incorporated Seawater Surface. Relaxation processes in the surface films lead to surface viscoelasticity and may affect the shape of isotherms, and consequently the recovered film parameters. The surface pressure-time responses of a natural marine surface film to a rapid ($\Delta t = 0.3\,s$) step relative surface area deformation $\Delta A/A$ (= 0.16), for a sample collected at Jelitkowo on

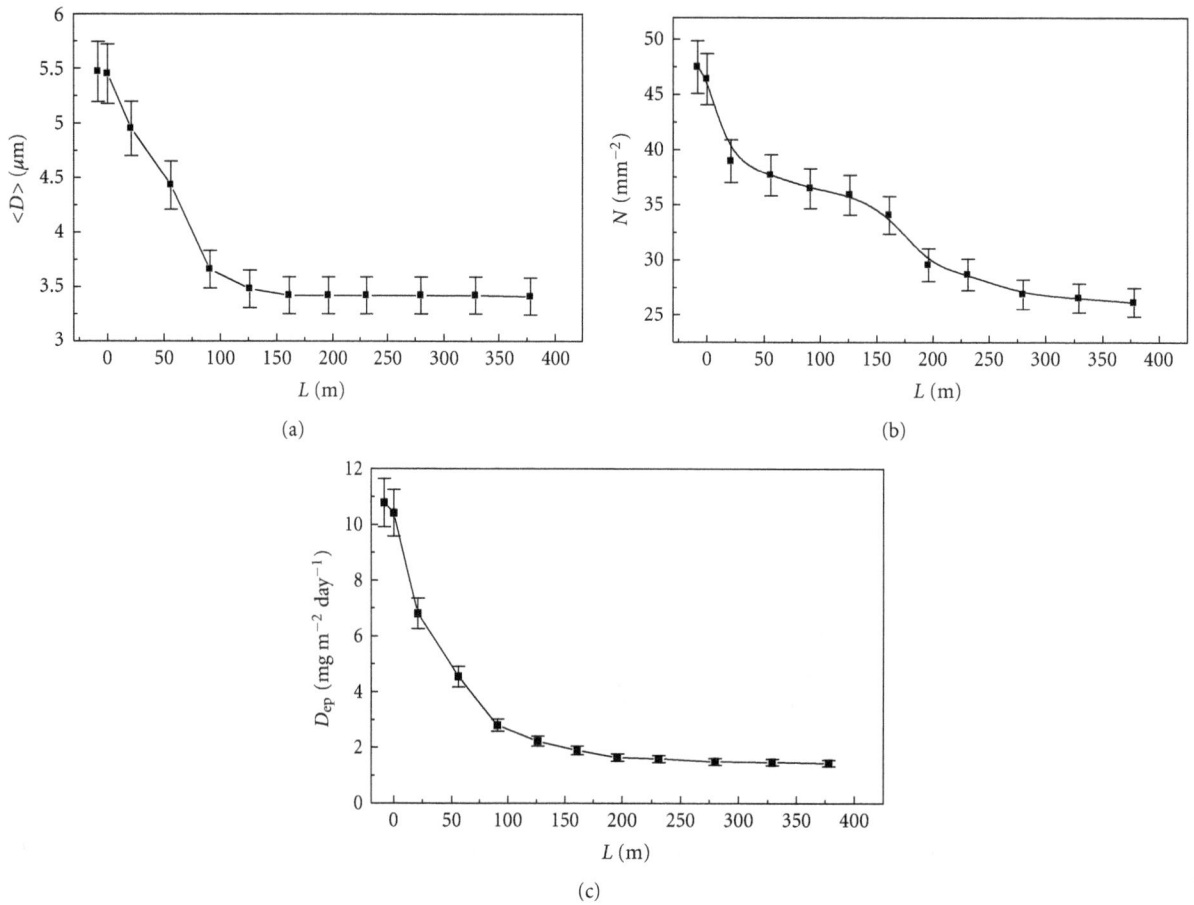

FIGURE 9: Mean natural dust diameter spatial (a), particle number flux (b), and mean daily deposition rate (c) profiles, registered along the Sopot pier on september 26, 2006, for a 24-hour period. $V_{10} = 4.0 \pm 1\,\mathrm{m\,s^{-1}}$; $T_{\mathrm{air}} = 21.5°C$.

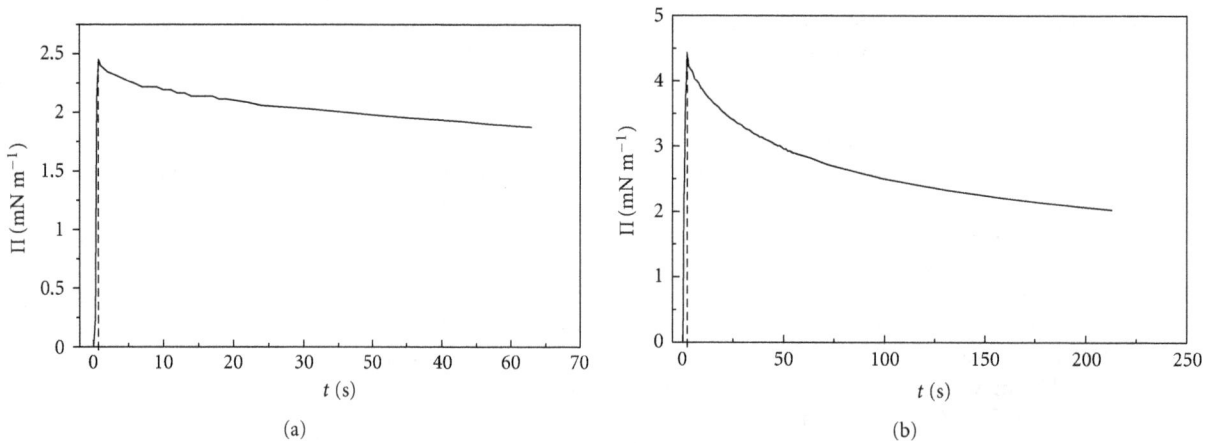

FIGURE 10: Results of stress-relaxation experiments: $\Pi(t)$ decay dependence of surface pressure versus time after a step rapid ($\Delta t = 0.3\,\mathrm{s}$) relative film area $\Delta A/A(= 0.16)$ compression, for a marine film sample collected at Jelitkowo on 11.12.2006 (sample state: $T = 19°C$, pH = 8.2) as a reference particle-free surface (a), and for composite surface (b), that is, monoparticulate talc layer covered the same seawater interface ($\Delta A/A = 0.17$; $\Delta t = 1.2\,\mathrm{s}$).

TABLE 4: Rheokinetic parameters of dilational surface viscoelasticity obtained in the stress-relaxation experiment performed on solid monoparticulate layers deposited at the sea water interface. A: particles-free surfactant natural film; S: mineral model particle film spread at sea water collected at Jelitkowo (December 11, 2006). Standard deviation is given in brackets.

No.	Sampling site date	$\Delta A/A$	Δt (s)	τ_1 (s)	τ_2 (s)	τ_3 (s)	E_{isoth} (mN m^{-1})	E_d (mN m^{-1})	E_i (mN m^{-1})	$|E|$ (mN m^{-1})	φ (°)
(1)	Brzeźno[A] July 12, 1999	0.13 (0.01)	0.4	2.0 (0.2)	21.9 (2.6)		29.12 (3.49)	12.03 (1.44)	3.72 (0.44)	12.59 (1.51)	17.2 (2.1)
(2)	Jelitkowo[A] November 29, 1997	0.12 (0.01)	0.7	2.4 (0.3)	12.2 (1.5)		25.81 (3.09)	6.98 (0.83)	2.45 (0.29)	7.40 (0.88)	19.3 (2.3)
(3)	Orłowo[A] July 15, 1999	0.08 (0.01)	0.3	1.6 (0.2)	10.1 (1.2)		24.48 (2.93)	17.58 (2.11)	5.31 (0.63)	18.37 (2.20)	16.8 (2.0)
(4)	Sopot[A] September 23, 1998	0.08 (0.01)	0.4	2.0 (0.2)	10.2 (1.2)		23.12 (2.77)	13.55 (1.62)	4.18 (0.50)	14.18 (1.70)	17.2 (2.1)
(5)	Jelitkowo[S] December 11, 2006	0.16 (0.01)	0.3	2.6 (0.3)	13.1 (1.5)	68.5 (8.2)	11.81 (1.42)	8.12 (0.97)	2.14 (0.25)	8.39 (1.03)	14.7 (1.7)
(6)	Talc[S]	0.17 (0.01)	3.0	9.9 (1.2)	30.5 (3.7)		20.18 (2.41)	11.77 (1.40)	4.08 (0.48)	12.45 (1.48)	19.9 (2.4)
(7)	Microsphere[S]	0.17 (0.01)	1.3	9.3 (1.1)	21.6 (2.6)		21.46 (2.57)	15.78 (1.89)	3.57 (0.43)	16.18 (1.93)	15.2 (1.8)
(8)	Combustion dust[S]	0.19 (0.01)	4.0	7.4 (0.9)	21.0 (2.5)		26.79 (3.21)	13.66 (1.64)	7.18 (0.85)	15.43 (1.86)	27.7 (3.3)

December 11, 2006, and for the same sample of seawater with the floating talc monoparticulate layer (Δt = 1.2 s); $\Delta A/A$ (= 0.17), are presented in Figures 10(a) and 10(b). The rate of the relaxation processes can be analyzed with the first-order equation (2). The equation applied to the Π-t plot from Figure 10 yielded two linear regions. The composite seawater film studies revealed a two- (three-) step relaxation process with the characteristic times τ_1 = (2.6–9.9) s, τ_2 = (13.1–30.5) s, and τ_3 = 68.5 s (for the talc-covered surface). So far, for natural surfactant seawater films, relaxation times were found to be in rather narrow ranges τ_1 = (1.3–2.9) s and τ_2 = (10.1–25.6) s [2]. The obtained relaxation times for the composite surface films were systematically longer, as a consequence of more complicated (complex) film structure leading to lengthening the time scale of film conformation changes at the interfacial system under surface stress.

The characteristic relaxation times τ, together with the applied step deformation time Δt, the relative area change $\Delta A/A$, and the remaining surface viscoelasticity parameters (real and imaginary parts of the dilational modulus E), its loss angle φ (from (2)–(3)), and the static isothermal modulus E_{isoth} (1) are collected in Table 4. The obtained rheokinetic parameters demonstrated that we are concerned with the complex structurally films having a relatively significant imaginary part of the complex E ($E_i/|E| \approx 0.22$–0.46) with values of the loss angle φ (15.2–27.7°) taking particularly high values for the most hydrophobic particles—combustion dust. In contrast, sea surface natural films (supposed to be solid particle free) demonstrated the purely elastic behavior ($E_d \sim |E|$; $E_d \gg E_i$) with the loss angles φ ranging from 14.7 to 19.3°, and $E_i/|E| \approx 0.25$–0.33. At compression rates ($[\Delta A/A]/\Delta t$) applied in these studies, the dilational viscoelasticity modulus can not be approximated by E_{isoth} (comparable values of E_{isoth} and $|E|$). It should be pointed out that a dilational modulus determined by quasi-isotherm measurements must be higher than a modulus obtained under nonequilibrium conditions, and the latter values are different from those determined by stepwise compression. In general, for structurally complex natural sea surface films, the surface rheology parameters may significantly differ from each other if derived from static and dynamic studies. Viscoelasticity of the particle-incorporated surfaces revealed time scales of the relaxation processes corresponding to the deformation frequency range 0.9–5.3 Hz relevant to water wave damping in the capillary-short gravity frequency region [34]. It should be noted that the wind waves Marangoni damping effect can not be explained quantitatively only by the dilational moduli obtained in this work; other aerodynamic parameters of the air/film-covered seawater interactions remain to be taken into account [35].

5. Conclusion

Surface pressure (Π) versus surface area (A) isotherms determined for monoparticulate solid layers at the air-water interface allowed the effect of hydrophobicity on the particle-particle and particle-subphase interactions to be quantified in terms of the removal energy E_r, contact

cross-sectional areas CCSA, and collapse energies E_c. Dust material wettability in contact with the seawater affected by the surfactant surface adsorption, resulting from the autophobing phenomenon leading to CA↑, was evaluated by contact angle techniques (wicking layer and microscope) applicable to solids appearing in a powdered form. The collected data revealed different wettability of the mineral material with water CA $\approx$ 47.3° to 106°, irregularly shaped particles with the mean diameters 5.8–7.6 μm, with density of 2.65 g cm^{-3} appearing at the mean daily deposition rates (2.30–10.82 mg m^{-2} day^{-1}) typical for clean marine coastal regions. Surface dilational modulus of composite surfaces E_{com} obtained using a theoretical approach, in which the solid particle deposition features (surface particle concentration, mean diameter, and solid/water phase contact angles) originated from the dry dust deposition flux laboratory and field studies agreed well with the direct isotherm and stress-relaxation particle-incorporated surface film Langmuir trough measurements. The presence of the hydrophobic particles in sea surface film caused an increase of the apparent modulus by a factor K = ($E_{\text{com}}/E_{\text{free}}$ = 1.29–1.58). The monoparticulate solid particle layers at the seawater-air, studied in the stress-relaxation experiments, revealed a three-step relaxation phenomenon, and demonstrated a viscoelastic surface behavior. In contrast, natural marine surface films (supposed to be particle-free) were found to be purely elastic ($E_d \gg E_i$; φ = 14.7–19.3°).

So far, an underestimated role played by solid dust incorporated in interfacial region of natural sea surface films may lead to a significant modification of the static and dynamic surface rheology with further implications to the dynamic exchange processes taking place between sea and atmosphere.

Acknowledgments

The study was carried out in the framework of scientific activity of the University of Gdansk (supported from DS/5200-4-0024-12).

References

[1] Z. A. Mazurek, J. S. Pogorzelski, and K. Boniewicz-Szmyt, "Evolution of natural sea surface film structure as a tool for organic matter dynamics tracing," *Journal of Marine Systems*, vol. 74, pp. S52–S64, 2008.

[2] S. J. Pogorzelski and A. D. Kogut, "Structural and thermodynamic signatures of marine microlayer surfactant films," *Journal of Sea Research*, vol. 49, no. 4, pp. 347–356, 2003.

[3] J. Lucassen, "Dynamic dilational properties of composite surfaces," *Colloids and Surfaces*, vol. 65, no. 2-3, pp. 139–149, 1992.

[4] Z. Hórvölgyi, M. Máté, A. Dániel, and J. Szalma, "Wetting behaviour of silanized glass microspheres at water-air interfaces: a Wilhelmy film balance study," *Colloids and Surfaces A*, vol. 156, no. 1–3, pp. 501–507, 1999.

[5] A. W. Adamson and A. Gast, *Physical Chemistry of Surfaces*, John Wiley & Sons, New York, NY, USA, 2nd edition, 1997.

[6] E. Kim, D. Kalman, and T. Larson, "Dry deposition of large, airborne particles onto a surrogate surface," *Atmospheric Environment*, vol. 34, no. 15, pp. 2387–2397, 2000.

[7] Z. G. Cui, B. P. Binks, and J. H. Clint, "Determination of contact angles on microporous particles using the thin-layer wicking technique," *Langmuir*, vol. 21, no. 18, pp. 8319–8325, 2005.

[8] A. W. Adamson, *Physical Chemistry of Surfaces*, John Wiley & Sons, New York, NY, USA, 8th edition, 1982.

[9] T. Kato, K. Iriyama, and T. Araki, "The time of observation of π-A isotherms III. Studies on the morphology of arachidic acid monolayers, observed by transmission electron microscopy of replica samples of one-layer Langmuir-Blodgett films using plasma-polymerization," *Thin Solid Films*, vol. 210-211, no. 1, pp. 79–81, 1992.

[10] F. Ravera, M. Ferrari, E. Santini, and L. Liggieri, "Influence of surface processes on the dilational visco-elasticity of surfactant solutions," *Advances in Colloid and Interface Science*, vol. 117, no. 1–3, pp. 75–100, 2005.

[11] J. van Hunsel and P. Joos, "Study of the dynamic interfacial tension at the oil/water interface," *Colloid & Polymer Science*, vol. 267, no. 11, pp. 1026–1035, 1989.

[12] M. R. Rodríguez Niño, P. J. Wilde, D. C. Clark, and J. M. Rodríguez Patino, "Surface dilational properties of protein and lipid films at the air-water interface," *Langmuir*, vol. 14, no. 8, pp. 2160–2166, 1998.

[13] Y. Jayalakshmi, L. Ozanne, and D. Langevin, "Viscoelasticity of Surfactant Monolayers," *Journal of Colloid And Interface Science*, vol. 170, no. 2, pp. 358–366, 1995.

[14] J. H. Clint and S. E. Taylor, "Particle size and interparticle forces of overbased detergents: a Langmuir trough study," *Colloids and Surfaces*, vol. 65, no. 1, pp. 61–67, 1992.

[15] M. Máté, J. H. Fendler, J. J. Ramsden, J. Szalma, and Z. Hórvölgyi, "Eliminating surface pressure gradient effects in contact angle determination of nano- and microparticles using a film balance," *Langmuir*, vol. 14, no. 22, pp. 6501–6504, 1998.

[16] B. S. Murray, "Equilibrium and dynamic surface pressure-area measurements on protein films at air-water and oil-water interfaces," *Colloids and Surfaces A*, vol. 125, no. 1, pp. 73–83, 1997.

[17] M. A. Rodríguez-Valverde, M. A. Cabrerizo-Vílchez, P. Rosales-López, A. Páez-Dueas, and R. Hidalgo-Álvarez, "Contact angle measurements on two (wood and stone) non-ideal surfaces," *Colloids and Surfaces A*, vol. 206, no. 1–3, pp. 485–495, 2002.

[18] C. J. van Oss, R. F. Giese, Z. Li et al., "Determination of contact angles and pore sizes of porous media by column and thin layer wicking," *Journal of Adhesion Science and Technology*, vol. 6, pp. 413–428, 1992.

[19] W. Wu, R. F. Giese, and C. J. Van Oss, "Change in surface properties of solids caused by grinding," *Powder Technology*, vol. 89, no. 2, pp. 129–132, 1996.

[20] G. Tolnai, A. Agod, M. Kabai-Faix, A. L. Kovács, J. J. Ramsden, and Z. Hórvölgyi, "Evidence for secondary minimum flocculation of Stöber silica nanoparticles at the air-water interface: film balance investigations and computer simulations," *Journal of Physical Chemistry B*, vol. 107, no. 40, pp. 11109–11116, 2003.

[21] A. Ulman, *An Introduction to Ultrathin Organic Films*, Academic Press, Boston, Mass, USA, 1991.

[22] S. Yariv, "Wettability of clay minerals," in *Modern Approaches to Wettability: Theory and Applications*, M. E. Schrader and G. Loeb, Eds., pp. 279–326, Plenum Press, New York, NY, USA, 1992.

[23] S. Caquineau, A. Gaudichet, L. Gomes, M. C. Magonthier, and B. Chatenet, "Saharan dust: clay ratio as a relevant tracer to assess the origin of soil-derived aerosols," *Geophysical Research Letters*, vol. 25, no. 7, pp. 983–986, 1998.

[24] A. Avila, I. Queralt-Mitjans, and M. Alarcón, "Mineralogical composition of African dust delivered by red rains over northeastern Spain," *Journal of Geophysical Research D*, vol. 102, no. 18, pp. 21977–21996, 1997.

[25] J. Keller and R. Lamprecht, "Road dust as an indicator for air pollution transport and deposition: an application of SPOT imagery," *Remote Sensing of Environment*, vol. 54, no. 1, pp. 1–12, 1995.

[26] H. W. Vallack and D. E. Shillito, "Suggested guidelines for deposited ambient dust," *Atmospheric Environment*, vol. 32, no. 16, pp. 2737–2744, 1998.

[27] B. Forster, X. Baide, and S. Xingwai, "Modelling suspended particle distribution in near coastal waters using satellite remotely-sensed data," *International Journal of Remote Sensing*, vol. 15, no. 6, pp. 1207–1219, 1994.

[28] P. Gentien, M. Lunven, M. Lehaître, and J. L. Duvent, "In-situ depth profiling of particle sizes," *Deep-Sea Research I*, vol. 42, no. 8, pp. 1297–1312, 1995.

[29] S. Chen and D. Eisma, "Fractal geometry of in situ flocs in the estuarine and coastal environments," *Netherlands Journal of Sea Research*, vol. 32, no. 2, pp. 173–182, 1995.

[30] D. J. Law, A. J. Bale, and S. E. Jones, "Adaptation of focused beam reflectance measurement to in-situ particle sizing in estuaries and coastal waters," *Marine Geology*, vol. 140, no. 1-2, pp. 47–59, 1997.

[31] R. Arimoto, B. J. Ray, N. F. Lewis, U. Tomza, and R. A. Duce, "Mass-particle size distributions of atmospheric dust and the dry deposition of dust to the remote ocean," *Journal of Geophysical Research D*, vol. 102, no. 13, pp. 15867–15874, 1997.

[32] S. A. Slinn and W. G. N. Slinn, "Predictions for particle deposition on natural waters," *Atmospheric Environment A*, vol. 14, no. 9, pp. 1013–1016, 1980.

[33] H. Sievering, "Small-particle dry deposition on natural waters: how large the uncertainty?" *Atmospheric Environment*, vol. 18, no. 10, pp. 2271–2272, 1984.

[34] W. Alpers and H. Huhnerfuss, "The damping of ocean waves by surface films: a new look at an old problem," *Journal of Geophysical Research*, vol. 94, pp. 6251–6265, 1989.

[35] G. Franceschetti, A. Iodice, D. Riccio, G. Ruello, and R. Siviero, "SAR raw signal simulation of oil slicks in ocean environments," *IEEE Transactions on Geoscience and Remote Sensing*, vol. 40, no. 9, pp. 1935–1949, 2002.

16

Polycyclic Aromatic Hydrocarbons in Various Species of Fishes from Mumbai Harbour, India, and Their Dietary Intake Concentration to Human

V. Dhananjayan[1, 2] and S. Muralidharan[1]

[1] Division of Ecotoxicology, Sálim Ali Centre for Ornithology and Natural History, Coimbatore 641108, India
[2] Regional Occupational Health Center, ICMR, Kannamangala Post, Poojanahalli Road, Devenahalli TK, Bangalore 562110, India

Correspondence should be addressed to V. Dhananjayan, dhananjayan_v@yahoo.com

Academic Editor: Swadhin Behera

Polycyclic aromatic hydrocarbons (PAHs) are ubiquitous environmental contaminants which have caused worldwide concerns as toxic pollutant. This study reports the concentrations of 15 PAHs in 5 species of fish samples collected along the harbour line, Mumbai, between 2006 and 2008. Among 5 species of fish investigated, Mandeli, *Coilia dussimieri*, detected the maximum concentration of PAHs ($P < 0.05$) followed by Doma, *Otolithes ruber*. The concentration of total and carcinogenic PAHs ranged from 17.43 to 70.44 ng/g wet wt. and 9.49 to 31.23 ng/g wet wt, respectively, among the species tested. The lower-molecular-weight PAHs were detected at highest levels. Estimated intakes of PAHs by fish consumption for the general population were ranged between 1.77 and 10.70 ng/kg body weight/day. Mandeli contributed to the highest intakes of PAHs. The toxic equivalents (TEQs) of PAHs were calculated using a TEQ proposed in literature, and the intake ranged from 8.39 to 15.78 pg TEQ/kg body weight/d. The estimated excess cancer risk value (2.37×10^{-7}–1.43×10^{-6}) from fish consumption for the general population exceeded the guideline value (1.0×10^{-6}) for potential cancer risk.

1. Introduction

Polycyclic aromatic hydrocarbons (PAHs) are ubiquitous anthropogenic pollutants that can be biologically amplified to high concentrations in food webs. Due to their lipophilicity, persistence, and high toxicity, these residues are readily accumulated in the tissues of nontarget living organisms where they may cause detrimental effects. PAHs are toxic, carcinogenic, and mutagenic to all organisms, including humans [1, 2]. The metabolites of PAHs may bind to proteins and DNA, which causes biochemical disruption and cell damage in animals and cancer in human [2]. The main sources of these contaminants in the environment include forest fire, natural petroleum seeps, combustion of fossil fuels, coal burning, and use of oil for cooking and heating [3, 4]. Other sources include domestic and industrial waste waters and sewage. As a consequence, environmental contamination by PAHs has steadily increased in recent years [5].

Dietary intake has been reported as an important route for human exposure to PAHs, except for smokers and occupationally exposed populations [6, 7]. Pollution by persistent chemicals is potentially harmful to the organisms at higher tropic levels in the food chain. The marine organisms like fish are able to accumulate severalfold higher concentration of PAHs than the surrounding water [8–10]. Fish is a major source of proteins and healthy lipids for people. In particular, the long-chain omega-3 fatty acids have been shown to have numerous beneficial roles in the human health [11]. Despite the human benefits of a fish diet, an issue of concern related to frequent fish consumption is the potential risk arising from exposure to toxic chemicals [12, 13]. In a recent year, a number of epidemiologic studies have reported that a large portion of human cancers, such as lung and prostate cancers, are attributable to dietary sources [14, 15]. Certain groups of population may have higher risks from dietary exposure of PAHs than the general populations [16].

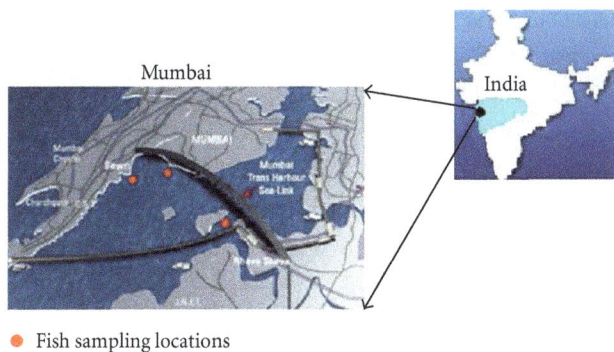

Mumbai

India

● Fish sampling locations

FIGURE 1: Study area showing fish samples collection locations at
Mumbai harbor line, Maharashtra, India.

In India there are many studies on the presence of
petroleum hydrocarbons in the Goa coastal water [17],
north-west coastal water [18, 19], fish and prawn from north
west coast of India [20], bivalve in east coast of India [21],
freshwater and fish [22, 23], soil and sediment [24], and
marine environment of Mumbai [25]. However, there is no
information available concerning dietary intake of PAHs and
their risk from fish consumption. The preset study was aimed
at assessment of PAHs in five species of fish collected along
the harbor line and estimate the cancer risk of PAHs through
fish consumption using the risk assessment guideline of the
United States Environmental Protection Agency [26].

2. Materials and Methods

2.1. Study Area. Mumbai is one of the major cities in India
which is located along the western coast of the country. City
with a human population density of 25,000 persons/km^{-2}
generates 2.2×10^6 m^3 d^{-1} of domestic sewage out of which
about 2×10^6 m^3 d^{-1} enters marine waters including creeks
and bays, largely untreated [27]. It has great diversification of
industries in metropolitan region. About 8% of industries in
the country are located around Mumbai in the upstream. A
variety of industries, including refineries and petrochemical
complexes, from this area release their effluents largely
untreated into the sea. There are number of ports wherein
the ship and cargo handling activities contribute to marine
pollution. Sewri-Mahul and Nhava mudflats about 1000 ha
have been identified as an important bird area (IBA) [28].
Sewri-Mahul mudflats (19°01′00″N, 72°52′60″E) (Figure 1)
extent over an area of 10 km long and 3 km wide is
dominated by mangroves all along the coast. The Sewri Bay
is situated just off the wide mouth of the Thane Creek along
the northern periphery of Mumbai's eastern harbour. These
locations were selected for fish collection.

2.2. Sample Collection. Five species of fish samples were
collected between 2006 and 2008. Species which were
commonly available, namely *Eleutheronema tetradactylum,
Coilia dussumieri, Otolithes ruber, Sardinella longiceps,* and
Mystus seenghala in the study locations were collected with

the help of local fisherman of the region, and the mor-
phometry measurements were taken immediately (Table 1).
On collection, fish samples were stored in pollution-free
sealed polythene covers and transported to the laboratory at
SACON, Coimbatore, in ice box and stored at $-20°C$ in the
deep freezer until analysis.

2.3. Sample Processing. Fish samples were taken out from
the deep freezer, thawed, and well cleaned in tap water
to remove any external dirt. Dissection was performed on
thawed fish, using solvent rinsed instruments and glass
dishes. The scales were sloughed off and muscle tissues
were dissected between the pectoral fin and vent of the
fish, minced into smaller pieces, and a subsample was taken
from the homogenate. Fish parts were then placed in solvent
rinsed glass jars with solvent rinsed aluminium foil lined
lids. Information regarding sampled location, species, lipid
content, length, and weight is presented in Table 1. Ten gram
of the sample was weighed using a top loading electronic
balance (Mettler AE420) and ground with anhydrous sodium
sulphate (40 g), and the mixture was packed in a thimble
(Whatman) and desiccated overnight prior to extraction.
The desiccated thimble was loaded in a Soxhlet apparatus
and extracted with dichloromethane (DCM) for 7 hrs. The
solvent was reduced (11 mL), and an aliquot (1 mL) was
taken for lipid estimation. Another aliquot (10 mL) was
subjected to removal of other contaminants by passing the
sample through a glass column packed with florisil and
eluted with 100 mL of DCM and hexane mixture (2 : 8). Then
the extracts were concentrated using rotary flask evaporation
to a final volume of 1 mL in acetonitrile and filtered using
0.45 μm syringe filter units. The eluant was blown using
nitrogen, redissolved in 2 mL CAN, and transferred into
HPLC autosampler vials for PAH analysis.

2.4. Lipid Estimation. One mL aliquot of sample was subjected to gravimetric determination of percent lipid content.

2.5. Estimation of PAHs. Samples were analyzed in the laboratory at Sálim Ali Centre for Ornithology and Natural History (SACON), Coimbatore, India. All the samples were
quantified for 15 components of PAHs (naphthalene, acenaphthene, fluorene, phenanthrene, anthracene, fluoranthene, pyrene, benz[*a*]anthracene, chrysene, benzo[*b*]fluorant-
hene, benzo[*k*]fluoranthene, benzo[*a*]pyrene, dibenzo[*a,h*]
anthracene, benzo[*g,h,i*]perylene, and indeno[1,2,3-*cd*]py-
rene) using HPLC with programmable fluorescence detection at excited and emission wavelengths of 260 and 500 nm,
respectively. About 20 μL of sample was injected through
an autosampler into C18 column (Zorbax 4.6 × 250 mm)
of 5 μm particle size. The temperature of the column was
maintained at 20°C. Water/acetonitrile (ACN) was used as
mobile phase with a flow of 1 mL/min. The initial content
of ACN was 50% and then increased into 60% (0–3 min)
and 95% (3–14 min). These levels were held constant for
24 minutes until the end of the analysis. Recoveries of the
compounds from fortified samples (50 ppb) ranged from
78% to 94%, and the concentrations were not corrected for
percent recovery. Concentrations of PAHs are reported on

TABLE 1: List of fish species included in the present study.

S. no.	Vernacular name (Marathi)	Scientific name	Place of collection	n	Length (cm) mean ± SD	Weight (g) mean ± SD
1	Doma	*Otolithes ruber*	S, M, N	22	17.9 ± 3.53	66.6 ± 37.5
2	Mandeli	*Coilia dussumieri*	S, M, N	77	15.4 ± 1.73	11.5 ± 3.22
3	Mathi	*Sardinella longiceps*	S, M	24	21.5 ± 1.00	102 ± 5.42
4	Ravas	*Eleutheronema Tetradactylum*	S, M	8	21.6 ± 8.02	138 ± 16.1
5	Singala	*Mystus seenghala*	M, N	11	17.3 ± 4.58	36.7 ± 6.14

n = Number of samples collected, S = Sewri, M = Mahul, N = Nhava.

a wet weight basis. Analyses were run in batches of 10 samples plus four quality controls (QCs) including one reagent blank, one matrix blank, one QC check sample, and one random sample in duplicate. The minimum detection limit for all the compounds analysed was 0.5 ng/g wet wt.

2.6. Estimation of Dietary Intake of PAHs through Fish. Dietary intake concentration was calculated by multiplying the PAH concentration measured in each species of fish by the per capita consumption. The World Health Organization (WHO) has recommended a minimum 11 kg fish consumption per capita per annum in India [29, 30].

2.7. Estimation of TEQ. Some PAHs are aryl-hydrocarbon receptors agonists and potent inducers of ethoxyresorufin-O-deethylase activity [31]. Hence, to accurately evaluate risk from intake of dioxin-like contaminants, the TEQ concentrations of PAHs, as well as polychlorinated dibenzo-*p*-dioxins and dibenzofurans (PCDD/Fs) and dioxin-like polychlorinated biphenyls (PCBs), should be considered. The H4IIE-specific potencies, relative to tetrachlorinated dibenzo-*p*-dioxin (TCDD), have been reported of same PAH compounds used in this study [32]. Relative potencies of BaA, Chr, BbF, BkF, BaP, InP, and DbA are 0.000025, 0.0002, 0.00253, 0.00478, 0.000354, 0.0011, and 0.00203, respectively [33], and these values were used to estimate TEQs contributed by PAHs (TEQ-$_{PAHs}$).

2.8. Determination of Cancer Risk Factor. The public concern regarding exposure to PAHs is associated with its potential carcinogenicity in humans [14, 15]. The potential health risks of ingesting fish contaminated with carcinogenic contaminates were evaluated for the Mumbai population using the risk assessment guideline of the USEPA [26]. The mean dietary intakes of the seven PAHs are considered probable human carcinogens by the US EPA, and hence they are considered. The general equation for estimating exposure, through ingestion of fish is as follows:

$$\text{Excess cancer risk} = \frac{\text{EI} \cdot \text{ED} \cdot \text{CSF}}{\text{BW} \cdot \text{AT}}, \tag{1}$$

where EI is estimated intake (mg/kg/d), ED is exposure duration (years; adults = 30 years), CSF is the oral cancer slope factor ((mg/kg/d)$^{-1}$), BW is human body weight (assuming 60 kg weight), and AT is the average time for carcinogens (years, assuming 70 years for adults). The CSF

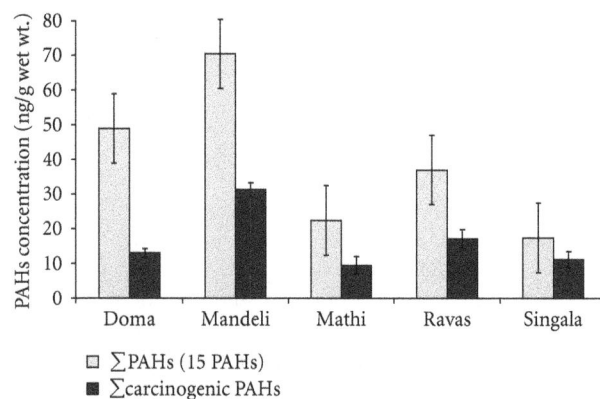

FIGURE 2: Concentration of carcinogenic PAHs and total PAHs among various species of fish along harbor line, Mumbai, India.

data for individual PAHs, obtained from the integrated risk information system reported by the USEPA (2004), are BaA (0.73), Chr (0.0073), Bbf (0.73), BkF (0.073), BaP (7.3), InP (0.73), and DbA (7.3).

2.9. Statistics. All the data were log transformed to get normal distribution. One Way Analysis of Variance (ANOVA) was performed to assess the variation among species. Means were compared using the Bonferroni multiple comparison test. The significant level was $P < 0.05$. All the calculations were done using statistical software, SPSS student version 10.

3. Results and Discussion

3.1. Concentration of PAHs among Fish Species. Concentration of lipids estimated and individuals components of PAHs among fish species in harbor line, Mumbai, collected between 2006 and 2008 are listed in Table 2. The lipid content of fish samples ranged from 3.3% to 4.4% on wet weight basis. The concentration of total PAHs ($\sum$PAH (the sum of 15 PAHs)) and carcinogenic PAHs ($\sum$CPAH (the sum of BaA, BbF, BkF, BaP, InP, and DbA)) are presented in Figure 2. The levels of $\sum$PAH and $\sum$CPAH ranged from 17.43 to 70.44 ng/g wet wt. and 9.49 to 31.23 ng/g wet wt, respectively. The maximum concentration of $\sum$PAH in marine fish species was found in *Coilia dussumieri* (70.44 ng/g wet wt.) ($P < 0.05$). Other species such as *Otolithes ruber, Eleutheronema tetradactylum* and *Mystus seenghala* detected relatively

Polycyclic Aromatic Hydrocarbons in Various Species of Fishes from Mumbai Harbour, India, and Their
Dietary Intake Concentration to Human

175

TABLE 2: Concentration (mean $\pm$ SD) of PAHs among fish from the Mumbai transharbour, Maharashtra, India.

Fishes $\Rightarrow$ PAHs $\Downarrow$	Doma ($n = 22$)	Mandeli* ($n = 77$)	Mathi ($n = 24$)	Ravas ($n = 8$)	Singala ($n = 11$)
Lipid content (%)	3.9	4.4	3.3	3.7	4.1
Naphthalene	19.94 $\pm$ 6.12	43.14 $\pm$ 10.9	17.67 $\pm$ 6.78	14.04 $\pm$ 6.36	25.7 $\pm$ 8.22
Acenaphthene	2.98 $\pm$ 1.68	5.48 $\pm$ 3.24	1.30 $\pm$ 1.06	6.13 $\pm$ 4.16	ND
Fluorene	5.36 $\pm$ 2.91	4.84 $\pm$ 3.01	4.10 $\pm$ 2.39	ND	6.70 $\pm$ 3.47
Phenanthrene	0.61 $\pm$ 0.52	0.99 $\pm$ 0.52	ND	ND	ND
Anthracene	ND	ND	ND	1.89 $\pm$ 0.98	ND
Fluoranthene	2.17 $\pm$ 1.05	2.10 $\pm$ 1.72	3.20 $\pm$ 1.94	5.20 $\pm$ 2.47	3.24 $\pm$ 1.69
Pyrene	1.64 $\pm$ 1.37	ND	ND	ND	ND
Benz[a]anthracene	ND	ND	ND	ND	ND
Chrysene	ND	8.85 $\pm$ 3.56	ND	2.79 $\pm$ 1.11	ND
Benzo[b]fluoranthene	1.20 $\pm$ 0.93	3.71 $\pm$ 1.85	ND	3.46 $\pm$ 1.08	ND
Benzo[k]fluoranthene	2.72 $\pm$ 1.24	ND	3.54 $\pm$ 2.88	ND	5.67 $\pm$ 3.78
Benz[a]pyrene	ND	ND	ND	1.25 $\pm$ 0.87	ND
Dibenzo[a,h]anthracene	4.44 $\pm$ 3.21	5.53 $\pm$ 2.40	ND	3.70 $\pm$ 2.07	1.27 $\pm$ 0.78
Benzo[g,h,i]perylene	2.06 $\pm$ 1.92	3.61 $\pm$ 1.99	4.61 $\pm$ 2.31	3.15 $\pm$ 2.10	1.83 $\pm$ 0.67
Indeno[1,2,3-cd]pyrene	ND	3.23 $\pm$ 2.13	ND	ND	1.73 $\pm$ 1.02

*ANOVA, $P < 0.05$. ND = not detected (below detectable limits).

equal concentration, whereas the minimum concentration was detected in *Sardinella longiceps*.

The total PAHs concentration reported in fish samples of the present study appears to be higher than the concentration reported in edible fishes (0.207–3.365 ppm) of the Gomti river, Lucknow, India [23]. The $\sum$PAHs concentration detected in the fishes of present study is comparable with study reported in the muscles of the fish, from fish pound of the Pearl river delta (49.59 ng/g wet wt) [34] and lower than the levels in the Mai Po Marshes Nature Reserve of Hong Kong (497 ng/g wet wt.) [35]. However, higher levels reported in the fish muscles from the Red Sea Coast (12.29 ng/g wet wt.) [36]. Among the various components of PAHs, naphthalene was the most frequently detected as reported in other studies [36, 37], followed by fluorene and acenaphthene. Benzo[b]anthracene could not be detected in any of the samples, while anthracene, pyrene, and benzo[a]pyrene were present only in one sample each. The absence or rather low detection of certain PAHs in the fish samples may be attributed to their rapid depuration or biotransformation [37]. The accumulation and depuration of PAHs in fish can be influenced by various factors including route and duration of exposure, lipid content of tissues, environmental factors, differences in species, age, and sex, and exposure to other xenobiotics [38].

3.2. Dietary Intake of PAHs through Fish Consumption. Calculated dietary intake concentration of PAHs through consumption of marine fish to human is presented in Figure 3. The average intake of PAHs through fish consumption was estimated to be 1.77–10.7 ng/kg body weight/d. Among fish species of fish analysed, Mandeli contributed to the highest intake of total PAHs. Estimated intakes through consumption of other species are less than 7 ng/kg body

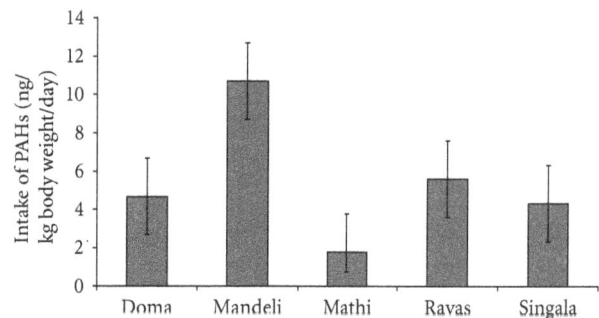

FIGURE 3: Calculated dietary intake concentration of PAHs through consumption of fishes from the Mumbai horbour line, Maharashtra, India.

weight/d. Only a few studies have examined dietary intakes of PAHs through fish consumption worldwide. The amount of average dietary intake of PAHs in humans through fishes estimated in Mumbai is far less compared to that in other countries. The estimated intake of PAHs from fish consumption reported for the general populations in other counties such as Spain (626–712 ng/d) [39], Kuwait (231 ng/d) [40], and Korea (13.8–16.7 ng/kg body weight/d [41]. This result seems to be associated with the consumption rate and accumulation level of PAHs in the fish available in Mumbai.

3.3. Contribution of PAHs to Total Toxic Equivalence ($\sum$TEQs). The estimated intake of TEQ-$_{PAHs}$ from fish consumption for the general population ranged between 8.3 and 15.78 pg TEQ/kg body weight/day (Figure 4); these levels were higher than the levels reported in Korea through consumption of seafood [41]. PAHs concentration was proposed

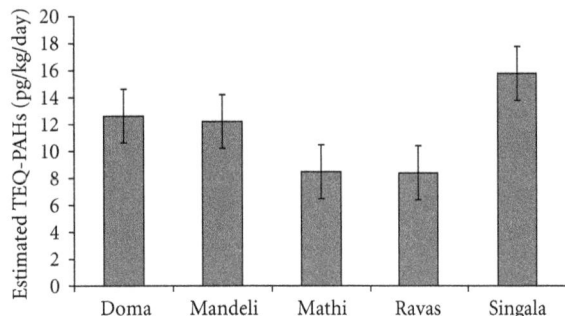

FIGURE 4: Estimated TEQ-PAHs concentration among various species of fishes from the Mumbai horbour line, Maharashtra, India.

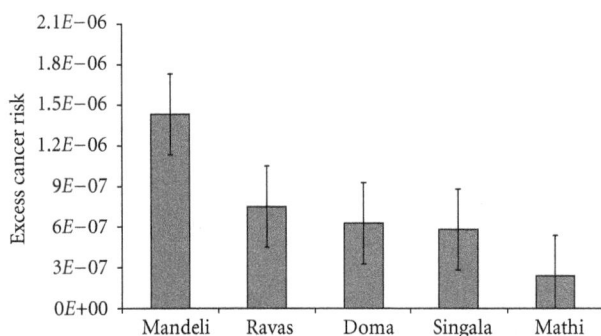

FIGURE 5: Calculated excess cancer risk associated with the consumption of marine fish at the Mumbai horbour line, Maharashtra, India.

as one of the toxic chemicals in food, consumer products, and the environment [32] for estimating the tolerable daily intake of 2 pg TEQ/kg body weight/day by the United and European Commission [42]. Based on these results, TEQ-PAHs estimated in the present study was the highest contributor to total TEQ intake from fish consumption, indicating that PAHs in fish should be considered in risk assessment along with PCDD/Fs and dioxin-like PCBs.

3.4. Assessment of Excess Cancer Risk. Excess cancer risk values, estimated from individual fish consumption for the general population are presented in Figure 5. Among the fish species of fish estimated, Mandeli showed relatively higher risk values than those shown by other specie because of their high concentration of carcinogenic PAHs. The excess cancer risk values estimated for Mandeli exceeded the cancer risk guideline value (1×10^{-6}) [26]. This result indicates that adverse effects from PAHs in fish could only be caused by lifetime consumption of fish.

The concentration of PAHs was measured in commonly consumed types of fishes available in the study locations in Mumbai. The PAH levels in Mumbai sea were moderate compared with those found in other countries. Dominant compounds of PAHs were the lower-molecular-weight aromatics, such as naphthalene and fluorine. PAH intakes by way of fish consumption by the general population were estimated. Intake of the TEQ values calculated from

PAH concentrations showed that PAHs were the highest contributor to total TEQ intake. The estimated excess cancer risk values from fish consumption for the general population exceeded the guideline value for potential cancer risk. The results of the resent study emphasize the importance of systematically monitoring PAH levels and fish intake and comparing them with published guidelines to protect human health.

Acknowledgments

The authors sincerely thank the Maharashtra State Road Development Corporation (MSRDC), India for financial assistance. They are grateful to Drs. V. S. Vijayan, S. N. Prasad, Lalitha Vijayan, R. Jayakumar, and Miss V. R. Peter, SACON for their support. They appreciate S. Patturajan and Muragesan for their assistance in all our laboratory works.

References

[1] D. E. Nacci, M. Kohan, M. Pelletier, and E. George, "Effects of benzo[a]pyrene exposure on a fish population resistant to the toxic effects of dioxin-like compounds," *Aquatic Toxicology*, vol. 57, no. 4, pp. 203–215, 2002.

[2] B. Armstrong, E. Hutchinson, J. Unwin, and T. Fletcher, "Lung cancer risk after exposure to polycyclic aromatic hydrocarbons: a review and meta-analysis," *Environmental Health Perspectives*, vol. 112, no. 9, pp. 970–978, 2004.

[3] E. R. Christensen and P. A. Bzdusek, "PAHs in sediments of the Black River and the Ashtabula River, Ohio: source apportionment by factor analysis," *Water Research*, vol. 39, no. 4, pp. 511–524, 2005.

[4] H.-B. Moon, K. Kannan, S.-J. Lee, and G. Ok, "Atmospheric deposition of polycyclic aromatic hydrocarbons in an urban and a suburban area of Korea from 2002 to 2004," *Archives of Environmental Contamination and Toxicology*, vol. 51, no. 4, pp. 494–502, 2006.

[5] P. C. Van Metre, B. J. Mahler, and E. T. Furlong, "Urban sprawl leaves its PAH signature," *Environmental Science and Technology*, vol. 34, no. 19, pp. 4064–4070, 2000.

[6] G. Scherer, S. Frank, K. Riedel, I. Meger-Kossien, and T. Renner, "Biomonitoring of exposure to polycyclic aromatic hydrocarbons of nonoccupationally exposed persons," *Cancer Epidemiology Biomarkers and Prevention*, vol. 9, no. 4, pp. 373–380, 2000.

[7] G. Falcó, J. L. Domingo, J. M. Llobet, A. Teixidó, C. Casas, and L. Müller, "Polycyclic aromatic hydrocarbons in foods: human exposure through the diet in Catalonia, Spain," *Journal of Food Protection*, vol. 66, no. 12, pp. 2325–2331, 2003.

[8] R. J. Law and J. Hellou, "Contamination of fish and shellfish following oil spill incidents," *Environmental Geosciences*, vol. 6, no. 2, pp. 90–98, 1999.

[9] I. Vives, J. O. Grimalt, P. Fernández, and B. Rosseland, "Polycyclic aromatic hydrocarbons in fish from remote and high mountain lakes in Europe and Greenland," *Science of the Total Environment*, vol. 324, no. 1–3, pp. 67–77, 2004.

[10] B. Johnson-Restrepo, J. Olivero-Verbel, S. Lu et al., "Polycyclic aromatic hydrocarbons and their hydroxylated metabolites in fish bile and sediments from coastal waters of Colombia," *Environmental Pollution*, vol. 151, no. 3, pp. 452–459, 2008.

Polycyclic Aromatic Hydrocarbons in Various Species of Fishes from Mumbai Harbour, India, and Their
Dietary Intake Concentration to Human

177

[11] H. M. Ismail, "The role of omega-3 fatty acids in cardiac protection: an overview," *Frontiers in Bioscience*, vol. 10, pp. 1079–1088, 2005.

[12] J. L. Domingo, A. Bocio, G. Falcó, and J. M. Llobet, "Benefits and risks of fish consumption. Part I. A quantitative analysis of the intake of omega-3 fatty acids and chemical contaminants," *Toxicology*, vol. 230, no. 2-3, pp. 219–226, 2007.

[13] I. Sioen, J. Van Camp, F. Verdonck et al., "Probabilistic intake assessment of multiple compounds as a tool to quantify the nutritional-toxicological conflict related to seafood consumption," *Chemosphere*, vol. 71, no. 6, pp. 1056–1066, 2008.

[14] G. L. Ambrosini, L. Fritschi, N. H. de Klerk, D. Mackerras, and J. Leavy, "Dietary patterns identified using factor analysis and prostate cancer risk: a case control study in Western Australia," *Annals of Epidemiology*, vol. 18, no. 5, pp. 364–370, 2008.

[15] M. Shen, R. S. Chapman, X. He et al., "Dietary factors, food contamination and lung cancer risk in Xuanwei, China," *Lung Cancer*, vol. 61, no. 3, pp. 275–282, 2008.

[16] R. Martí-Cid, J. M. Llobet, V. Castell, and J. L. Domingo, "Evolution of the dietary exposure to polycyclic aromatic hydrocarbons in Catalonia, Spain," *Food and Chemical Toxicology*, vol. 46, no. 9, pp. 3163–3171, 2008.

[17] S. P. Fondekar, R. S. Topgi, and R. J. Noronha, "Distribution of petroleum hydrocarbons in Goa coastal waters," *Indian Journal of Marine Sciences*, vol. 9, no. 4, pp. 286–288, 1980.

[18] A. N. Kadam and V. P. Bhangale, "Petroleum hydrocarbons in northwest coastal waters of India," *Indian Journal of Marine Sciences*, vol. 22, no. 3, pp. 227–228, 1993.

[19] M. S. Shailaja, R. Rajamanickam, and S. Wahidulla, "Increased formation of carcinogenic PAH metabolites in fish promoted by nitrite," *Environmental Pollution*, vol. 143, no. 1, pp. 174–177, 2006.

[20] P. Mehta, A. N. Kadam, S. N. Gajbhiye, and B. N. Desai, "Petroleum hydrocarbon concentration in selected species of fish and prawn form northwest coast of India," *Indian Journal of Marine Science*, vol. 23, no. 2, pp. 123–125, 1994.

[21] P. C. Mohan and R. R. Prakash, "Concentration of petroleum hydrocarbons in bivalve Mytilopsis sallei and in the habour waters of Visakhapatnam, east coast of India," *Indian Journal of Marine Sciences*, vol. 27, no. 3-4, pp. 496–498, 1998.

[22] A. Malik, K. P. Singh, D. Mohan, and D. K. Patel, "Distribution of polycyclic aromatic hydrocarbons in Gomti river system, India," *Bulletin of Environmental Contamination and Toxicology*, vol. 72, no. 6, pp. 1211–1218, 2004.

[23] A. Malik, P. Ojha, and K. P. Singh, "Distribution of polycyclic aromatic hydrocarbons in edible fish from Gomti river, India," *Bulletin of Environmental Contamination and Toxicology*, vol. 80, no. 2, pp. 134–138, 2008.

[24] S. A. Ingole, S. S. Dhaktode, and A. N. Kadam, "Determination of petroleum hydrocarbons in sediment samples from Bombay harbour, Dharamtar creek and Amba river estuary," *Indian Journal of Environmental Protection*, vol. 9, no. 2, pp. 118–123, 1989.

[25] M. K. Chouksey, A. N. Kadam, and M. D. Zingde, "Petroleum hydrocarbon residues in the marine environment of Bassein-Mumbai," *Marine Pollution Bulletin*, vol. 49, no. 7-8, pp. 637–647, 2004.

[26] U.S. EPA, "Volume I. Human Health Evaluation Manual (HHEM) (Part A, Baseline Risk Assessment). Interim Final," Office of Emergency and Remedial Response, Washington, DC, USA, EPA/540/1-89/002. NTIS PB90-15558, 1989.

[27] M. D. Zingde, "Marine pollution—what are we heading for?" in *Ocean Science: Trends and Future Directions*, pp. 229–246, Indian National Science Academy, New Delhi, India, 1999.

[28] M. Z. Islam and A. R. Rahmani, *Important Bird Areas in India: Priority Sites for Conservation*, Indian Bird Conservation Network, Bombay Natural History Society and Bird life International, Oxford University Press, Oxford, UK, 2004.

[29] P. V. Dehadrai, "Aquaculture and environment," in *Souvenir, National Aquaculture Week*, pp. 13–16, Aquaculture Foundation of India, Chennai, India, 1997.

[30] A. Kumari, R. K. Slnha, K. Gopal, and K. Prasad, "Dietary intake of persistent organochlorine residues through gangetic fishes in India," *International Journal of Ecology and Environmental Sciences*, vol. 27, no. 2, pp. 117–120, 2001.

[31] D. L. Villeneuve, J. S. Khim, K. Kannan, and J. P. Giesy, "Relative potencies of individual polycyclic aromatic hydrocarbons to induce dioxinlike and estrogenic responses in three cell lines," *Environmental Toxicology*, vol. 17, no. 2, pp. 128–137, 2002.

[32] Committee on Toxicity (CoT) of Chemicals in Food, Consumer Products, and theEnvironment, and Food standards Agency/Department of Health, Brussel, Belgium, 2001.

[33] K. L. Willett, P. R. Gardinali, J. L. Sericano, T. L. Wade, and S. H. Safe, "Characterization of the H4IIE rat hepatoma cell bioassay for evaluation of environmental samples containing polynuclear aromatic hydrocarbons (PAHs)," *Archives of Environmental Contamination and Toxicology*, vol. 32, no. 4, pp. 442–448, 1997.

[34] K. Y. Kong, K. C. Cheung, C. K. C. Wong, and M. H. Wong, "The residual dynamic of polycyclic aromatic hydrocarbons and organochlorine pesticides in fishponds of the Pearl River delta, South China," *Water Research*, vol. 39, no. 9, pp. 1831–1843, 2005.

[35] Y. Liang, M. F. Tse, L. Young, and M. H. Wong, "Distribution patterns of polycyclic aromatic hydrocarbons (PAHs) in the sediments and fish at Mai Po Marshes Nature Reserve, Hong Kong," *Water Research*, vol. 41, no. 6, pp. 1303–1311, 2007.

[36] A. A. Z. DouAbul, H. M. A. Heba, and K. H. Fareed, "Polynuclear Aromatic Hydrocarbons (PAHs) in fish from the Red Sea Coast of Yemen," *Hydrobiologia*, vol. 352, no. 1–3, pp. 251–262, 1997.

[37] S. C. Deb, T. Araki, and T. Fukushima, "Polycyclic aromatic hydrocarbons in fish organs," *Marine Pollution Bulletin*, vol. 40, no. 10, pp. 882–885, 2000.

[38] U. Varanasi, J. E. Stein, and M. Nishimoto, "Chemical carcinogenesis in feral fish: uptake, activation, and detoxication of organic xenobiotics," *Environmental Health Perspectives*, vol. 71, pp. 155–170, 1987.

[39] G. Falcó, A. Bocio, J. M. Llobet, and J. L. Domingo, "Health risks of dietary intake of environmental pollutants by elite sportsmen and sportswomen," *Food and Chemical Toxicology*, vol. 43, no. 12, pp. 1713–1721, 2005.

[40] T. Saeed, S. Al-Yakoob, H. Al-Hashash, and M. Al-Bahloul, "Preliminary exposure assessment for Kuwaiti consumers to polycyclic aromatic hydrocarbons in seafood," *Environment International*, vol. 21, no. 3, pp. 255–263, 1995.

[41] H.-B. Moon, H.-S. Kim, M. Choi, and H.-G. Choi, "Intake and potential health risk of polycyclic aromatic hydrocarbons associated with seafood consumption in Korea from 2005 to 2007," *Archives of Environmental Contamination and Toxicology*, vol. 58, no. 1, pp. 214–221, 2010.

[42] Scientific Committee on Food, "Opinion of the Scientific Committee on Food on the risk assessment of dioxins and dioxin-like PCBs in food," CS/CNTM/DIOXIN/20 FINAL, Brussels, Belgium, 2001, http://ec.europa.eu/food/fs/sc/scf/out90_en.pdf.

17

Numerical Modeling of the Interaction of Solitary Waves and Submerged Breakwaters with Sharp Vertical Edges Using One-Dimensional Beji & Nadaoka Extended Boussinesq Equations

Mohammad H. Jabbari,[1] **Parviz Ghadimi,**[1] **Ali Masoudi,**[1] **and Mohammad R. Baradaran**[2]

[1] *Department of Marine Technology, Amirkabir University of Technology, Hafez Avenue No. 424,*
P.O. Box 15875-4413, Tehran, Iran
[2] *Civil Engineering Group, Islamic Azad University, Maymand, Iran*

Correspondence should be addressed to Parviz Ghadimi; pghadimi@aut.ac.ir

Academic Editor: Grant Bigg

Using one-dimensional Beji & Nadaoka extended Boussinesq equation, a numerical study of solitary waves over submerged breakwaters has been conducted. Two different obstacles of rectangular as well as circular geometries over the seabed inside a channel have been considered in view of solitary waves passing by. Since these bars possess sharp vertical edges, they cannot directly be modeled by Boussinesq equations. Thus, sharply sloped lines over a short span have replaced the vertical sides, and the interactions of waves including reflection, transmission, and dispersion over the seabed with circular and rectangular shapes during the propagation have been investigated. In this numerical simulation, finite element scheme has been used for spatial discretization. Linear elements along with linear interpolation functions have been utilized for velocity components and the water surface elevation. For time integration, a fourth-order Adams-Bashforth-Moulton predictor-corrector method has been applied. Results indicate that neglecting the vertical edges and ignoring the vortex shedding would have minimal effect on the propagating waves and reflected waves with weak nonlinearity.

1. Introduction

Boussinesq type equations are among the most practical mathematical models used in offshore engineering. These equations include nonlinear terms as well as dispersion terms. Thus, they are one of the most robust tools for hydrodynamic study of nearshore waves. During the years from 1871 to 1872, Boussinesq introduced these equations by adding dispersion effects to the shallow water equations originally known as Saint Venant. These equations have a hyperbolic structure with derivatives of high order in order to numerically model the dispersion-based physics. Peregrine [1] introduced what is known as the basic type of Boussinesq equations. Using Boussinesq equations for inviscid fluids, continuity equation with integral representation and applying respective boundary conditions, the basic Peregrine-Boussinesq equations for

long waves over variable seabeds can be derived. Many efforts have been made for the development of Boussinesq equations. These efforts have been made with the aim of enhancing dispersion (ratio of water depth to wave length) characteristics of the equations in order to preserve their validity for deep waters applications. As one of the first attempts, Witting [2] used the momentum equations on the integrated depth in one dimension so he could present practical equations based on the velocity terms that were defined on the free surface. In the governing equations, Taylor series were used to represent velocity components. Coefficients of the expansion were derived in a way to achieve the best agreement with the linear dispersive properties. Later on, Madsen et al. [3] developed the dispersive properties of these equations to include deep water conditions by obtaining resulting terms based on the assumption of long waves in shallow waters and adding them to the classic equations of Abbott et al. [4]. Other forms of the extended Boussinesq equations were derived by Nwogu [5] where the velocity components in an arbitrary distance from the calm free surface were used. These equations were derived directly from 2-dimensional continuity and Euler equations on a seabed with variable depth. Another type of extended Boussinesq equations was obtained by Beji and Nadaoka [6] using an algebraic manipulation of the classical Peregrine equations. Extended Boussinesq equations carry similar nonlinear and dispersive properties. This means that, while their algebraic representation differs from each other, the dispersive properties of these equations could be indicated to be equivalent after performing a respective change of variable.

Various physical phenomena have been simulated so far using the extended Boussinesq equations for many practical applications. Simulation of undular bore that was carried out by Peregrine [7] can be noted as the first physical modeling using Boussinesq equations in which a second-order accurate finite difference method with one corrector step for the free surface was employed. Almost twenty years later, Abbott et al. [4] introduced the first physical model for the regular wave shoaling on a sloping beach using the Boussinesq equations. They used a finite difference scheme for spatial terms of the equation in order to cancel out the nonphysical dispersion sources. Later, Schäffer et al. [8] and Nwogu [5] modeled the regular and irregular waves propagation on the sloping bottom using the extended Boussinesq equations. In the first attempt, Schäffer et al. [8] applied the breaking effects to the equations when the wave nonlinearity was being increased and the wave was getting closer to breaking up conditions. Moreover, experiments of Beji and Battjes [9] that were carried out for the regular wave propagation over a submerged trapezoidal bar are among the most applicable practical experiments in this field. They used the Boussinesq equations of Abbott et al. [4] while introducing a predictor-corrector scheme in order to numerically model their experiments.

Using a finite difference method, Dingemans [10] compared the classical and extended Boussinesq equations with each other in the case of regular wave propagation over a submerged breakwater. In the meantime, with the introduction of a fourth-order predictor-corrector method for time integration and spatial discretization of extended Boussinesq equations of Nwogu [5], Wei and Kirby [11] made efforts to decrease the truncation errors to achieve a fourth-order accurate scheme. They applied a multitude of physical models such as random wave shoaling on a slope, 2-dimensional sloshing wave evolution in a basin, and regular wave propagation over shoal in order to verify the accuracy of their numerical method. Ohyama et al. [12] introduced a finite difference method for Nwogu's Boussinesq equation [5] and modeled the regular wave propagation over a bar. In another attempt, Wei et al. [13] applied Nwogu's Boussinesq equation [5] for the simulation of solitary wave propagation on a sloping beach. Ambrosi and Quartapelle [14] applied the modified Taylor-Galerkin finite element method to the classical Peregrine equations in order to model the solitary wave propagation and also its interaction with a cylinder. Using Galerkin finite

element method with a complex temporal scheme (known as Sprint), Walkley and Berzins [15] used the Nwogu's extended Boussinesq equations for the numerical simulation of regular wave propagation over a submerged breakwater and the 1-dimensional solitary wave propagation over a beach attached to the coastal wall. Wave propagation in portal regions over regular geometries was also modeled by Li et al. [16] using Beji and Nadaoka extended Boussinesq equations while applying Galerkin Finite Element Method. Walkley and Berzins [17] extended their numerical method to consider 2-dimensional case of wave propagation in portal regions over irregular geometries. In the past decade, Sørensen et al. [18] simulated the 2-dimensional physical phenomenon of surf zone using unstructured meshes. Nonlinear oscillations and harbor resonance problem were investigated by Woo and Liu [19] using a finite element method on Nwogu's extended Boussinesq equations. In another attempt, the problem of sloshing in a closed tank was modeled by Lin and Man [20] using Nwogu's extended Boussinesq equations with the application of a finite difference method. In the recent years, Ghadimi et al. [21] simulated solitary wave shoaling over coastal beds using Beji and Nadaoka's extended Boussinesq equations and with the application of a finite element method. They [21] presented a reasonable estimation of wave breaking using their numerical technique. More recently, Liu et al. [22] modeled solitary wave run-up in a cylinder group using a finite element method approach on Beji and Nadaoka's extended Boussinesq equations.

Although there seemed to be a wide range of applications for the extended Boussinesq equations with the above mentioned studies being only a brief overview on their implementations, the use of these equations for modeling submerged bars and breakwaters is rarely reported as in Ting et al. [23] and Yao et al. [24].

In the present study using the FEM approach of Ghadimi et al. [21] for the numerical solution of Beji and Nadaoka's extended Boussinesq equations, interaction of solitary wave with submerged bars for nondiffracting waves condition is investigated. Since the geometry of the submerged bars includes sharp vertical edges, it has been shown that these geometries cannot be directly implemented using the current numerical method. Therefore, a set of geometrical techniques are studied and then applied for the problem at hand.

2. Governing Equations

In the present work, Beji and Nadaoka's extended Boussinesq equations are employed. These equations include various terms of free surface, $\eta(x, y, t)$, integrated velocity vector in depth, $\mathbf{u} = (u, v)$, depth of the seabed profile which is measured from the calm free-surface level, $H(x, y)$, and gravity acceleration, g, which are presented as follows:

$$\eta_t + \nabla \cdot \left[(H + \eta) \mathbf{u} \right] = 0,$$

$$\mathbf{u}_t + (\mathbf{u} \cdot \nabla) \mathbf{u} + g\nabla\eta$$

$$= (1 + \beta) \frac{H}{2} \nabla \left[\nabla \cdot (H\mathbf{u}_t) \right] + \beta \frac{gH}{2} \nabla \left[\nabla \cdot (H\nabla\eta) \right]$$

$$- (1 + \beta) \frac{H^2}{6} \nabla \left[\nabla \cdot (\mathbf{u}_t) \right] - \beta \frac{gH^2}{6} \nabla \left[\nabla^2\eta \right],$$

$$(1)$$

where β is a free variable for which the value of 1/5 is proposed by Beji and Nadaoka [6]. Nonlinearity and dispersive properties are introduced by parameters ε and μ, respectively. As the wave approaches the beach, the ratio of wave amplitude to water depth (a/H), which is the source of nonlinearity, would increase and the ratio of water depth to wave length (H/λ) will decrease. For Beji and Nadaoka's extended Boussinesq equations with the consideration of the above mentioned value for β, the error of the dispersion relation with $\mu \approx 0.5$ would be less than 0.5 percent. It is worth mentioning that the accumulative error of the equations for $\mu \approx 0.17$ would be less than 2 percent.

3. Numerical Method

The FEM technique of Ghadimi et al. [25] is employed for the numerical solution of the governing equations in 1-dimensional form. The Galerkin FEM using linearly interpolated elements and functions is used for the discretization of the spatial terms. In order to couple the discretized form of the temporal and spatial terms, a fourth-order Adams-Bashforth-Moulton predictor-corrector method has been applied. Since the present method is 2nd-order accurate in terms of spatial discretization, Von-Neumann stability analysis has been employed. Therefore, the aforementioned scheme would be stable for $C_r = \sqrt{gH}(\Delta t/\Delta x) < 0.5$, where C_r is the courant number.

3.1. Initial and Boundary Conditions.
In the present study, the computational domain is a rectangular channel with the length and depth of l and H, respectively. A solitary wave of amplitude of a is propagated from the left side of this channel at the time $t = 0$. This solitary wave is defined with profile velocity given as follows:

$$\eta(x,t) = a \operatorname{sech}^2\left[\sqrt{\frac{3a}{4H^3}}(x-ct)\right],$$

$$u(x,t) = \sqrt{gH}\frac{A}{H}\frac{1}{\cosh\left(\sqrt{3a/4H^3}(x-ct)\right)^2}, \quad (2)$$

$$c = \sqrt{g(H+a)},$$

in which c is the propagation velocity of the solitary wave. A fully reflecting boundary condition is imposed on the exit boundary, and the normal velocity on the boundary is assumed to be zero:

$$u \cdot \mathbf{n} = 0 \quad \text{or} \quad u = 0. \quad (3)$$

Here, $\mathbf{n}$ is the unit normal vector on the reflecting boundary.

4. Solitary Wave Propagation over the Submerged Bars

In the present study, the propagation of solitary wave over two types of submerged bars is simulated. The first simulation considers a, submerged bar. This geometry has vertical edges and is impossible to be implemented in the model directly using the present method. Therefore, a set of polygonal approximations with inclined sides have been employed instead of the vertical edges, and the results are presented. The second submerged bar is an attached step to the wall which also has vertical edges and, therefore, requires further modifications.

4.1. Solitary Wave Propagation over a Semicircular Bar.
For the case of solitary wave propagation over a semicircular submerged bar, the experimental investigations of Cooker et al. [26] are employed. They performed a set of experiments to investigate the interaction of solitary waves with amplitude a over a semicircular submerged bar of radius R and introduced a relation between the reflected and transmitted wave for breaking and nonbreaking conditions. Since a breaking model is not employed in the present method, the submerged bar of radius $R = 0.6$ m and a solitary wave of amplitude $a = 0.311$ m are used for modeling the interaction of solitary wave with the seabed depicted in Figure 1 (in the deep side of a tank $H = 1$ m).

The scale of the wave nonlinearity just before reaching the submerged bar in the fixed-depth channel is $\varepsilon = a/H = 0.311$ which is far from the breaking conditions. When the wave reaches the semicircular submerged bar due to the decrease in depth, it is expected for the wave amplitude to elevate. In the region close to the semicircular bar, the nonlinearity based on the experimental results would be equal to $\varepsilon = 0.4$. Once again, this case is safely away from the breaking conditions, and during its propagation, the solitary wave will theoretically exhibit a nonbreaking procedure. Due to the decrease in water depth which occurs when the wave passes over the submerged bar crest, wave would be dispersed, and the wave crest, which was sharpened due to shoaling, requires strict considerations. As stated before, the semicircular submerged bar shown in Figure 1 cannot be numerically modeled using the present FEM approach. Since the computational domain is discretized using finite elements, the neighboring nodes of an element laying on the convex semicircular surface would have discrete depths and as the wave gets sharpened and the free surface gets accelerated, a discrete profile would be obtained over time from the numerical results (Figure 2). As a remedy to this numerical phenomenon, it is required to approximate the convex surface with a regular or irregular polygon (triangle or quadrilateral) in order to have more than two adjusting nodes on each side of the slope. In fact, there should exist a minimum of one intermediate node on each slope so that elements on that side would be discretized more than once (Figure 3).

In the present study, three gauges have been employed in order to record the time series of the propagating solitary wave with two of them being placed before the submerged bar with distances of 0.9 m and 2.5 m from the bar, while another one is placed 0.4 m behind the bar. Four numerical experiments have been arranged with four different polygonal approximations of the semicircular bar (Figure 4), and the resulting time series of the shoaling numerical solution are compared with the gauge results. In the present numerical method, the computational domain inside the channel has a length of $l = 40$ m and is discretized using 1000 linear finite

Numerical Modeling of the Interaction of Solitary Waves and Submerged Breakwaters with Sharp Vertical Edges Using
One-Dimensional Beji & Nadaoka Extended Boussinesq Equations

181

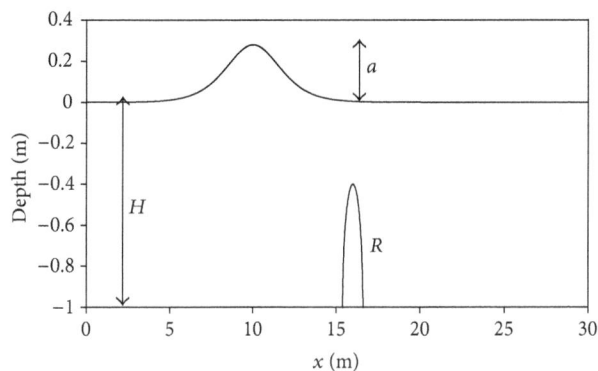

FIGURE 1: Physical modeling of the interaction of a solitary wave and a semicircular submerged bar.

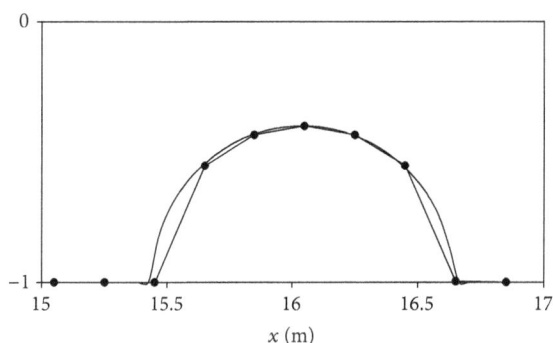

FIGURE 2: A semicircular bar where the convex surface is discretized using a general approach.

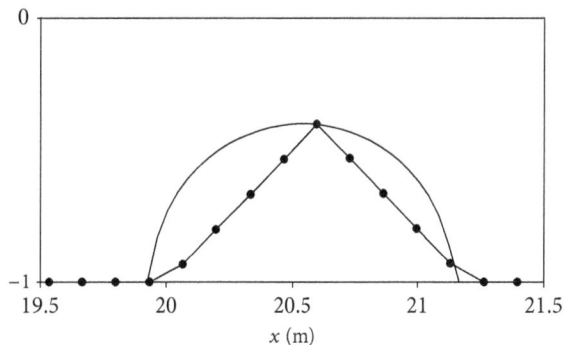

FIGURE 3: Approximation of nodes height on a semicircular bar so that a minimum of two elements are defined on each slope.

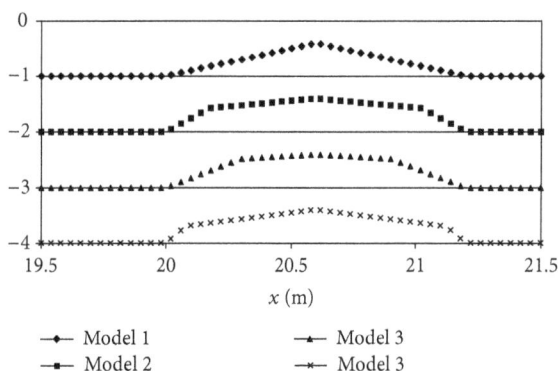

FIGURE 4: Four different approximations of the semicircular bar.

elements. The time step for the numerical solution is set equal to $\Delta t = 0.01$.

Figure 4 represents four different models employed for the approximation of semicircular submerged bar that are used in the present numerical simulation. The approximated geometries are set in a way that they are all inscribed in a semicircle with a radius of 0.6 m. Model 1 represents an inscribed triangle, and model 2 is a regular semioctagon in which the intersection of the middle sides forms a 45 degrees angle with respect to the center of the geometry. Other models include geometries with a 30- and 60-degree angle between the intersections of the middle sides. Therefore, the distributions of the resulting finite element nodes on each side are shown in Figure 4. The calculated time series for the models introduced earlier are shown in Figures 5–8.

As depicted in these figures, numerical and experimental results have a slight difference between them in terms of phase angle which could be attributed to the irrotational flow assumption for the Boussinesq governing equations and also the proposed approximation of the semicircular submerged bar. Numerical results recorded by the first gauge (from left) give two peaks in the time series for all four models. The first peak is due to the approaching wave crest while the second one denotes the reflecting wave crest. Based on the results of the experiments, a distinct crest for the reflecting wave cannot be distinguished, while both experimental and numerical

results demonstrate a similar time range for the reflecting wave. The second gauge shows the moment at which the wave passes the first slope (left side) of the submerged bar and therefore wave shoaling occurs. An overestimation is observed in the results of the second gauge which shows its relevance to the results of the first gauge.

The celerity of the reflecting wave using the Boussinesq model is lower than that of the empirical theories and this causes an overestimation in the free surface elevation that was recorded by the second gauge. In the meantime, a third gauge records wave dispersion and shows two peaks. After crossing the submerged bar, the wave starts to disperse and therefore the second reflection occurs; the idea that is, supported by the second peak recorded in the time series. Contrary to the first gauge, the second peak recorded by the third gauge cannot be observed. In fact, the second reflection occurs sooner in the Boussinesq model. After crossing the submerged bar, the wave amplitude using fourth approximation model (Model 4) is modeled more precisely. It is shown that only a triangular approximation of the semicircular submerged bar would give favorable results for the free surface elevation over the bar when the Beji and Nadaoka extended Boussinesq equations are applied. This is due to the fact that the wave-front would be accelerated while crossing over the bar which would cause instabilities for the case of the models 2, 3, and 4.

This phenomenon can be observed as nonphysical dispersions that are shown on Figures 6, 7, and 8. That is why

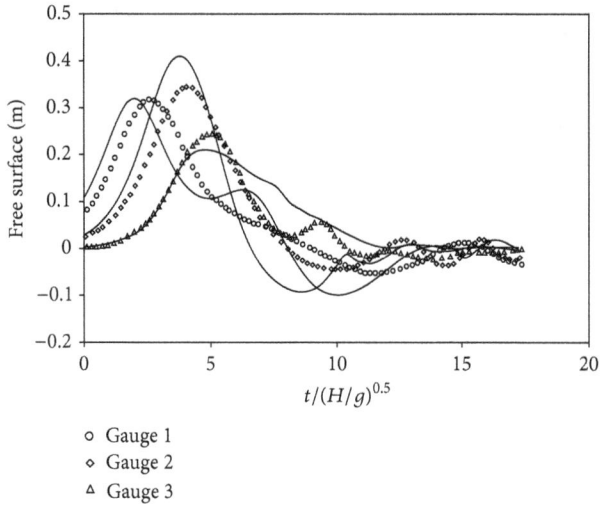

- o Gauge 1
- ◊ Gauge 2
- △ Gauge 3

FIGURE 5: Time series of the gauges compared to the experimental results for Model 1.

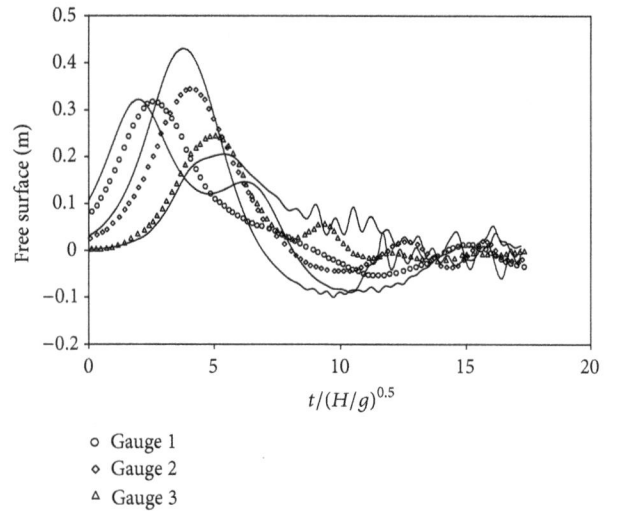

- o Gauge 1
- ◊ Gauge 2
- △ Gauge 3

FIGURE 7: Time series of the gauges compared to the experimental results for Model 3.

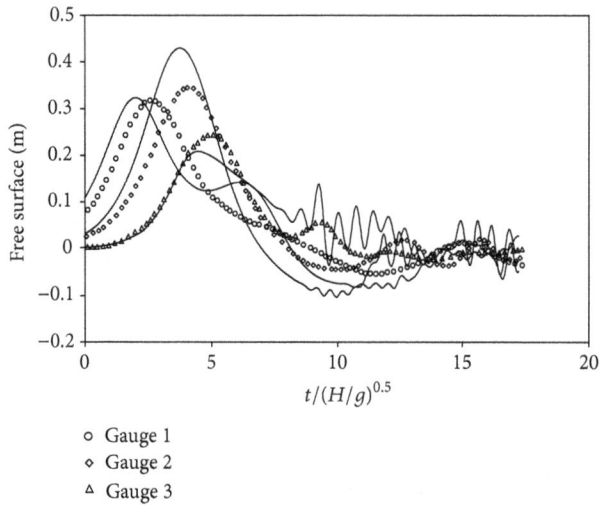

- o Gauge 1
- ◊ Gauge 2
- △ Gauge 3

FIGURE 6: Time series of the gauges compared to the experimental results for Model 2.

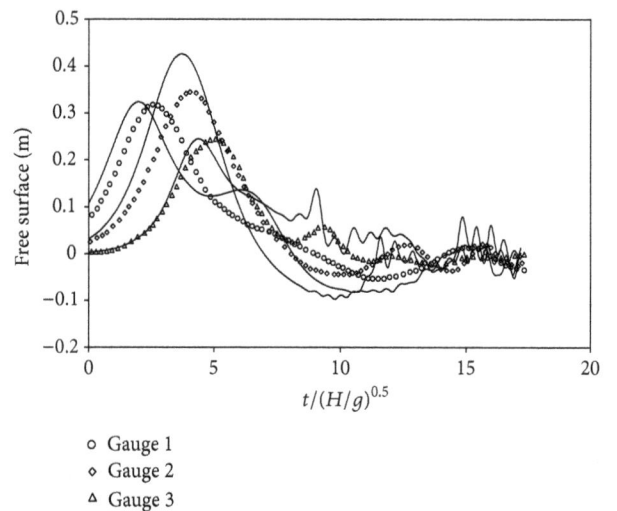

- o Gauge 1
- ◊ Gauge 2
- △ Gauge 3

FIGURE 8: Time series of the gauges compared to the experimental results for Model 4.

some deviations are observed between the present results and the ones obtained from the experiments. Mentioned cases are the main reason of difference in numerical and experimental results. It is worth mentioning that the main goal of the present study was to have a suitable numerical modeling of the physical problem at hand using the Boussinesq equations due to the fact that their computational speed exceeds their rival alternatives in fluid motion equations. Time history of the propagating wave over Model 1 (Triangular Approximation) is shown in Figure 9 for the time intervals of 0.6 seconds.

4.2. Solitary Wave Propagation over a Step Attached to the Coastal Wall. Solitary waves are stable during their propagation over a fixed depth, and this characteristic makes them more destructive. Therefore, a set of vertical concrete caissons is generally used in front of the coastal walls in order to act as a breakwater at times of low tide. These costal steps would

become unstable when collide with these kinds of solitary waves and would be destructive over time. On the other hand, their presence in the coastal line for the prevention of the wave impacts on the coastal wall is of high importance. Therefore in this part of the present study, propagating solitary wave over a step, that is, attached to the coastal wall, is modeled using the Beji and Nadaoka extended Boussinesq equations. Experimental results of Grilli et al. [27] are used for the validation purposes. In their experimental model [27], they introduced a solitary wave with nondimensional amplitude of $\varepsilon = a/H = 0.33$ in the deep side of a tank ($H = 1$ m) while the wave would approach a step with a depth of $h_1 = 0.67$ m. The wave profile is recorded over propagation using different gauges. Using the present numerical model, the introduced problem is simulated, and the free surface

Numerical Modeling of the Interaction of Solitary Waves and Submerged Breakwaters with Sharp Vertical Edges Using One-Dimensional Beji & Nadaoka Extended Boussinesq Equations

183

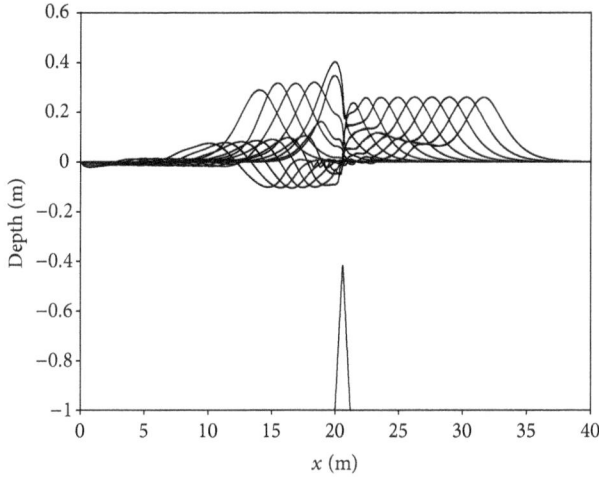

FIGURE 9: Solitary wave propagation time history for Model 1.

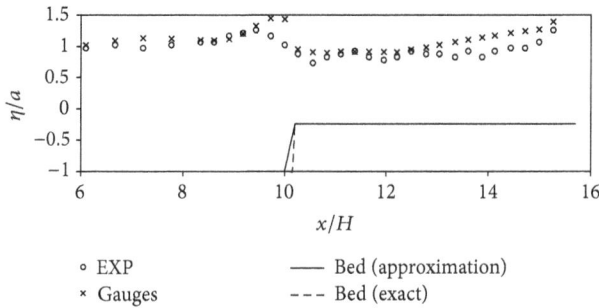

| ○ EXP | —— Bed (approximation) |
| × Gauges | - - - Bed (exact) |

FIGURE 10: Free surface elevation over a step.

elevations are compared with the experimental results in Figure 10.

Due to the fact that the vertical face of the step in this model causes a discontinuity between the two different depths, a gradual slope is used in a way that enables the required continuity of the finite elements in this region. The obtained numerical solution is in a good agreement with the experimental results for the respective gauges.

5. Conclusion

In the present paper, Beji and Nadaoka extended Boussinesq equations have been applied for numerical modeling of solitary waves over submerged bars with sharp vertical edges using finite element method. A semicircular submerged bar along with a stepped seabed was implemented. Since these bars have vertical edges, a discontinuity occurs in depth which makes it impossible to have a direct implementation of the present numerical method. Thus, a gradual slope was employed instead of the vertical edges. In the first type of submerged bar, inscribed polygons in the semicircle were used instead of the semicircular submerged obstacle. Irregular geometries were chosen in such a way that the intersection of the middle sides makes a 30-, 45- and 60-degree angle for various models, respectively. Numerical

results for all four models demonstrate that in these experimental cases, the Boussinesq model simulates the collision causing the first reflection of solitary wave to occur a bit sooner and the second reflecting wave appears a bit later. However, due to being more efficient compared to other equations used in simulating fluid motions, the present Boussinesq model shows to be a suitable criterion for the present experimented phenomenon. Due to the fact that the wave front gets sharpened after crossing over the submerged bar, implementation of sequential polygonal slopes would cause instabilities for the dispersed wave. As a result, only the triangular approximation of the semicircular bar is proved to be acceptable in the present numerical model. Moreover, the solitary wave propagation over a step attached to the coastal wall was investigated, and favorable agreement of the numerical results with the experimental data was achieved. It is worth mentioning that, due to the discontinuity problem that was discussed earlier, the step was replaced by a gradual slope in the second test case.

References

[1] D. H. Peregrine, "Calculations of the development of an undular bore," *Journal of Fluid Mechanics*, vol. 25, no. 2, pp. 321–330, 1966.

[2] J. M. Witting, "A unified model for the evolution nonlinear water waves," *Journal of Computational Physics*, vol. 56, no. 2, pp. 203–236, 1984.

[3] P. A. Madsen, R. Murray, and O. R. Sørensen, "A new form of the Boussinesq equations with improved linear dispersive properties," *Coastal Engineering*, vol. 15, pp. 371–388, 1991.

[4] M. B. Abbott, A. D. McCowan, and I. R. Warren, "Accuracy of short-wave numerical models," *Journal of Hydraulic Engineering*, vol. 110, no. 10, pp. 1287–1301, 1984.

[5] O. Nwogu, "Alternative form of Boussinesq equations for near-shore wave propagation," *Journal of Waterway, Port, Coastal & Ocean Engineering*, vol. 119, no. 6, pp. 618–638, 1993.

[6] S. Beji and K. Nadaoka, "A formal derivation and numerical modelling of the improved boussinesq equations for varying depth," *Ocean Engineering*, vol. 23, no. 8, pp. 691–704, 1996.

[7] D. H. Peregrine, "Long waves on a beach," *Journal of Fluid Mechanics*, vol. 27, pp. 815–827, 1967.

[8] H. A. Schäffer, P. A. Madsen, and R. Deigaard, "A Boussinesq model for waves breaking in shallow water," *Coastal Engineering*, vol. 20, no. 3-4, pp. 185–202, 1993.

[9] S. Beji and J. A. Battjes, "Numerical simulation of nonlinear wave propagation over a bar," *Coastal Engineering*, vol. 23, no. 1-2, pp. 1–16, 1994.

[10] M. W. Dingemans, "Comparison of computations with Boussinesq-like models and laboratory measurements," Technical Report H1684.12, MAST G8M Coastal Morphodynamics Research Programme, 1994.

[11] G. Wei and J. T. Kirby, "Time-dependent numerical code for extended Boussinesq equations," *Journal of Waterway, Port, Coastal & Ocean Engineering*, vol. 121, no. 5, pp. 251–261, 1995.

[12] T. Ohyama, W. Kioka, and A. Tada, "Applicability of numerical models to nonlinear dispersive waves," *Coastal Engineering*, vol. 24, no. 3-4, pp. 297–313, 1995.

[13] Ge Wei, J. T. Kirby, S. T. Grilli, and R. Subramanya, "A fully nonlinear Boussinesq model for surface waves. Part 1. Highly

nonlinear unsteady waves," *Journal of Fluid Mechanics*, vol. 294, pp. 71–92, 1995.

[14] D. Ambrosi and L. Quartapelle, "A Taylor-Galerkin method for simulating nonlinear dispersive water waves," *Journal of Computational Physics*, vol. 146, no. 2, pp. 546–569, 1998.

[15] M. Walkley and M. Berzins, "A finite element method for the one-dimensional extended Boussinesq equations," *International Journal For Numerical Methods in Fluids*, vol. 29, pp. 143–157, 1999.

[16] Y. S. Li, S. X. Liu, Y. X. Yu, and G. Z. Lai, "Numerical modeling of Boussinesq equations by finite element method," *Coastal Engineering*, vol. 37, no. 2, pp. 97–122, 1999.

[17] M. Walkley and M. Berzins, "A finite element method for the two-dimensional extended Boussinesq equations," *International Journal for Numerical Methods in Fluids*, vol. 39, no. 10, pp. 865–886, 2002.

[18] O. R. Sørensen, H. A. Schäffer, and L. Sørensen, "Boussinesq type modeling using an unstructured finite element technique," *Coastal Engineering*, vol. 50, pp. 181–198, 2004.

[19] S. B. Woo and P. L. F. Liu, "Finite-element model for modified Boussinesq equations. II: applications to nonlinear harbor oscillations," *Journal of Waterway, Port, Coastal and Ocean Engineering*, vol. 130, no. 1, pp. 17–28, 2004.

[20] P. Lin and C. Man, "A staggered-grid numerical algorithm for the extended Boussinesq equations," *Applied Mathematical Modelling*, vol. 31, no. 2, pp. 349–368, 2007.

[21] P. Ghadimi, M. H. Jabbari, and A. Reisinezhad, "Calculation of solitary wave shoaling on plane beaches by extended Boussinesq equations," *Engineering Applications of Computational Fluid Mechanics*, vol. 2, no. 2, pp. 1–29, 2012.

[22] S. Liu, Z. Sun, Li, and j, "An unstructured FEM model based on Boussinesq equations and its application to the calculation of multidirectional wave run-up in a cylinder group," *Applied Mathematical Modelling*, vol. 36, no. 9, pp. 4146–4164, 2011.

[23] C. L. Ting, M. C. Lin, and C. M. Hsu, "Spatial variations of waves propagating over a submerged rectangular obstacle," *Ocean Engineering*, vol. 32, no. 11-12, pp. 1448–1464, 2005.

[24] Y. Yao, Z. Huang, S. G. Monismith, and E. Y. M. Lo, "1DH Boussinesq modeling of wave transformation over fringing reefs," *Ocean Engineering*, vol. 47, pp. 30–42, 2012.

[25] P. Ghadimi, M. H. Jabbari, and A. Reisinezhad, "Finite element modeling fo one-dimensional boussinesq-equations," *International Journal of Modeling, Simulation, and Scientific Computing*, vol. 2, no. 2, pp. 207–235, 2011.

[26] M. J. Cooker, D. H. Peregrine, C. Vidal, and J. W. Dold, "Interaction between a solitary wave and a submerged semicircular cylinder," *Journal of Fluid Mechanics*, vol. 215, pp. 1–22, 1990.

[27] S. T. Grilli, M. A. Losada, and F. Martin, "Wave impact forces on mixed breakwaters," *Coastal Engineering*, vol. 88, pp. 1–14, 1992.

18

Bioaccumulation of Polycyclic Aromatic Hydrocarbons and Mercury in Oysters (*Crassostrea rhizophorae*) from Two Brazilian Estuarine Zones

Ronaldo J. Torres,[1] **Augusto Cesar,**[2] **Camilo D. S. Pereira,**[2] **Rodrigo B. Choueri,**[2]
Denis M. S. Abessa,[3] **Marcos R. L. do Nascimento,**[4] **Pedro S. Fadini,**[1] **and Antonio A. Mozeto**[1]

[1] *Laboratório de Biogeoquímica Ambiental (LBGqA)–Núcleo de Estudos, Diagnósticos e Intervenções Ambientais (NEDIA),*
  Departamento de Química, UFSCar, Rod. Washington Luiz, 13565-905 São Carlos, SP, Brazil
[2] *Centro de Ciências do Mar e Meio Ambiente, Baixada Santista UNIFESP, Avenida Dona Costa, 95, 11060-001 Santos, SP, Brazil*
[3] *Campus Experimental do Litoral Paulista, Universidade Estadual Paulista, Pça. Infante D. Henrique, s/no, 11330-900 São Vicente,*
  SP, Brazil
[4] *Laboratório de Poços de Caldas (LAPOC), CNEN, Comissão Nacional de Energia Nuclear, Rodovia Andradas, km 13,*
  37701-970 Poços de Caldas, MG, Brazil

Correspondence should be addressed to Ronaldo J. Torres, rjtorres2000@yahoo.com.br

Academic Editor: Roberto Danovaro

Nowadays, organisms are increasingly being used in biomonitoring to assess bioavailability and bioaccumulation of contaminants. This approach can use both native and transplanted organisms in order to accomplish this task. In Brazil, most of the studies related to bioaccumulation of contaminants in oysters deal with metals. The present work employs this kind of test in Brazilian coastal estuaries (Santos and Paranaguá) to evaluate total mercury and polycyclic aromatic hydrocarbon contamination in sediments and oysters (native and caged *Crassostrea rhizophorae*). The methodologies employed were based on known USEPA methods. Results have shown a significant contamination in Santos sediments and consequent bioavailability of organisms. Paranaguá sediments presented lower contamination in sediments, but native oysters were able to accumulate total Hg. The experiments done with caged oysters did not show significant bioaccumulation of Hg and PAHs in the Paranaguá site, but proved to be an excellent tool to assess bioavailability in the Santos estuary since they were able to bioaccumulate up to 1,600% of total PAH in the samples from the inner part of this estuary when compared to control organisms. Multivariate statistical analyses employed to these results have separated the sites evaluated and the most contaminated samples from the least contaminated.

1. Introduction

Some organic and inorganic chemical contaminants have the capacity of persisting in the environment, bioaccumulate in tissues, and are toxic to organisms. The main classes of elements and compounds that belong to this category are some metals such as mercury, cadmium, and lead as well as those denominated POPs (persistent organic pollutants) such as pesticides, dioxins, polychlorinated biphenyls (PCBs), and polycyclic aromatic hydrocarbons (PAHs) [1, 2].

Once in the environment, the contaminants interact with sediments, water column, and organisms; such interactions are controlled by several physical and chemical processes, and the final result may be the chemical release, immobilization, or their transformation into more reactive forms or subproducts, which are more effectively available to organisms [3]. Bioavailability is also governed by kinetics and partitioning of the contaminant in the environment [4].

Bioaccumulation is the process by which a chemical is absorbed by an organism exposed to it. It is a net result of competing processes of absorption, ingestion, digestion, and excretion [5] and involves also the endogenous processes of biological depuration. Bioaccumulation studies can also be interpreted in human health risk analysis when the target organisms are cultivated or used for human consumption. PAH bioaccumulation is dependent on fatty and lipid-rich tissues and organs [6].

Since the mid-1970s, environmental assessment programs have utilized bivalve mollusks as monitoring agents of chemical contaminants in marine areas [7]. Because of their sedentary habit and ability to bioaccumulate pollutants, mussels and oysters have been employed as sentinel organisms in environmental quality monitoring studies in coastal regions. The assessment of oysters and mussels as sentinel organisms has also been used in Brazil in recent years [8–11].

The estuarine systems of Santos and Paranaguá are major port areas in southeastern Brazil. Studies undergone in these areas identified contamination and the bioavailability of contaminants in water and sediments from these systems through chemical, toxicological, and ecological assessments, as consequences of anthropic activities in this region [12–15]. One particular study [16] has evidenced bioaccumulation of contaminants in fish, crustaceans, and mollusks from the Santos Estuarine System. However, the relationship between sediment contamination and bioaccumulation was not effectively assessed. Some of these previous studies have shown high levels of mercury and PAH, especially in the Santos estuary [12, 13, 15, 16].

The central hypothesis of this study is that a number of anthropic activities, namely, industrial and municipal effluents from the study area as well as the dredging process and ship movement across the estuaries, can contaminate the organisms that live in these regions with metals such as mercury and hydrocarbons. Chemical analysis of the sediments and native and caged organisms may give the answers for the possibility of bioaccumulation and the risks posed to humans if these organisms are ingested. Whole integration of the results can establish a synoptic application of the assessment of environmental contamination, toxicity, and bioaccumulation. Assessment of other chemicals will be held for future studies.

2. Materials and Methods

Paranaguá Estuarine System (PES) and Santos Estuarine System (SES) (Figures 1(a) and 1(b), resp.) are two of the most important port areas in Brazil. PES is influenced by the port and seasonal tourism in an almost untouched coastal area. SES has heavy industries in it (petrochemicals, metallurgy, and fertilizes among others) and a large urban concentration. Both areas comprise ecologically significant ecosystems, such as mangroves and Atlantic rain forest; in addition, artisanal fisheries and familiar aquaculture activities are performed in many places all over the PES. Whilst for SES there is vast information on sediment contamination [15, 16], sediment toxicity [17, 18], and benthic communities' alterations [12], for PES little information about sediment contamination, toxicity, or benthic communities is available [14].

In both estuaries, four sampling stations were chosen (Figures 1(a) and 1(b)) in which sediment was collected through a van Veen dredge. Native oysters (*Crassostrea rhizophorae*) (Guilding, 1828) (Bivalvia: Ostreidae) were collected where possible, and one cage containing about 30

oysters from the same species (purchased from a farm in a reference area located in Guaratuba bay, about 40 km southwards PES) was placed at each point for bioaccumulation assessment. The sediments and native oysters were placed in plastic bags for analysis of mercury and in aluminum containers for PAH quantification, and they were kept frozen (−20°C) prior to analysis. Two distinct periods were chosen for these experiments which were a dry season (August 2008) and a rainy season (May 2009). All samplings, handlings and storage of the sediments and organisms were conducted according to USEPA guidance [19].

After 30 days of exposure, the oysters were removed from the cage and transported to laboratory where the soft tissues were detached from the shells and placed in plastic bags or aluminum containers and finally frozen prior to laboratory analysis for Hg and PAH.

For the analysis of Hg, the sediment was dried in the oven at 60°C for 24 hours and grounded and homogenized. Soft tissues were digested *in natura* and part of them was dried for moisture content since results are usually expressed in dry weight. All laboratory analyses followed methods 245.5 and 245.6 from [19] where the sample is weighted and digested with a mixture of nitric and sulfuric acids, potassium permanganate, potassium persulfate with heating, and hydroxylamine hydrochloride, and stannous chloride was added prior to instrumental analysis by cold vapor atomic absorption spectrophotometry (CV-AAS-Varian, FS-220). Quantitation limit was $0.2 \, \mu g \, kg^{-1}$ for both sediments and organisms.

Sixteen priority PAH listed by USEPA [20] were extracted in both sediments and oysters according to method 3550b [21]. The individual compounds analyzed are naphthalene, acenaphthylene, acenaphthene, fluorene, phenanthrene, anthracene, fluoranthene, pyrene, benzo[a]anthracene, chrysene, benzo[b]fluoranthene, benzo[k]fluoranthene, benzo[a]pyrene, indeno[1,2,3-cd]pyrene, dibenzo[a,h]anthracene, and benzo[ghi]perylene. In this method, parts of the sediments and tissues were dried to obtain moisture content and other parts were digested *in natura* by ultrasound extraction with a mixture of hexane/acetone. The extract was submitted to a clean-up process in a glass column with silica gel and alumina eluted with a mixture of dichloromethane/hexane. Instrumental determination was done according to method 8270 [19, 21] by gas chromatography coupled to a mass spectrometer by Shimadzu, model QP2010. Quantitation limits have been established in the range from $0.081 \, \mu g \, kg^{-1}$ to $0.936 \, \mu g \, kg^{-1}$.

For quality assurance/quality control (QA/QC) procedures, blank extractions were done and replicate analysis was performed for all samples for Hg and PAH. For oyster analysis, in order to ensure enough weight for analysis, composite samples with two or three organisms were analyzed. Percent deviation between replicates was always below 5%. Also, percent recoveries were analyzed. Recovery was calculated by spiking some of the samples with known concentrations of the specific standard solutions. Method recoveries were evaluated by analysis of the certified sediment NIST-1944 (New York/New Jersey Waterways) and tissue NIST-2974a ("Organics in Freeze-Dried Mussel Tissue"). Recoveries were

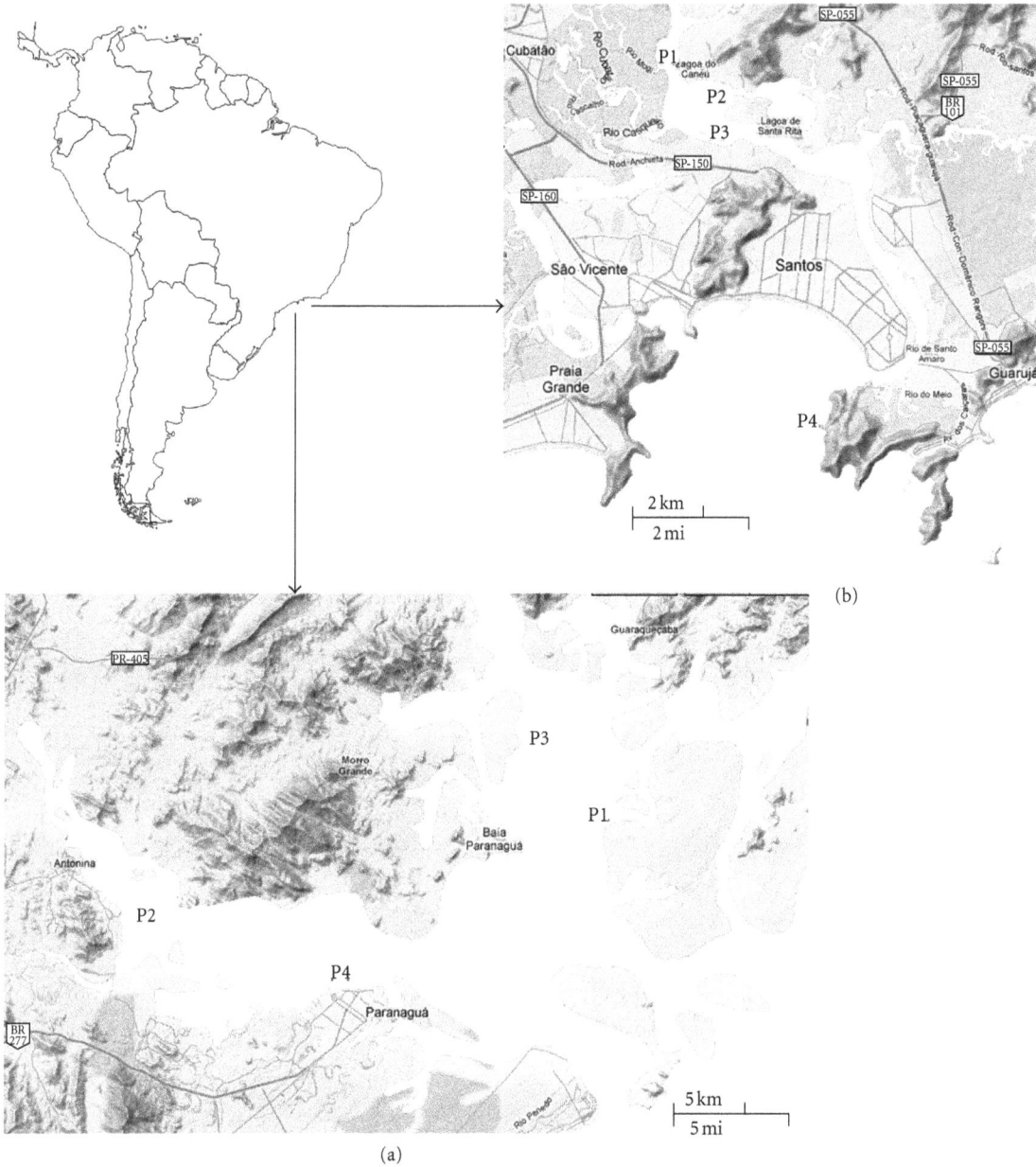

FIGURE 1: Distribution of sampling sites in (a) Paranaguá Estuarine System (PES-25°16′S–25°34′S; 48°17′W–48°42′W): P1-Guapicú; P2-Porto de Antonina; P3-Gamelas; P4-Porto de Paranaguá and B) Santos Estuarine System (SES-23°45′S–24°5′S; 46°15′W–46°30′W): P1-Piaçaguera; P2-Bagres Island; P3-Alamoa; P4-Palmas Island.

of 115% for Hg and varied from 68.3% to 118.8% for PAH. All the glasswares were washed with Extran by Merck, rinsed in acetone and methanol PA (Merck or Synth), and oven-dried at 105°C for POP analysis and also set in 20% nitric acid bath for Hg analysis. All the reagents used for extractions were HPLC grade and supplied by Baker, Merck, or Mallinckrodt.

Chemical results were compared with their respective controls using Student's t-test applied to compare chemical composition and comparisons of these contaminants in the study site. Data of both estuaries were analyzed separately by multivariate analyses, that is, principal component analysis (PCA) for pattern recognition. These analyses were done by the use of the program PAST 2.03 from the University of Oslo.

3. Results and Discussion

Table 1 presents a synthesis of all results acquired in this project. Even though two different seasons were assessed, results are presented as simple means of the values obtained because these values were significantly equal when compared

TABLE 1: Concentration of mercury and PAH in sediments and oysters from Santos and Paranaguá estuarine systems (mean of dry season—august 2008—and rainy season—may 2009).

Local/matrix	Total Hg (mg kg^{-1})	Bioaccumulation*%	Total PAH (μg kg^{-1})	Bioaccumulation*%
Sediments-Paranaguá				
P1 Guapicú	0.09	—	10.01	—
P2 Antonina Port	0.11	—	9.59	—
P3 Gamelas	0.13	—	16.43	—
P4 Paranaguá Port	0.15	—	21.18	—
Sediments-Santos				
P1 Piaçaguera Channel	1.92	—	1.406,13	—
P2 Bagres Island	1.97	—	431.71	—
P3 Alemoa	1.60	—	133.72	—
P4 Palmas Island	0.36	—	36.48	—
Native oysters-Paranaguá				
P1 Guapicú	0.35	—	32.20	—
P2 Antonina Port	0.22	—	31.42	—
P3 Gamelas	0.14	—	34.69	—
P4 Paranaguá Port	NA	—	NA	—
Native oysters-Santos				
P1 Piaçaguera Channel	NA	—	NA	—
P2 Bagres Island	0.32	—	182.99	—
P3 Alemoa	0.37	—	121.18	—
P4 Palmas Island	0.32	—	51.72	—
Caged oysters				
Control oyster (T-zero)	0.08	—	32.52	—
Transplanted oysters, after 30 d-Paranaguá				
P1 Guapicú	0.07	−18.9	42.18	29.7
P2 Antonina Port	0.09	10.7	35.83	10.2
P3 Gamelas	0.08	3.1	33.41	2.7
P4 Paranaguá Port	0.03	−66.0	53.70	65.2
Transplanted oysters, after 30 d-Santos				
P1 Piaçaguera Channel	0.12	44.0	584.33	1.697.0
P2 Bagres Island	0.20	150.9	234.32	620.6
P3 Alemoa	0.15	86.2	295.31	808.2
P4 Palmas Island	0.08	0.6	50.54	55.4

* Caged oysters percent biaccumulation in the 30 day experiment. NA: Not available.

to each other and when Student's t-test was applied. The analysis of the sediments has shown that SES is more contaminated by Hg and PAH than PES. Assessment of contaminants in the sediments of PES, the highest Hg concentration, was in the port area P4 which violated the SQG (sediment quality guideline) and ERL (effect range low) which is of 0.15 mg kg^{-1} [15]. Conceptually, the ERL is the 10th percentile of a series of data of ascending levels of contaminants and their related toxicity effects. This guideline is also present in the Brazilian Resolution CONAMA 344/04 as Level 1; such resolution regards the chemical composition of sediments to be dredged and disposed in jurisdictional Brazilian waters, and Level 1 is a threshold effect level, that is, the lowest level, below which adverse effects rarely occur [22–24].

Observing the results of sediments from SES, it was possible to see that sediments from the inner portions of the estuary (Piaçaguera, Bagres Island, and Alemoa Terminal) presented levels of Hg much above the SQG that is called ERM (Effect Range Medium) which is of 0.72 mg kg^{-1} and it is the 50th percentile (medium) of a series of data of ascending levels of contaminants and their related toxicity effects [15]. This is representative of concentrations above which toxic effects to the benthic communities frequently occur. It is also related in the above mentioned CONAMA resolution 344/04 as Level 2 and if this sediment was to be dredged, special care must be assigned to handling and disposal of this material. Only sediments from P4 (das Palmas Island) had concentrations below ERM, but still above ERL. This point is very close to an old disposal site for the dredging activities that took place in Santos in the past [16] and also suffers influence from the port area, especially when dredging activities take place. This site is influenced

Bioaccumulation of Polycyclic Aromatic Hydrocarbons and Mercury in Oysters (Crassostrea rhizophorae) from Two Brazilian Estuarine Zones

189

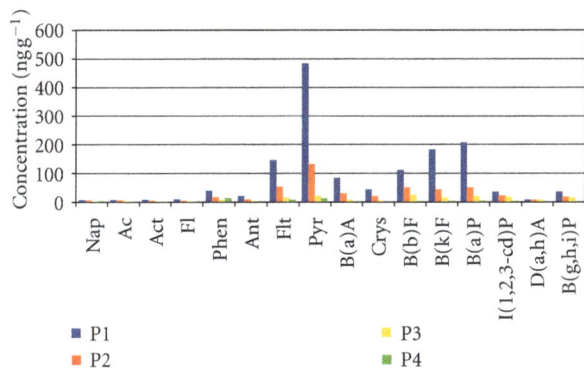

FIGURE 2: Distribution of individual PAHs in sediments from SES. Nap—naphthalene, Ac—acenaphthylene, Act—acenaphthene, Fl—fluorene, Phen—phenanthrene, Ant—anthracene, Flt—fluoranthene, Pyr—pyrene, B(a)a—benzo[a]anthracene, Crys—chrysene, B(b)F—benzo[b]fluoranthene, B(k)F—benzo[k]fluoranthene, B(a)P—benzo[a]pyrene, I(1,2,3-cd)P—indeno[1,2,3-cd]pyrene, D(a,h)A—dibenzo[a,h]anthracene, and B(g,h,i)P—benzo[ghi]perylene.

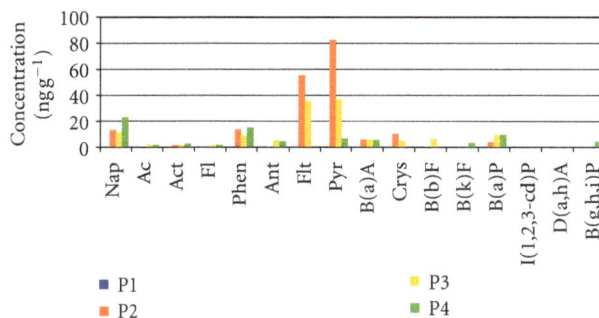

FIGURE 3: Distribution of individual PAHs in native oysters in SES. Nap—naphthalene, Ac—acenaphthylene, Act—acenaphthene, Fl—fluorene, Phen—phenanthrene, Ant—anthracene, Flt—fluoranthene, Pyr—pyrene, B(a)a—benzo[a]anthracene, Crys—chrysene, B(b)F—benzo[b]fluoranthene, B(k)F—benzo[k]fluoranthene, B(a)P—benzo[a]pyrene, I(1,2,3-cd)P—indeno[1,2,3-cd]pyrene, D(a,h)A—dibenzo[a,h]anthracene, and B(g,h,i)P—benzo[ghi]perylene.

also by the underwater sewage effluent from the city of Santos that is dumped in the bay and carried to das Palmas Island by weather currents from the SW during storm events [17].

With respect to the PAH, PES has shown very low concentrations in the sediments and little difference among sampling sites. According to ERL/ERM sediment quality values, no harm to the sediment organisms would be caused by the PAH in PES. On the other hand, in SES, violation of ERL for several individual PAHs (known to be responsible for causing cancer in humans [20]) such as pyrene, benzo(a)anthracene, and benzo(a)pirene are not difficult to occur [12, 13, 15, 17]. It is possible to note that there is a gradient of contamination that is higher in the inner part of Santos estuary, a fact that was observed in previous studies such as [25–27]. Historically, SES has presented high concentration of metals and hydrocarbons due to the industrial complexity of Cubatão, which includes a petroleum refinery plant and several fertilizers and chemical companies and due to the port area of Santos [12, 16, 17, 25, 26]. Figure 2 shows the distribution of individual PAHs in sediments of SES.

Assessing the native oysters collected from these two estuaries, it is noteworthy that in the two most contaminated points (Paranaguá Port in PES—P4—and Piaçaguera Channel in SES—P1) there were no organisms to be collected. Concentrations of Hg in soft tissues of collected native oysters were quite homogeneous among sampling stations in both PES and SES. The not so clear correlation between levels of Hg in sediments and organisms in PES may be a result of the absence of conditions for the efficient mercury methylation as a result, for example, of the inorganic mercury complexation by the organic matter, an interaction that decreases the methylation rate. On the other hand, even if methylated, the formed methylmercury can be complexed with bisulfide, decreasing its bioavailability [5, 28]. This scenery propitiates a complex context where appropriate conditions

for mercury methylation, like organic matter presence, that determines ideal redox conditions for the formation of the organic metallic product and sulfate reducing bacteria activity can, paradoxically, inhibit methylation as a result of the Hg^{2+} complexation or indirectly be responsible for the complexation of methylmercury by the sulfide species. These aspects of the mercury cycle along with oyster feeding habits, hypothesize reasons that explain why Hg in sediments either from PES or SES were not completely bioavailable, or, in case it was available, in PES Hg levels in organisms where higher than in the sediments, whereas in SES the levels of Hg in the sediments were higher than in soft tissues of the native organisms (exception for P4, where Hg concentration was very close in the organisms and the sediments).

Similarly to Hg, PAH levels in native oysters from PES were homogeneous along the estuarine system, reinforcing that the levels of PAH in sediments do not pose risk to C. rhizophorae. Levels in sediments were lower than in native organisms for all PES sampling points. In SES, PAH concentration in native organisms varied according to the concentration in sediments; in points P2 and P3, but in point P4, PAH levels in the native oysters were higher than in the sediments. These results show a higher concentration of PAH in native oysters from PES and SES compared to the concentration of up to $8.0\,mg\,kg^{-1}$ in native oysters from around the world, reported by Neff [5]. Figure 3 presents the distribution of the concentrations of individual PAHs in native oysters of SES.

Comparing these results to those of the experiment with caged oysters, it is possible to observe that the concentrations in native oysters from PES were higher than those of the transplanted organisms for Hg, but almost the same for PAH. The longer exposure period that the native organisms have in relation to the caged organisms explains the higher Hg concentrations in native oysters comparatively to transplanted organisms. On the other hand, the low levels of PAH in PES sediments did not produce important disparities between the concentration of this contaminant in native and

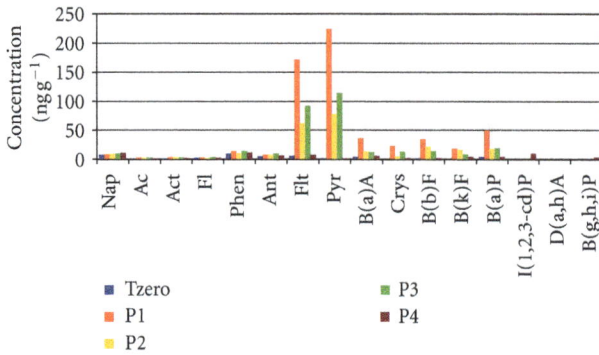

FIGURE 4: Distribution of individual PAHs in control (T-zero) and transplanted oysters in SES. Nap—naphthalene, Ac—acenaphthylene, Act—acenaphthene, Fl—fluorene, Phen—phenanthrene, Ant—anthracene, Flt—fluoranthene, Pyr—pyrene, B(a)a—benzo[a]anthracene, Crys—chrysene, B(b)F—benzo[b]fluoranthene, B(k)F—benzo[k]fluoranthene, B(a)P—benzo[a]pyrene, I(1,2,3-cd)P—indeno[1,2,3-cd]pyrene, D(a,h)A—dibenzo[a,h]anthracene, and B(g,h,i)P—benzo[ghi]perylene.

transplanted oysters as did in SES, despite the difference on time of exposure.

When the oyster tissues from the caged experiment were analyzed, a comparison of the results of the transplanted organisms to the results of the control oysters (also called T-zero), it was possible to observe the kinetics of bioaccumulation of these contaminants. In PES, concentration of Hg and PAH was quite the same as the control organisms and in some cases even lower. This strongly indicates that contaminants in the sediments of PES may not be bioavailable to the organisms. van Straalen [29] establishes that measurements of chemicals in biological matrices can be used to estimate ecological risk not by total concentration in the environment, but by estimating the bioavailable fraction of these contaminants.

Assessing the data from organisms transplanted to SES and comparing to control organisms, it is possible to observe that concentrations of these contaminants were higher in Piaçaguera (P1), Bagres Island (P2), and Alemoa (P3) and very close to T-zero in Palmas Island (P4). This proves that there is an unquestionable bioavailability of contaminants from sediments to the water column and to the transplanted oysters. Hg has shown from 44 to 150% increase in concentration, while PAH reached more than 1,600% increase in Piaçaguera (P1), more than 600% in Bagres Island (P2), and more than 800% in Alemoa (P3). The evaluations of contaminant residues in organisms are real indicators of contaminant bioavailability as have been shown by a number of studies [29]. Figure 4 shows individual PAHs distribution in these transplanted organisms.

It is possible to note a direct relationship in the concentrations of all PAHs, especially fluoranthene, pyrene, benzo(a)anthracene, benzo(b)fluoranthene, benzo(k)fluoranthene, and benzo(a)pyrene from the sediments, native and caged oysters as can be seen in Figures 2, 3, and 4.

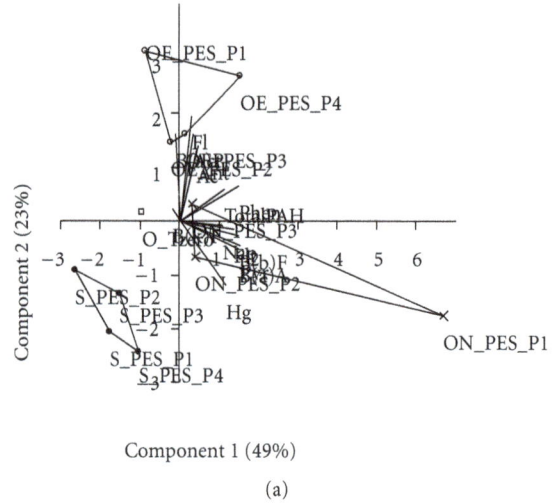

(a)

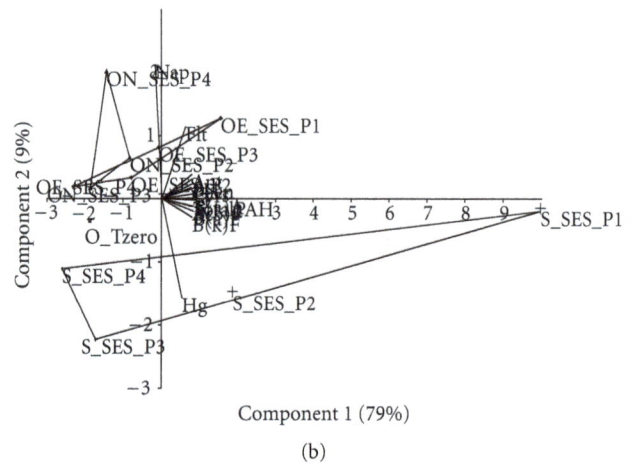

(b)

FIGURE 5: Principal component analysis of sediments, native oysters, and transplanted oysters. (a) Paranaguá, (b) Santos. OE—oyster in Experiment; ON—native oyster; S—sediments; SES—Santos Estuarine System; PES—Paranaguá Estuarine System. Nap—naphthalene, Ac—acenaphthylene, Act—acenaphthene, Fl—fluorene, Phen—phenanthrene, Ant—anthracene, Flt—fluoranthene, Pyr—pyrene, B(a)a—benzo[a]anthracene, Crys—chrysene, B(b)F—benzo[b]fluoranthene, B(k)F—benzo[k]fluoranthene, B(a)P—benzo[a]pyrene, I(1,2,3-cd)P—indeno[1,2,3-cd]pyrene, D(a,h)A—dibenzo[a,h]anthracene, and B(g,h,i)P—benzo[ghi]perylene.

Principal component analysis of the whole data set did not show a clear separation of the groups, but when processed each site alone it shows well-defined groupings of noncontaminated sediments from PES along with a separate group of oysters from the experiment and other groups of native oysters (Figure 5(a)) (groupings are presented as polygons, which represent the compartments—sediments, native oysters, and transplanted oysters). The first two components account for 72% of the significance, but they are very close to each other.

Analyzing the PCA for SES (Figure 5(b)), the first principal component represents 79% of the data and the second component represents 9% giving a total of 88%. Vectors

Bioaccumulation of Polycyclic Aromatic Hydrocarbons and Mercury in Oysters (Crassostrea rhizophorae) from Two Brazilian Estuarine Zones

191

representing Hg, individual PAHs, and total PAH are mostly projected in principal component 1 with only a small contribution related to component 2 which is mostly represented by Hg and two low molecular weight individual PAHs (naphthalene and fluoranthene). Contaminated sediments from SES are clearly separated from the organisms and sediments from Piaçaguera (P1) and Ilha dos Bagres (P2) which are in a different part of the PCA. Oysters grouping did not show a clear distinction between native and caged oysters except for the oysters transplanted to Piaçaguera (P1) and the native oysters from das Palmas Island (P4). It is possible to observe almost the same pattern in the sediments and caged oysters pointing to pyrogenic PAH of high molecular weight probably desorbed from the sediments from P1. Native oysters from P4 show some accumulation of some petrogenic low molecular weight individual PAHs present in this place as oil and gasoline spills from small cruise and fishing boats in the marina from the fisherman's club present in the island.

4. Conclusions

The aim of this study was to assess sediment quality and bioavailability of contaminants in the estuarine systems of Santos and Paranaguá in southeastern Brazil. Chemical analysis has shown that sediments and native oysters in PES are not significantly impacted, while in the inner part of SES it presented high concentrations of Hg and PAH. Also, the most contaminated sites have shown an absence of native organisms, while caged oysters presented varying levels of bioaccumulation.

From an overall evaluation of these results, it is possible to conclude that the use of transplanted oysters is a wonderful tool not only for bioaccumulation studies, but also for discriminating sites with different levels of contamination. Also, caged oysters, in this case Crassostrea rhizophorae, proved to be good indicating organisms for bioavailability and monitoring programs since they bioaccumulated significant amounts of contaminants in Santos estuary.

Acknowledgments

The authors would like to thank Brazilian Education Ministry—Coordenação de Aperfeiçoamento de Pessoal e Ensino Superior (CAPES no. 099/06) and Spanish Education Ministry—Ministerio de Educación—Dirección General de Universidades (MECD/DGU no. PHB2005-0100-PC) for the International Cooperation Program Brazil-Spain in financing this study.

References

[1] M. K. Hill, Understanding Environmental Pollution, Cambridge University Press, 3rd edition, 2010.

[2] United States Environmental Protection Agency, "Bioaccumulation testing and interpretation for the purpose of sediment quality assessment. Status and needs," Office of Water (4305). Office of Solid Waste (5307W). EPA-823-R-00-001. p. 111, 2000.

[3] National Research Council, Bioavailability of Contaminants in Soils and Sediments: Processes, Tools, and Applications, National Academies Press, 2003.

[4] C. J. Leeuwen and T. G. Vermeire, Risk Assessment of Chemicals: An Introduction, Springer, Dordrecht, The Netherlands, 2nd edition, 2007.

[5] J. M. Neff, Bioaccumulation in Marine Organisms. Effect of Contaminants From Oil Well Produced Water, Elsevier, Amsterdam, The Netherlands, 2002.

[6] E. Cortazar, L. Bartolomé, S. Arrasate et al., "Distribution and bioaccumulation of PAHs in the UNESCO protected natural reserve of Urdaibai, Bay of Biscay," Chemosphere, vol. 72, no. 10, pp. 1467–1474, 2008.

[7] International Mussel Watch, Initial Implementation Phase. Draft Final Report, National Oceanic and Atmospheric Administration, 1993.

[8] M. C. R. Amaral, M. F. Rebelo, J. P. M. Torres, and W. C. Pfeiffer, "Bioaccumulation and depuration of Zn and Cd in mangrove oysters (Crassostrea rhizophorae, Guilding, 1828) transplanted to and from a contaminated tropical coastal lagoon," Marine Environmental Research, vol. 59, no. 4, pp. 277–285, 2005.

[9] M. Costa, E. Paiva, and I. Moreira, "Total mercury in Perna perna mussels from Guanabara Bay—10 years later," Science of the Total Environment, vol. 261, no. 1–3, pp. 69–73, 2000.

[10] M. F. Rebelo, M. C. R. Amaral, and W. C. Pfeiffer, "High Zn and Cd accumulation in the oyster Crassostrea rhizophorae, and its relevance as a sentinel species," Marine Pollution Bulletin, vol. 46, no. 10, pp. 1354–1358, 2003.

[11] C. A. R. Silva, B. D. Smith, and P. S. Rainbow, "Comparative biomonitors of coastal trace metal contamination in tropical South America (N. Brazil)," Marine Environmental Research, vol. 61, no. 4, pp. 439–455, 2006.

[12] D. M. S. Abessa, R. S. Carr, E. C. P. M. Sousa et al., "Integrative ecotoxicological assessment of a complex tropical estuarine system," in Marine Pollution: New Research, Nova Science, New York, NY, USA, 2008.

[13] R. B. Choueri, A. Cesar, D. M. S. Abessa et al., "Development of site-specific sediment quality guidelines for North and South Atlantic littoral zones: comparison against national and international sediment quality benchmarks," Journal of Hazardous Materials, vol. 170, no. 1, pp. 320–331, 2009.

[14] R. B. Choueri, A. Cesar, R. J. Torres et al., "Integrated sediment quality assessment in Paranaguá Estuarine System, Southern Brazil," Ecotoxicology and Environmental Safety, vol. 72, no. 7, pp. 1824–1831, 2009.

[15] R. J. Torres, D. M. S. Abessa, F. C. Santos et al., "Effects of dredging operations on sediment quality: contaminant mobilization in dredged sediments from the Port of Santos, SP, Brazil," Journal of Soils and Sediments, vol. 9, no. 5, pp. 420–432, 2009.

[16] M. L. Lamparelli, M. P. Costa, V. A. Prósperi et al., Sistema Estuarino de Santos e São Vicente, Relatório Técnico CETESB, São Paulo, Brazil, 2001.

[17] D. M. S. Abessa, R. S. Carr, B. R. F. Rachid, E. C. P. M. Sousa, M. A. Hortelani, and J. E. Sarkis, "Influence of a Brazilian sewage outfall on the toxicity and contamination of adjacent sediments," Marine Pollution Bulletin, vol. 50, no. 8, pp. 875–885, 2005.

[18] A. Cesar, C. D. S. Pereira, A. R. Santos et al., "Ecotoxicology assessment of sediments from Santos and Sao Vicente Estuarine System Brazilian Journal of Oceanography, vol. 54, no. 1, pp. 55–63, 2006.

[19] United States Environmental Protection Agency, "A compendium of chemical, physical and biological methods for assessing and monitoring the remediation of contaminated sediment sites," No. 68-W-99-033, p. 291, 2003.

[20] United States Environmental Protection Agency, "Procedures for the derivation of equilibrium partitioning sediment benchmarks (ESBs) for the protection of benthic organisms: PAH Mixtures," EPA-822-R-02-013, p. 175, 2003.

[21] United States Environmental Protection Agency, *Test Methods For Evaluating Solid Waste (SW-846)*, US Environmental Protection Agency, Office of Solid Waste, Economic, Methods, and Risk Analysis Division, 1996.

[22] United States Army Corps of Engineers, "Use of sediment quality guidelines (SQGs) in dredged material management," Dredging Research Technical Note EEDP-04-29, p. 14, 1998.

[23] S. L. Simpson, G. E. Batley, A. A. Chariton et al., *Handbook for Sediment Quality Assessment*, CSIRO Publishing, Bangor, Australia, 2006.

[24] R. J. Wenning, G. E. Batley, C. G. Ingersoll, and D. W. Moore, *Use of Sediment Quality Guidelines and Related Tools for the Assessment of Contaminated Sediments*, Society of Environmental Toxicology and Chemistry, 2005.

[25] P. M. Medeiros and M. Caruso Bícego, "Investigation of natural and anthropogenic hydrocarbon inputs in sediments using geochemical markers. I. Santos, SP—Brazil," *Marine Pollution Bulletin*, vol. 49, no. 9-10, pp. 761–769, 2004.

[26] C. C. Martins, M. C. Bicego, and R. C. Montone, "Hidrocarbonetos marcadores geoquímicos em testemunhos de sedimentos do sistema estuarino de Santos e São Vicente, SP," in *Proceedings of the 28a Reunião Anual da Sociedade Brasileira de Química*, Poços de Caldas, Brazil, 2005.

[27] A. A. Mozeto, R. J. Torres, F. C. Santos, M. DelGrande, D. M. S. Abessa, and M. R. L. Nascimento, "Effects of dredging activities on contaminated sediment quality: contaminant mobilization in dredging sediments of the Port of Santos, SP, Brazil," in *Proceedings of the 17th SETAC Europe Annual Meeting*, p. 72, Porto, Portugal, May 2007.

[28] N. M. Lawson and R. P. Mason, "Accumulation of mercury in estuarine food chains," *Biogeochemistry*, vol. 40, no. 2-3, pp. 235–247, 1998.

[29] N. M. van Straalen, "Chapter 18 Contaminant concentrations in organisms as indicators of bioavailability: a review of kinetic theory and the use of target species in biomonitoring," *Developments in Soil Science*, vol. 32, pp. 449–477, 2008.

Permissions

The contributors of this book come from diverse backgrounds, making this book a truly international effort. This book will bring forth new frontiers with its revolutionizing research information and detailed analysis of the nascent developments around the world.

We would like to thank all the contributing authors for lending their expertise to make the book truly unique. They have played a crucial role in the development of this book. Without their invaluable contributions this book wouldn't have been possible. They have made vital efforts to compile up to date information on the varied aspects of this subject to make this book a valuable addition to the collection of many professionals and students.

This book was conceptualized with the vision of imparting up-to-date information and advanced data in this field. To ensure the same, a matchless editorial board was set up. Every individual on the board went through rigorous rounds of assessment to prove their worth. After which they invested a large part of their time researching and compiling the most relevant data for our readers. Conferences and sessions were held from time to time between the editorial board and the contributing authors to present the data in the most comprehensible form. The editorial team has worked tirelessly to provide valuable and valid information to help people across the globe.

Every chapter published in this book has been scrutinized by our experts. Their significance has been extensively debated. The topics covered herein carry significant findings which will fuel the growth of the discipline. They may even be implemented as practical applications or may be referred to as a beginning point for another development. Chapters in this book were first published by Hindawi Publishing Corporation; hereby published with permission under the Creative Commons Attribution License or equivalent.

The editorial board has been involved in producing this book since its inception. They have spent rigorous hours researching and exploring the diverse topics which have resulted in the successful publishing of this book. They have passed on their knowledge of decades through this book. To expedite this challenging task, the publisher supported the team at every step. A small team of assistant editors was also appointed to further simplify the editing procedure and attain best results for the readers.

Our editorial team has been hand-picked from every corner of the world. Their multi-ethnicity adds dynamic inputs to the discussions which result in innovative outcomes. These outcomes are then further discussed with the researchers and contributors who give their valuable feedback and opinion regarding the same. The feedback is then collaborated with the researches and they are edited in a comprehensive manner to aid the understanding of the subject.

Apart from the editorial board, the designing team has also invested a significant amount of their time in understanding the subject and creating the most relevant covers. They scrutinized every image to scout for the most suitable representation of the subject and create an appropriate cover for the book.

The publishing team has been involved in this book since its early stages. They were actively engaged in every process, be it collecting the data, connecting with the contributors or procuring relevant information. The team has been an ardent support to the editorial, designing and production team. Their endless efforts to recruit the best for this project, has resulted in the accomplishment of this book. They are a veteran in the field of academics and their pool of knowledge is as vast as their experience in printing. Their expertise and guidance has proved useful at every step. Their uncompromising quality standards have made this book an exceptional effort. Their encouragement from time to time has been an inspiration for everyone.

The publisher and the editorial board hope that this book will prove to be a valuable piece of knowledge for researchers, students, practitioners and scholars across the globe.

List of Contributors

K. B. Padmakumar and V. N. Sanjeevan
Centre for Marine Living Resources and Ecology, Ministry of Earth Sciences, Kochi 37, Kerala, India

N. R. Menon
School of Marine Sciences, Cochin University of Science and Technology and Nansen Environmental Research Center, Kochi 16, Kerala, India

Kalpesh Patil, M. C. Deo and Subimal Ghosh
Indian Institute of Technology, Bombay, Mumbai 400 076, India

M. Ravichandran
Indian National Centre for Ocean Information Services, Hyderabad 500090, India

Eduardo G. G. de Farias and Joao A. Lorenzzetti
Divisao de Sensoriamento Remoto, Instituto Nacional de Pesquisas Espaciais (INPE), CP 515 12227-010 Sao Jose dos Campos SP, Brazil

Bertrand Chapron
Laboratoire d Oceanographie Spatiale, Institut Franc ais de Recherche pour l Exploitation de la Mer (IFREMER), 70 29280 Plouzane, France

J. P. Le Roux
Geology Department, Faculty of Physical and Mathematical Sciences, University of Chile/Andean Geothermal Center of Excellence, Post-office Box 13158, Santiago, Chile

Suhas Shetye, Babula Jena and Rahul Mohan
National Centre for Antarctic & Ocean Research, Headland Sada, Goa 403 804, India

Maruthadu Sudhakar
Ministry of Earth Sciences, Prithvi Bhavan, New Delhi 110 003, India

João A. Lorenzzetti and Fabian G. Dias
Divisao de Sensoriamento Remoto, Instituto Nacional de Pesquisas Espaciais (INPE), CP 515, 12227-010 Sao Jose dos Campos, SP, Brazil

V. Dhananjayan
Industrial Hygiene and Toxicology Division, Regional Occupational Health Centre (Southern), ICMR, Kannamangala PO, Bangalore 562 110, India

S. Muralidharan and Vinny R. Peter
Division of Ecotoxicology, Salim Ali Centre for Ornithology and Natural History, Coimbatore 641 108, India

Fred M. Vukovich
FMV Atmospheric and Marine Consultants, 8033 Hawkshead Rd, Wake Forest, NC 27587, USA

C. P. Simha, P. C. S. Devara and S. K. Saha
Physical Meteorology and Aerology Division, Indian Institute of Tropical Meteorology, Dr. Homi Bhabha Road, Pashan, Pune 411 008, India

K. N. Babu and A. K. Shukla
DPD, Oceansat-II UP (OCM2 Validation), Space Applications Centre, ISRO, Ahmedabad 380 015, India

Jamil Tajam and Mohd Lias Kamal
Centre of Ocean Research, Conservation & Advances (ORCA), Division of Research, Industrial Linkage, Community Network & Alumni, Universiti Teknologi MARA, 02600 Arau, Perlis, Malaysia

Neha Nandkeolyar and G. Sandhya Kiran
Department of Botany, Faculty of Science, M.S. University, Baroda, Vadodara 390002, India

Ajai and Mini Raman
Marine Optics Division, Marine and Planetary Sciences Group, Space Application Centre, Ahmedabad 380015, India

R. K. Sarangi
Marine Geo and Planetary Sciences Group, Space Applications Centre (ISRO), Ahmedabad 380 015, India

C. G. Diedrich
Paleo Logic, Nansenstraße 8, D-33790 Halle, Germany

Mukesh Gupta, Randall K. Scharien and David G. Barber
Centre for Earth Observation Science, Department of Environment and Geography, Clayton H. Riddell Faculty of Environment, Earth, and Resources, University of Manitoba, Winnipeg, MB, Canada

Adriana Z. Mazurek
Department of Physics, Warsaw University of Life Sciences, Nowoursynowska 159, 02-776 Warsaw, Poland

Stanisław J. Pogorzelski
Institute of Experimental Physics, University of Gdansk, Wita Stwosza 57, 80-952 Gdansk, Poland

V. Dhananjayan
Division of Ecotoxicology, Salim Ali Centre for Ornithology and Natural History, Coimbatore 641108, India
Regional Occupational Health Center, ICMR, Kannamangala Post, Poojanahalli Road, Devenahalli TK, Bangalore 562110, India

S. Muralidharan
Division of Ecotoxicology, Salim Ali Centre for Ornithology and Natural History, Coimbatore 641108, India

Mohammad R. Baradaran
Civil Engineering Group, Islamic Azad University, Maymand, Iran

Mohammad H. Jabbari, Parviz Ghadimi and Ali Masoudi
Department of Marine Technology, Amirkabir University of Technology, Hafez Avenue No. 424, P.O. Box 15875-4413, Tehran, Iran

Mohammad R. Baradaran
Civil Engineering Group, Islamic Azad University, Maymand, Iran

Ronaldo J. Torres, Pedro S. Fadini and Antonio A. Mozeto
Laboratorio de Biogeoquımica Ambiental (LBGqA)–Nucleo de Estudos, Diagnosticos e Intervencoes Ambientais (NEDIA), Departamento de Quimica, UFSCar, Rod. Washington Luiz, 13565-905 Sao Carlos, SP, Brazil

Augusto Cesar, Camilo D. S. Pereira and Rodrigo B. Choueri
Centro de Ciencias do Mar e Meio Ambiente, Baixada Santista UNIFESP, Avenida Dona Costa, 95, 11060-001 Santos, SP, Brazil

Denis M. S. Abessa
Campus Experimental do Litoral Paulista, Universidade Estadual Paulista, Pca Infante D. Henrique, s/no, 11330-900 Sao Vicente, SP, Brazil

Marcos R. L. do Nascimento
Laboratorio de Pocos de Caldas (LAPOC), CNEN, Comissao Nacional de Energia Nuclear, Rodovia Andradas, km 13, 37701-970 Pocos de Caldas, MG, Brazil